Particle Size Distribution III

ACS SYMPOSIUM SERIES 693

Particle Size Distribution III

Assessment and Characterization

Theodore Provder, EDITOR
ICI Paints

Developed from a symposium sponsored by the Division
of Polymeric Materials: Science and Engineering
at the 212th National Meeting
of the American Chemical Society,
Orlando, Florida,
August 25–29, 1996

American Chemical Society, Washington, DC

Library of Congress Cataloging-in-Publication Data

Particle size distribution III : assessment and characterization / Theodore Provder, editor.

p. cm.—(ACS symposium series, ISSN 0097–6156; 693)

"Developed from a symposium sponsored by the Division of Polymeric Materials: Science and Engineering at the 212th National Meeting of the American Chemical Society, Orlando, Florida, August 25–29, 1996."

Includes bibliographical references and indexes.

ISBN 0–8412–3561–9

1. Particle size determination—Congresses.

I. Provder, Theodore, 1939– . II. American Chemical Society. Division of Polymeric Materials: Science and Engineering. III. American Chemical Society. Meeting (212th : 1996 : Orlando, Fla.) IV. Series.

TA418.8.P343 1998
620.1′.1299—dc21 98–16110
CIP

Distributed by Oxford University Press

PRINTED IN THE UNITED STATES OF AMERICA

Advisory Board

ACS Symposium Series

Foreword

THE ACS SYMPOSIUM SERIES was first published in 1974 to provide a mechanism for publishing symposia quickly in book form. The purpose of the series is to publish timely, comprehensive books developed from ACS sponsored symposia based on current scientific research. Occasionally, books are developed from symposia sponsored by other organizations when the topic is of keen interest to the chemistry audience.

Before agreeing to publish a book, the proposed table of contents is reviewed for appropriate and comprehensive coverage and for interest to the audience. Some papers may be excluded in order to better focus the book; others may be added to provide comprehensiveness. When appropriate, overview or introductory chapters are added. Drafts of chapters are peer-reviewed prior to final acceptance or rejection, and manuscripts are prepared in camera-ready format.

As a rule, only original research papers and original review papers are included in the volumes. Verbatim reproductions of previously published papers are not accepted.

ACS BOOKS DEPARTMENT

Contents

ELECTROPHORETIC AND ELECTROACOUSTIC SEPARATION AND ANALYSIS OF PARTICLES IN DILUTE AND CONCENTRATED REGIMES

INDEXES

Preface

The previous two books in this series (*1, 2*) emphasized the revitalization in particle size distribution analysis which was technologically driven by advances in electronics and computer technology coupled with market forces generated by customer needs for improved particle size measurements for the characterization of materials. Improved particle size measurements include the following attributes:

- user friendly, reliable, cost-effective and smart instruments
- high rate of sample measurement throughput
- wide dynamic particle size range
- improved resolution
- analysis of colloidal systems in concentrated regimes
- at-line and on-line measurement capability

The chapters in this book reflect the current activity in measurement techniques, methodology and application to a range of particulate systems and dispersions with topics divided into three distinct sections.

The first section deals with static and dynamic light scattering methods and applications. Improved and novel instrumentation described include high resolution CCD detectors, novel spectrometer designs, unique automatic dilution systems, use of fiber optic probes and advances in hybrid instrumentation. These hybrid instruments extend the measurement particle size range and include the following combinations: static light scattering coupled to photon correlation spectroscopy; photon correlation spectroscopy coupled to single particle optical sensing; and time of transition method coupled to dynamic image analysis. In this section it also is shown that light scattering methods have promise for at-line and on-line particle size distribution analysis in both dilute and concentrated regimes. These techniques are applied to a wide range of particulate systems.

The second section emphasizes current activity in the development and refinement of fractionation methods and their application to a range of colloidal systems. The fractionation methods include thermal field flow fractionation, electric field flow fractionation, capillary hydrodynamic fractionation, and sedimentation in a centrifugal force field.

The third section examines current activity in electrophoretic and electroacoustic separation, analysis and characterization of particles in dilute and concentrated regimes. Several recently developed novel instrumental techniques are described.

It is expected that the improvements in instrumentation capability and measurement science will continue and be manifested in commercially available instrumentation. I hope that this book and the first two books (*1, 2*) in this series will be useful guides to those first becoming acquainted with the field of particle size analysis and particle characterization and will spur further activity among experienced practitioners

Acknowledgments

I thank the authors for their effective oral and written communications and the reviewers for their helpful critiques and constructive comments.

Theodore Provder
ICI Paints
Strongsville, OH 44136

1. *Particle Size Distribution: Assessment and Characterization;* Provder, T., Ed.; ACS Symposium Series 332; American Chemical Society, Washington, DC, 1987.

2. *Particle Size Distribution II: Assessment and Characterization;* Provder, T., Ed.; ACS Symposium Series 472; American Chemical Society, Washington, DC, 1991.

Static and Dynamic Light-Scattering Methods and Applications

Chapter 1

Design of a Small Angle Spetrometer: Application to Milk Casein

M. Alexander[1], F. R. Hallett[1], and D. G. Dalgleish[2]

[1]Department of Physics and [2]Department of Food Science, University of Guelph, Guelph, Ontario N1G 2W1, Canada

A Small Angle Integrated Light Scattering (SAILS) Spectrometer was designed with the innovative use of a Charged Couple Device (CCD) camera. In contrast to previous spectrometers, SAILS has very few optical surfaces, minimizing the interference with the main beam and allowing the recovery of low angle scattering data.

The scattered light falls on a diffusive plate which is photographed by the CCD camera. The image on this plate produces a digitized two dimensional array, covering the scattering angles from 10 to 20 degrees. This data is then stored as an ASCII file for further analysis.

Due to this non-invasive technique and the speed of data collection (on the order of 40 sec), we are able to study dynamic phenomena such as casein micelle aggregation. As this process takes place, the intensity of the scattered light is recorded as a function of time. The size distribution of the scatterers can be obtained from the experimental data by a discrete inversion of the angle dependent scattering. Due to the narrow Q regime of SAILS we encountered some problems in retrieving this information. Nevertheless, we were successful in following the change in scattered intensity in times as short as a couple of minutes and we were able to see the reduction in light scattered as the glycoprotein moieties of the κ - casein were removed.

Ever since the invention of the laser in the 1960's, the theory of light scattering has been an expansive and growing field. Originally, the method of choice had been Dynamic Light Scattering or DLS (1-5) and for numerous years it has been used to determine the diffusion rates and sizes of macromolecules in suspension (6,7). Traditionally, DLS spectrometers made use of a photomultiplier tube to gather light intensity information. Though a reliable method, the drawback of DLS was the long time required to collect scattering data. This fact limits this technique to the study of systems with relatively slow dynamics.

With the advent of increasingly fast computers and more sensitive and rapid light detectors, a new method, called Integrated Light Scattering or ILS, became more and more the technique of choice (8-10).

ILS measures very different properties of macromolecules in solution. While DLS traces intensity fluctuations as a function of time for a certain, constant angle, ILS monitors the time averaged total scattered intensity as a function of angle.

Mie Theory

If we imagine a particle subjected to an incoming light wave, a dipole will be induced in it. This dipole will in turn re-create a scattered electric field with the same average frequency as the incoming one. A detector placed in its way will then measure an intensity related to the time averaged square of this electric field, and an angular dependence related to the shape and size of the scattering particle.

For scatterers whose sizes are small comparable to the incoming wavelength, the scattered intensity I_{scatt} (Q,R) of a spherical particle of radius R is given by

$$\frac{I_{scatt}(Q,R)}{I_o} = C R^6 P(Q,R) \tag{1}$$

where I_o is the incident intensity, C an experimentally determined constant, P(Q,R) is called the scattering factor and Q is the scattering vector given by

$$Q = \frac{4\pi n}{\lambda} \sin\frac{\theta}{2} \tag{2}$$

where n is the index of refraction of the medium, λ is the wavelength of the light and θ is the scattering angle.

In the case where the dimensions of the particle are comparable to the wavelength of the light, different parts of the particle will receive different phases of the incoming wave which will interfere with each other. In this case, the scattered intensity I_{scatt} (Q,R) is given by (11)

$$\frac{I_{scatt}(Q,R)}{I_o} = \frac{|F(\theta)|^2}{k^2 r^2} \tag{3}$$

where k is wave number, r is the distance from the scattering particle to point of measurement and $F(\theta)$ is called the amplitude function given by

$$F(\theta) = \sum_{n=1}^{\infty} \frac{2n+1}{n(n+1)} \left[a_n \pi_n(\cos\theta) + b_n \tau_n(\cos\theta) \right] \tag{4}$$

where

$$\pi_n(\cos\theta) = \frac{1}{\sin(\theta)} P_n^{\,1} \cos(\theta)$$
$$\tau_n(\cos\theta) = \frac{d}{d\theta} P_n^{\,1} \cos(\theta) \tag{6}$$

where $P_n^{\,1} \cos(\theta)$ are the associated Legendre polynomials.
The coefficients a_n and b_n are given by

$$a_n = \frac{\psi_n'(y)\, \psi_n(x) - m\, \psi_n(y)\, \psi_n'(x)}{\psi_n'(y)\, \xi(x) - m\, \psi_n(y)\, \xi(x)}$$
$$b_n = m \frac{\psi_n'(y)\, \psi_n(x) - \psi_n(y)\, \psi_n'(x)}{m\, \psi_n'(y)\, \xi(x) - \psi_n(y)\, \xi(x)} \tag{7}$$

where

$$\psi_n(z) = z j_n(z) \quad and \quad \xi(z) = z h_n^{(2)}(z) \tag{8}$$

and the primes indicate differentiation with respect to the argument in parenthesis. j_n (z) and h_n (z) are the Ricatti-Bessel functions (11), m is the ratio of the index of refraction of the particle over the index of refraction of the medium and the arguments are

$$x = ka \quad and \quad y = mka \tag{9}$$

where $k = 2\pi/\lambda$.

Small Angle Scattering

A more recent addition to the already vast field of light scattering has been the development of small angle light scattering (SALS). The range of scattering angles probed by traditional spectrometers lies somewhere between 20 and 340 degrees. It was not till the advent of fast, wide dynamic range and extremely sensitive detectors (12) that small angle light scattering became a viable technique.

The length scale being probed at small angles can range from several hundred Angstroms to several microns. Larger systems such as polymers, polymer solutions, gels, colloids, micelles, emulsions, etc, contain structures that fall into the so called mesoscopic region (13). It is in this region that the advantage of SAILS is observed. There is a significant amount of literature describing the application of small angle techniques (14-17). Koch (18) studies the physical properties of bacteria, Holoubek (19) applied low angle scattering to the study of anisotropic spherulites and Kubota (24) described the use of a Fourier lens to improve the signal to noise ratio of scattered light obtained by a diode array.

Another interesting application of SAILS is to studies of aggregation. These processes deal with larger structures, where sizes can be readily determined by SAILS. There is an interest in investigating the actual rate of formation of these superstructures. Since some phenomena such as crystallization or aggregation can occur in relatively short times, it is critical to have a fast recording detector. It is here that Charge Couple Devices or CCDs come to play an important role in small angle scattering.(20-24)

CCD detectors have a superb dynamic range, 14 to 16 bits, incredibly good sensitivity that allows them to detect under 10 photons striking the detection area, and very fast response rate. Their two dimensional nature permits scattering measurements in both horizontal and vertical scattering planes simultaneously and the minute size of the pixels (on the order of 10's of microns) allows large numbers of independent light scattering measurements to be made.

Casein Micelle Theory

One of the most common suspensions and yet one that is not completely understood is bovine milk which is used to manufacture many different food products.

Cow's milk is a colloidal system made up of five principal components which are (25)

I) Water 87.3 %
ii) Lactose 4.6 %
iii) Fat 3.9 %
iv) Proteins 3.3 %
v) Salts 0.7 %

The solvent medium is an aqueous solution of lactose in which the rest of the components mentioned above are dispersed. Starting with the most abundant, fat globules appear in sizes ranging from 0.5 to 10 microns. Each globule is surrounded by a thin protective membrane that prevents it from coagulating into one continuous phase (26).

The second dispersed element is protein which is present in two forms, caseins (80 % of total protein) and whey proteins. Casein micelles are aggregates of the caseins containing large numbers of protein molecules as well as calcium phosphate. Their known function is to carry calcium phosphate and protein in an easily digestive form from mother to nursling. They also create a coagulum in the stomach of the young that slowly releases nutrients down the digestive tract (26). There are four proteins, α_{s1}, α_{s2}, β, (mainly hydrophobic) and κ-caseins (amphiphilic) (27,28). At room temperature and in their native state, the micelles are fairly stable. In general, α_{s1}, α_{s2} and β are mainly found "inside" the globule, while the κ-casein covers the surface of the micelle, rendering it steric and charge stabilization (26).

By the addition of the proteolytic enzyme, rennet, the hydrophilic part of the κ-casein is split off and the steric stabilization is removed. This is one of the processes by which aggregation can occur. Figure 1 shows the process.

Design of SAILS

The layout of SAILS is very simple indeed. Figure 2 shows a schematic diagram of the apparatus.

The light source is a vertically polarized Helium Neon laser (λ=632.8nm) of 17 mWatts power (model LHRP - 1701, research Electro Optics Inc., Boulder, Colorado, USA) This beam is focussed onto a diffusive screen by an antireflection coated lens (KPX 12 AQ 14, Newport Corp., Mississauga, Ont. CANADA). It is further collimated by several pinholes to finally illuminate the sample which is encased in a cubic, flat faced quartz cuvette of 1cm light path.(Hellma Canada Lmt., Concord, Ont., CANADA) The scattered light radiated by the sample is projected onto a diffusive screen (Custom made, Oriel Corporation, Stratford, Ct. USA) placed 40 cm away, along the laser axis. The image on this plate is photographed by the Charged Couple Device (CCD) camera (specs in next section) placed 40 cm behind the screen, producing a two dimensional array, covering the scattering angles already mentioned.

The CCD Detector. The CCD used is a PI TEK 512x512 pixel (Princeton Instruments Inc., Trenton, NJ, USA), thermoelectrically cooled front illuminated, with 200msec array readout time. The setup closely imitates a standard 35 mm camera. The front lens captures the desired image and focuses it into the CCD array. In front of it lays the shutter which limits the exposure time of the pixels to the incoming light. The back enclosure contains the heat removal block. Most TE/CCD detectors must be cooled during operation, for the simple reason that dark charge is thermally induced. A Peltier effect thermoelectric cooler cools the CCD and a thermal sensing diode monitors the temperature to 0.05° C. The controller box of the CCD is a ST-130 Controller (Princeton Instruments) with a 16 bit resolution A/D, operating at 500kHz. The commercial software that runs the controller and CCD is WinView 1.2A (Princeton Instruments). All the information gathered by the device is sent to a PC as a bitmap file, which gets in turn, converted into an ASCII file for further manipulation.

Determination of Scattering Angles. In order to get an accurate measurement of

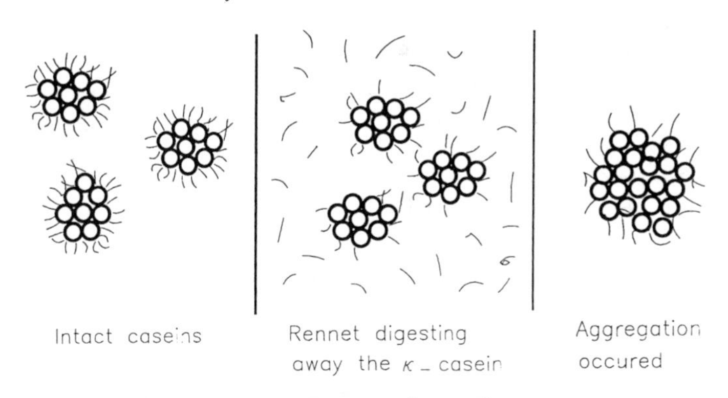

Figure 1. Process of renneting.

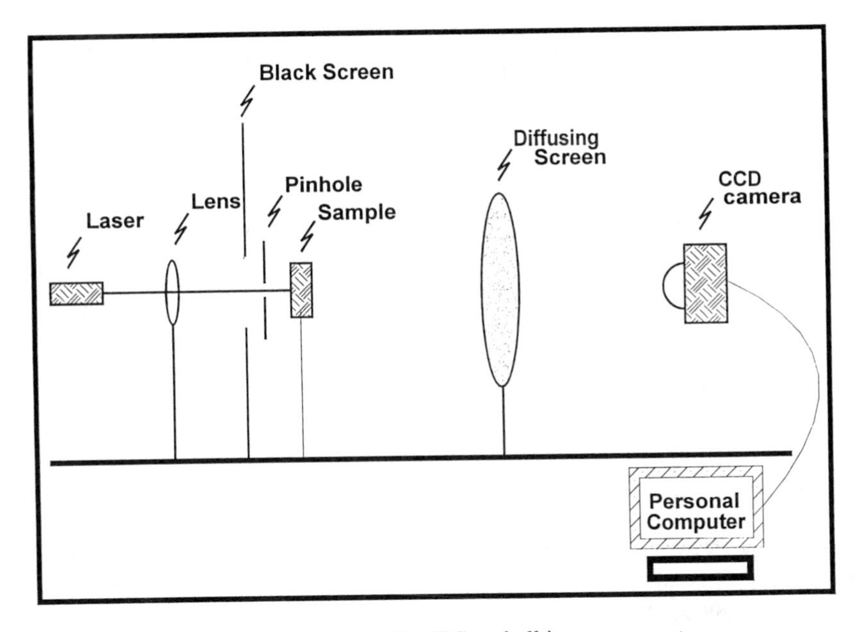

Figure 2. Schematic of SAILS and all its components.

the scattering angles, it is indispensable to assign a pixel number to every point on the diffusive screen. To achieve this, a transparency with a series of black concentric rings with radial increases of 1 cm was drawn and placed on the diffusive screen. An uniform bright light was shone on the "sample" side of the apparatus and the CCD was enabled. A series of pictures were taken of this static situation. The pixels that "see" the dark lines of the pattern would record a far lower number of counts than the rest of the screen. From this, an average pixel number was assigned to each black line. Simultaneously, we could geometrically calculate the angle in air that each line subtended with the main beam at the sample position. Again, to increase the amount of light gathered and improve statistics, three rows of pixels were assigned to each angle, the average pixel row found from this experiment, plus one row on each side.

Due to the geometry of SAILS, further correction factors are necessary. These are:

i) Path differences in the scattered rays reaching the CCD - As the scattering angle increases, so does the distance that the scattered rays have to travel. Because light intensity is inversely proportional to path length squared, there is a need to correct each angle for this difference.

ii) Refraction corrections - As explained above, the determination of the angles is done geometrically in air. As a ray of light enters the sample chamber, it travels through a water based medium (sample), quartz (sample holder) and finally air. Each one of these interfaces will bend the ray slightly. Since we are interested in scattering angles in the scattering medium it is necessary to transform these angles using Snell's law.

iii) Transmissivities corrections - A ray of light passing through a change of medium will be partially transmitted and partially reflected depending on angle of incidence. Though a minute source of error, the change of transmissivities with angle in the quartz face should be accounted for.

iv) The last set of corrections arises from inhomogeneities of the diffusive screen. Due to the nature of the screen, light rays might diffuse differently at different sites. To correct for this effect, a measurement of the intensity of the light given off by a 0.5 watt tungsten light bulb was measured and a last set of correction factors calculated. A detailed description of the calculations of the above mentioned corrections will be published elsewhere.

Sample Preparation And Procedure

Latex Spheres Experiment. The solid latex spheres used to test SAILS were 269, 343, 503 and 993 nm in diameter purchased from Duke Scientific Corporation (Palo Alto, California, cat #5100 A). For preparation, the container housing these spheres was gently inverted to obtain a homogeneous distribution in the solution. Next, they were placed in a low power ultrasonic bath for about 30 seconds to promote dispersion. At this point the spheres were ready to be used. By means of a micro pipet, 18, 16, 12 and 5 μlitres of the solution were diluted in 25 ml of distilled, de-ionized, dust free water. The samples were then carefully loaded into the sample chamber and placed in the spectrometer.

Milk Experiment. Casein micelles were prepared from reconstituted skim milk powder. 100 grams of powder were dissolved in 1000 ml of distilled, deionized filtered water and left overnight in the refrigerator. Sodium azide was added to inhibit bacterial growth. From a portion of this solution, milk ultra filtrate (UF) was prepared using an Amicon TCF high-shear ultrafiltration module, fitted with a Diaflo PM10 membrane-molecular weight cut-off 10,000. This UF was kept at 4°C. Simultaneously, another portion of the milk went through three successive centrifugations, at 6,000, 15,000 and 25,000 rpm. These were done at 20°C using a Beckman L8-70M ultracentrifuge. The last pellet of sedimented micelles was then redispersed in the previously made UF to a concentration of 125g wet micellar casein/liter to ensure stability and kept in the fridge overnight for a good dispersion. At the time of the experiment, the milk was further diluted, 100 μl in 3 ml of UF filtered through 0.22 μm pore diameter filters (Acrodisc, Prod# 4618, Gellman Sciences, Montreal, Quebec, Canada), to avoid multiple scattering.

For the aggregation experiment, the sample scattering was measured once at t = 0 sec, 7.4μl of rennet were then added *in-situ* and the time was recorded. The sample was observed for two hours, taking measurements every three minutes.

Results

Latex Spheres. Figure 3 shows the intensity vs angle plot of experimental data for the 269, 343, 503 and 993 nm diameter latex spheres. Superimposed are the fitted functions generated by the inversion program (29) using Mie theory for solid spheres as the theoretical fit.

The agreement between the theoretical and experimental scattering curves proves to be very good. The percentage errors for the 269, 343, 503 and 993 nm spheres are 6.2%, 3.7%, 1.8% and 0.3% respectively. It is clearly seen that as the size of the scatterers increases, the percentage error decreases. This is mainly due to the fact that at SAILS' small angle range, the scattering profiles, P(Q,R) for different small size scatterers are relatively flat and it is very hard for any inversion program to reliably obtain a mean radius from these scattering curves because of their relatively small information content. Nevertheless, SAILS did a very good at recovering latex spheres' sizes from their light scattering profile.

Casein Micelles. Figure 4 shows the experimental intensity vs angle curves for the casein micelles at different times after the addition of rennet. As expected, the slope of the curves increases as the size gets larger.
The next step is to extract the mean radii of these aggregates as a function of time, and gain information on the kinetics of the process. Due to the similarity of the curves shown if Figure 3 and the fact that the scattering curves of small particles, such as the micelles, are essentially flat over SAILS' scattering angles, an inversion of intensities could not be performed. However, this flatness means that the scattering factor, P(Q,R), is nearly constant over this range (8). Using this approximation Equation (1) simplifies to

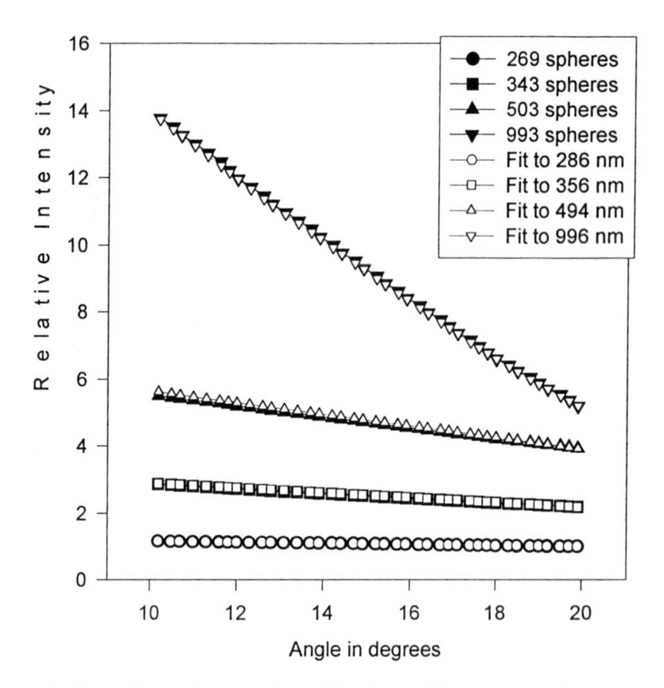

Figure 3. Intensity vs Angle plot of the four different sizes latex spheres. Superimposed are the fits obtained by the inversion program.

$$\frac{I_{scatt}(Q,R)}{I_o} = C\,R^6 \qquad (10)$$

From equation (10) it is trivial to extract the casein micelle radii as a function of time provided that the changes in scattering intensity with time are known and a starting radius is available. Using DLS, a starting radius of 95.96 nm was obtained.

Figure 5 plots the casein micelle radius, for a fixed scattering angle, as a function of time. This graph shows very clearly the increase in size as time lapses. It also shows the importance of a fast detection system since the changes are on the order of two minutes. It is interesting to note the small "dip" in the curve between 10 minutes after rennet till about 40 minutes. This is thought to be due to the stripping of the κ - casein before the micelles start to aggregate. The effective radius of the micelle probably diminishes since the protruding protein is no longer there. This "hair" is about 5 to 9 nm in length, therefore we can see that SAILS can detect minute changes in size.

Conclusion

We were fully successful in building a low angle, light scattering spectrometer with fast readout time to follow dynamic phenomena. We were able to collect scattering data for calibrating polystyrene spheres and retrieve their mean size with an inversion of I(Q) program. Due to the fast readout time of the CCD, we could follow the aggregation curve of casein micelles in milk. However, we could not invert the casein micelles' scattering curve mainly due to the difficulty in fitting to small sizes ($d < 300$ nm) where the scattering profile is nearly flat in the range of angles probed by SAILS.

Note

Since the above work was done, there was a major design change in the sample holder of SAILS that enabled us to obtain more accurate correction factors that would yield better sizes after an inversion. We are also considering changes in SAILS that would result in an increase of 50% in its Q range. This should enable us to size particles smaller than 300nm in radius. We are also investigating larger food systems, namely aggregation phenomenon in protein emulsions.

Literature Cited

1) T. Craig, T.J. Racey and F.R. Hallett, *The Application of laser Light Scattering to the Study of Biological Motion,* Plenum Publishing Corporation, 1983.

2) Hallett, F.R., Watton, J. And Krygsman, P., *Biophysical Journal*, **59**, 357, (1991).

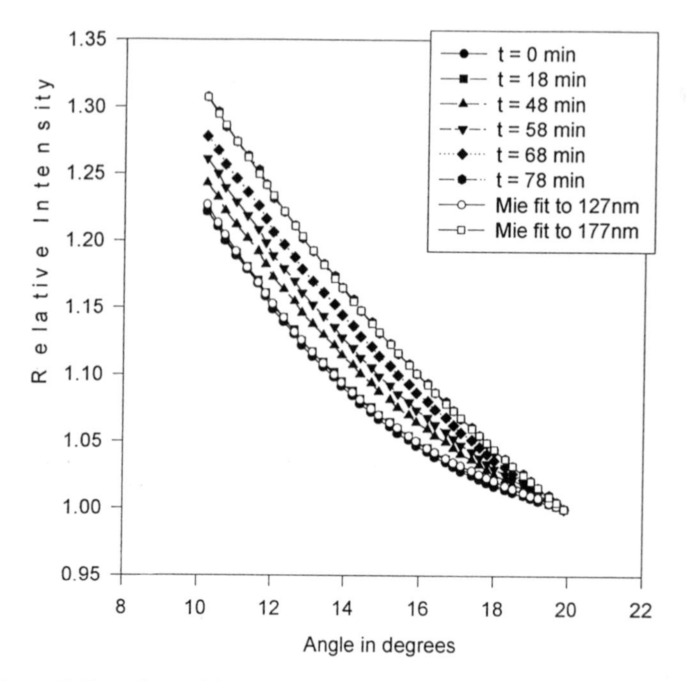

Figure 4. Superimposition of experimental and theoretical generated curves for caseins.

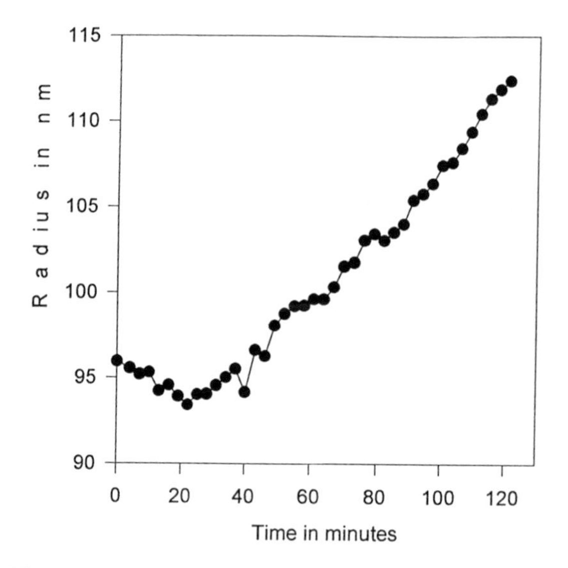

Figure 5. Plot of Radius vs Time of the casein micelles obtained from Equation 5.1 calculated at 20°. A decrease in size can be clearly seen early in the process.

3) Goll, J.H., Stock, G.B., *Biophysical Journal*, **19**, 265, (1977) .
4) Helmstedt, M., *Makromol. Chem.*, Macromol. Symp., **18**, 37, (1988).
5) Berne, B.J. and Pecora, R., *Dynamic Light Scattering*, Wiley Interscience, New York, (1976).
6) Craig, T., Blaber, A. And Hallett, F.R., *Biology of Reproduction*, **29**, 1189, (1983).
7) Hallett, F.R., Nickel, B., Craig, T., *Biopolymers*, **24**, 947, (1985).
8) Tanford, C., *Physical Chemistry of Macromolecules*, John Wiley and Sons, New York, (1961).
9) Kerker, M., *The Scattering of Light and Other Electromagnetic Radiation*, Academic Press, New York, (1969).
10) B. Chu, Z. Zhou, G. Wu and H. Farrell, Jr., *Journal of Colloid and Interface Science* 170, 102, (1995)
11) Van de Hulst, H.C., *Light Scattering by Small Particles*, John Wiley and Sons, New York (1957).
12) Blouke, M.M., *Laser Focus World*, **27**, A17 (1991).
13) de Gennes, P.G., *Scaling Concepts in Polymer Physics*, Cornell University Press, Ithaca, New Yok, (1979).
14) Henderson, S.J., *Biophys. Jour.*, **70**, 1618 (1996).
15) Rathborne, S.J., Haynes, C.A., Blanch, H.W. and Prausnitz, J.M., *Macromolecules,* **23**, 3944, (1990).
16) Meehan, E.J., Gyberg, A.E., *Applied Optics*, **12**, 3, (1973).
17) Kaye, W., *Analytical Chem.*, **45**, 2, (1973).
18) Koch, A.L., *Journal of Microbiological Methods*, **9**, 139 (1989).
19) Holoubek, J., *Journal of Polymer Science*, **32**, 351, (1994).
20) Tant, M.R., Culberson,W.T., *Polymer Engineer and Science*, **33**, 17, (1993).
21) Culberson, W.T., Tant, M.R., *J. of Applied Pol. Science*, **47**, 395, (1993).
22) Cumming, A., Wiltzius, P., Bates, F.S. and Rosedale, J.H., *Physical Review A*, **45**, 2, (1992).
23) Wong, A.P.Y., Wiltzius, P., *Rev. Sci. Instrum.*, **64**, 9, (1993).
24) Kubota, K., Kuwahara, N., *Jpn J. Appl. Phys.*, **31**, 11, (1992).
25) Mulder, S. Walstra, P., *The Milk Fat Globule*, Commonwealth Agricultural Bureau, U.K. (1974).
26) Holt, C., *Adv. Prot. Chem.*, **43**, 63, (1993).
27) Dalgleish, D.G., Hallett, F.R., *Food Research International*, **28**, 3, (1995).
28) Dalgleish, D.G., Horne, D.S., Law, A.J.R., *Biochim. Biophys. Acta*, **991**, 383, (1989).
29) Strawbridge, K.B. and Hallett, F.R., *Macromolecules*, **27**, 2283, (1994).

Chapter 2

Capturing Static Light Scattering Data Using a High-Resolution Charge-Coupled Device Detector

William B. Conklin, James P. Olivier, and Michael L. Strickland

Micromeritics Instrument Corporation, 1 Micromeritics Drive, Norcross, GA 30093

Static light scattering data traditionally have been captured using photodiode ring detectors and linear arrays. These devices impose limits on angular resolution of the data and provide only a one-dimensional view of scattered light intensity versus scattering angle. Use of a charge coupled device (CCD) enables the capture of two-dimensional, high resolution scattering patterns. The fine details in these scattering patterns closely match the theoretical scattering patterns for spherical particles derived from Mie theory. The two-dimensional array of pixels on the CCD is mapped to concentric conic sections corresponding to various scattering angles based on the precise location and angle of the light incident on the sample. Varying the angle of the incident light beam relative to the sample-CCD axis and remapping the pixels allows capture of scattered light over a wide range of angles.

Measuring particle size by static light scattering involves illuminating sample particles with monochromatic, collimated light and measuring the intensity of light scattered by the particles at various angles relative to the direction of the incident light. A particle scatters light in a pattern dependent on the size and shape of the particle, the refractive index of the particle material relative to the surrounding medium, and the wavelength of the incident light. The complete solution for scattering of a monochromatic plane wave of light incident on a homogeneous spherical particle is attributed to Gustav Mie (*1*). Useful presentations of the theory are provided by several authors (*2*)(*3*)(*4*).

Instruments for determining particle size by static light scattering commonly measure the intensity of the forward scattered light by focusing the light by means of a field lens onto a ring detector, which is a series of silicon photodiodes formed as sections of concentric circles (*5*)(*6*). The ring detector is at the focal plane of the field lens and mounted perpendicular to the incident light with the unscattered portion of the light passing through an aperture at the center of the concentric circles. Each photodiode section then captures light scattered through a particular solid angle range relative to the incident light, giving a measure of intensity of that light. The range of angles covered may be extended by placing a linear array of photodiodes at angles larger than those covered by the ring detector. The data from all detectors for intensity versus scattering angle are then compared to Mie theory or to Fraunhofer diffraction theory (*7*) to determine the particle size distribution whose theoretical scattering pattern most closely matches the observed pattern.

Use of ring detectors and linear arrays of photodiodes imposes practical limits on the measurement. The minimum spacing and size of sections of the ring detector limit angular resolution and the number of angles for which intensity data are captured. The outer segments of the ring detector are generally large, sacrificing angular resolution for a larger signal. Precise mechanical alignment of the center of the detector with the unscattered beam is also required.

Description of the Apparatus

In the present work a high resolution charge coupled device (CCD) is used to capture scattered light data. The CCD is a two-dimensional array of sensors or pixels that accumulate electrical charge in proportion to the quantity of light incident on the surface of the pixels. After exposure to light, the charge from each pixel is shifted row by row to the edge of the array, where each row of charges is shifted pixel by pixel into a signal amplifier, and the data are processed serially. The CCD presently used (Kodak model KAF-1300L, Eastman Kodak Company, Rochester, NY) has 1024 rows and 1280 columns for a total of 1,310,720 pixels, each 16 micrometers square.

A schematic of the apparatus is shown in Figure 1. A 30 mW, 687 nm wavelength solid state laser (Toshiba model TOLD 9150, Toshiba Corporation, Tokyo) is used as the light source. A beam splitter directs approximately 50 % of the light onto a photodiode. The output of the photodiode is integrated over the time the laser is on to measure the quantity of light
administered. The remainder of the light is focused onto the end of a 4 micrometer diameter, single mode optical fiber. Prior to the light entering the fiber an optical density 4.4 filter can be moved into the beam or removed, providing a 25000-to-1 reduction in light intensity. The output of the fiber is collimated using an achromatic lens to form a beam with a $1/e^2$ diameter of 11 mm. This beam passes through a glass flow cell containing sample particles dispersed in fluid, usually water. The collimator is mounted on a rotation mechanism whose axis of rotation passes through the cell, so that the collimated light can be presented to the cell at various angles up to 45 degrees from perpendicular to the cell walls. Beyond the cell is a plano-convex lens of 200 mm focal length and the CCD mounted at focal length from the lens. The lens focuses all parallel light emanating from the cell at a particular angle to a single point on the CCD, performing a Fourier transform on the light, so that the CCD receives a two-dimensional pattern of light distributed by scattering angle, with differences in origin within the illuminated volume removed. Scattering angle is defined as the angle between the direction of scattered light and the direction of the unscattered light.

Collection and Processing of Data

The CCD is placed so that the unscattered light is focused on the CCD near one edge. The precise location of the center of this focused beam is determined by locating the point of highest intensity on the CCD. This corresponds to 0 degrees scattering angle. Increasing scattering angles, symmetric about the axis of the incident light, approximate conic sections projected on the plane of the CCD, varying from true conic sections due to refraction of light at the fluid-to-glass and glass-to-air interfaces of the cell window. If the light axis were normal to the CCD and there were no refraction, each scattering angle would project as a circle on the CCD; for other light axis-to-CCD angles each scattering angle would project as an ellipse, a parabola, or a hyperbola depending on the light axis-to-CCD angle and the scattering angle. In practice, the projection of each scattering angle onto the CCD is calculated by ray tracing to account for refraction at the cell window. The surface of best focus of the lens is not a plane, but curves in toward the lens at increasing angles from the optic axis of the lens. The CCD is angled 6 degrees from normal to the lens optic axis to approximate the focal surface in that region, providing better focus overall across the CCD. This angle is also accounted for in projecting scattering angles onto the CCD.

All pixels lying between the projections of two scattering angles are associated, in that their aggregate output represents the quantity of light scattered into the solid angle bounded by the two scattering angles. The data from these pixels are averaged to determine the average quantity of light per unit area scattered in the associated solid angle. Before averaging the output of each pixel is corrected for dark current signal and variation in response to uniform illumination, including variation due to the distance from each pixel to the optical center of the lens. A series of solid scattering angles is chosen in approximately a logarithmic progression with the smallest width solid angles at the smallest scattering angles and increasing width with increasing scattering angle. The smallest solid scattering angles have a span of 0.0025 degrees, or about the width of a single pixel; the largest have a span of 0.1 degrees. The present work uses 465 solid scattering angles from 0.005 degrees to 36 degrees for samples in water.

Only a limited range of scattering angles project onto the CCD when the unscattered light is focused near one edge of the CCD. In the present system with an approximately 200 mm focal length lens and a 20.48 mm length CCD, the widest angle projecting onto the CCD is about 5.5 degrees. When data have been captured for scattering angles from 0 degrees to 5 degrees the light source is rotated so that the new axis of the unscattered light beam is 5 degrees from the previous unscattered light axis. In this configuration, light scattered through angles of approximately 5 degrees to 10 degrees project onto the CCD. Pixels are remapped to new solid scattering angles based on the new projections and data for the new scattering angles are captured. This process continues in 5 degree increments of incident light angle up to 45 degrees. Data captured at each incident light angle are corrected for reflectance and refraction when the light enters the cell, and for increasing extinction due to increasing path length of the light through the cell as the incident light angle increases. Due to refraction of light in the suspending water, data captured at an incident light angle of 45 degrees corresponds to scattering angles of 32 to 36 degrees.

The CCD is an integrating device, accumulating charge in proportion to the total amount of light incident on each pixel during an exposure. This is in contrast to photodiodes, which generate a continuous signal proportional to the intensity of light incident on the photodiode at any instant. Each pixel on the CCD can store up to approximately 150,000 electrons. Any light generating charge per pixel in excess of this cannot be measured. This upper limit and the lower limit on measurable signals due to electronic noise in the signal processing impose a dynamic range of about a factor of 10 on quantities of light that can be accurately measured during a single exposure. Since the scattering patterns to be measured can span 8 orders of magnitude in intensity, it is necessary to take a number of exposures using different quantities of light in order to bring all portions of the measured pattern on scale. On-scale data from each exposure must be ratioed to the quantity of light, or "light dose", incident on the sample during that exposure. The light dose is varied by changing the drive current of the laser, by pulsing the laser for precisely controlled lengths of time, and by inserting or removing the O.D. 4.4 optical filter. Multiple laser pulses per CCD exposure are also used to improve sampling of particles illuminated during a single CCD exposure. The light doses are measured with the photodiode and associated integrating electronics.

Data from the various exposures taken at a single incident light angle are scaled by light dose and combined to produce a single, wide intensity range composite of the scattering pattern at the angles observed. Then data from the various incident light angles are scaled for reflectance and extinction and combined to produce a single, virtually continuous wide angle composite of the sample scattering pattern from 0 degrees to 36 degrees scattering angle. A background pattern is similarly measured with no sample present, scaled to account for extinction, and subtracted from the sample composite. The result is the scattering pattern for the sample expressed as average intensity versus scattering angle.

Results

The apparatus was used to measure the scattering of monosize polystyrene spheres ranging from 0.1 micrometers to 1000 micrometers in diameter dispersed in water. The resulting scattering patterns match well with patterns derived from Mie theory. Figure 2 shows the pattern measured for 10.2 μm polystyrene / 2% divinyl benzene (SDVB) particles, standard deviation 0.092 μm, obtained from Bangs Laboratories, Inc., Carmel, IN. This pattern is compared with the Mie curve for a narrow size distribution of spheres around 10.2 μm diameter with refractive index 1.58. The curve shown is the average of curves for 10.145 μm and 10.292 μm diameter spheres calculated using a modified version of the algorithm published by Bohren and Huffman (*8*). Degree of polarization of the incident light is 0.778. The high resolution of the scattering angle axis allows the details of the first 11 lobes of the scattering pattern to be clearly seen. The angle and intensity of these lobes are very sensitive to both refractive index and particle size, as is illustrated by comparisons with the patterns for 10.0 μm refractive index 1.58 and 10.2 μm refractive index 1.60 in Figure 3.

Figures 4 through 7 show comparisons of measured scattered intensity with Mie theory for particle diameters of 1004 μm, 100 μm, 1.6 μm, and 0.3 μm, respectively. The samples were obtained from Duke Scientific Corporation, Palo Alto, CA. Results are in good agreement with the theoretical models.

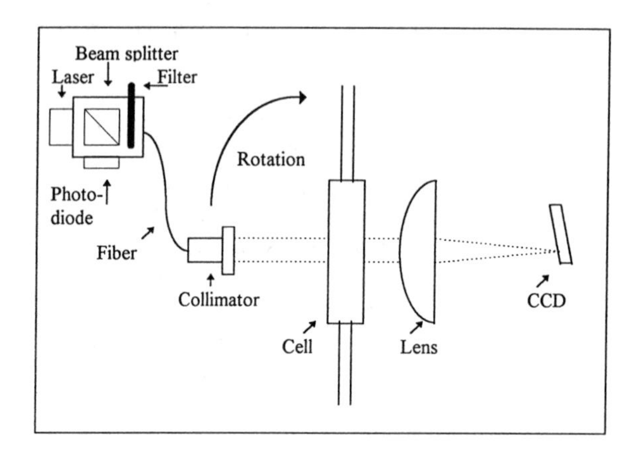

Figure 1. Schematic View of the Apparatus

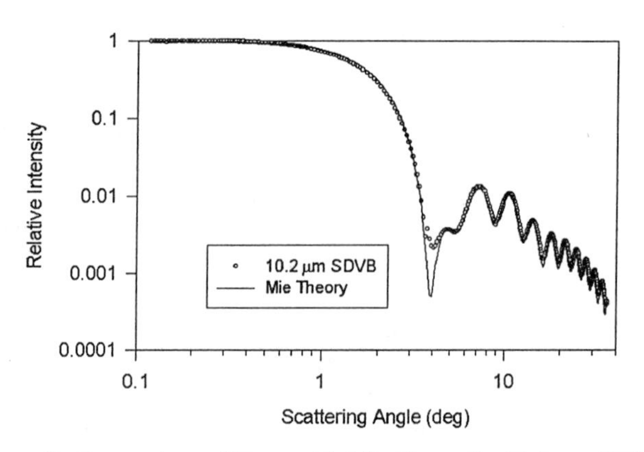

Figure 2. Comparison of Data with Mie Curve for 10.2 μm SDVB

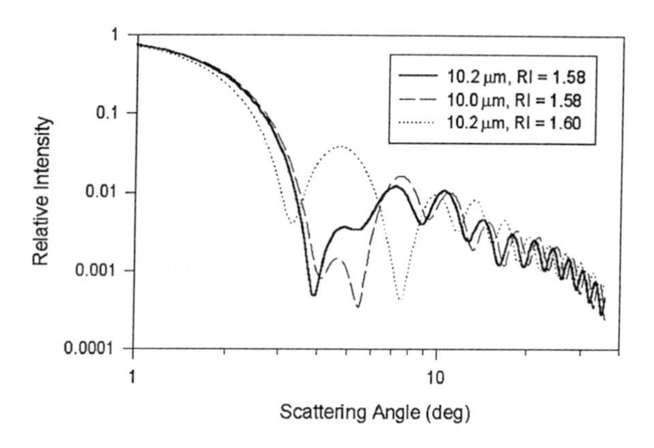

Figure 3. Comparison of Mie Curves for 10.2 µm RI 1.58, 10.0 µm RI 1.58, and 10.2 µm RI 1.60

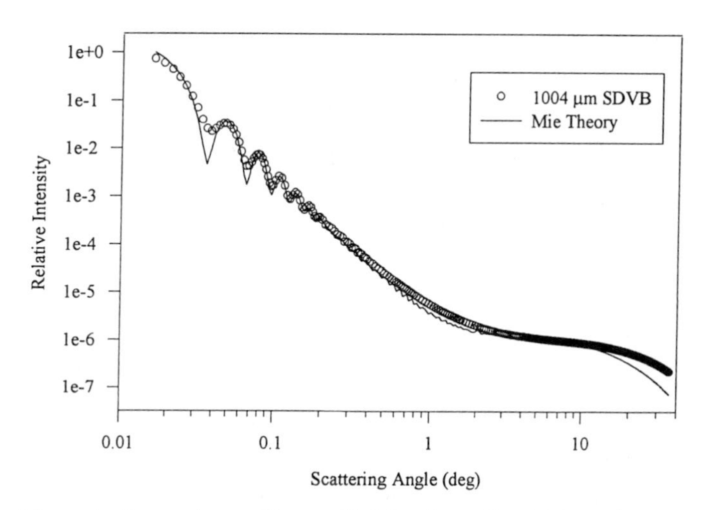

Figure 4. Comparison of Data with Mie Curve for 1004 µm SDVB

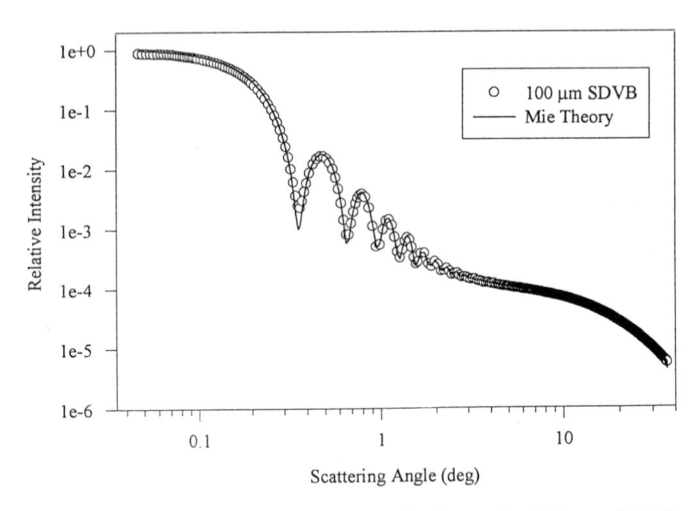

Figure 5. Comparison of Data with Mie Curve for 100 µm SDVB

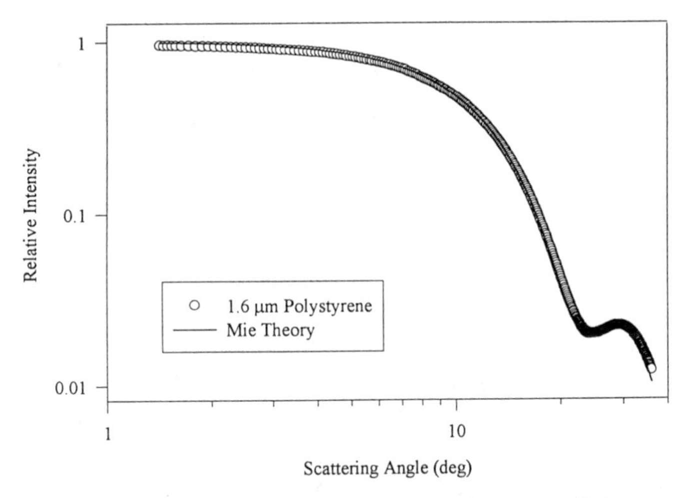

Figure 6. Comparison of Data with Mie Curve for 1.6 µm Polystyrene

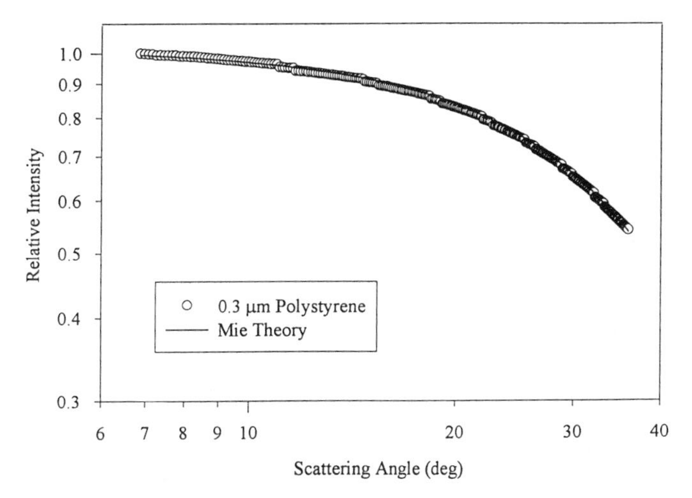

Figure 7. Comparison of Data with Mie Curve for 0.3 μm Polystyrene

Conclusion

The high resolution CCD appears to be well suited to measuring static light scattering patterns, showing fine detail in the patterns that agree closely with Mie theory. The ability to remap the pixels to correspond to solid scattering angles of arbitrary shape projection allows use of a single detector to collect data for a wide range of scattering angles, reducing the cost and complexity of the apparatus. The ability to measure the precise location of the unscattered beam on the CCD eliminates the need for fine mechanical alignment. A ring detector, by comparison, is fabricated with ring sections of fixed size and shape which must be precisely aligned concentric to the axis of the unscattered beam. The CCD has a more limited dynamic range than the silicon photodiodes in a typical ring detector; however use of controlled and measured light doses and the optical filter allow ensemble measurements covering a very wide dynamic range.

The current apparatus is useful for measuring scattering patterns for particle diameters of about 1000 micrometers to about 0.1 micrometer. Larger sizes could be measured by using a longer focal length field lens to provide greater scattering angle resolution. The small end of the size range is limited by a combination of the broadening of the scattering pattern captured at 0 to 36 degrees and the rapidly decreasing scattering power of particles with decreasing size in this regime.

Literature Cited

(1) Mie, G. *Ann. Phys.* **1908**; *vol 25*, pp 377-445.
(2) Van de Hulst, H.C. *Light Scattering by Small Particles*; Dover Publications: New York, NY, 1981; pp 114-130.
(3) Kerker, M. *The Scattering of Light and Other Magnetic Radiation*; Academic Press: San Diego, CA, 1969; pp 39-54.
(4) Bohren, C. F.; Huffman, D. R. *Absorption and Scattering of Light by Small Particles*; John Wiley & Sons: New York, NY, 1983; pp 82-129.
(5) Allen, Terence *Particle Size Measurement*, Fourth Edition; Chapman and Hall: London, UK, 1990, pp 715-718.
(6) Barth, H.G. *Modern Methods of Particle Size Analysis*; John Wiley & Sons: New York, NY, 1984; pp 135-209.
(7) Van de Hulst, H.C. *Light Scattering by Small Particles*; Dover Publications: New York, NY, 1981; pp 103-108.
(8) Bohren, C. F.; Huffman, D. R. *Absorption and Scattering of Light by Small Particles*; John Wiley & Sons: New York, NY, 1983; pp 477-482.

Chapter 3

A Novel Automatic Dilution System for On-Line Particle Size Analysis

P. Sacoto[1], F. Lanza[2], H. Suarez[1], and Luis H. Garcia-Rubio[1,3]

[1]Department of Chemical Engineering, University of South Florida, Tampa, FL 33620
[2]Universidad del Zulia, Maracaibo, Venezuela

On-line methods for the characterization of micron and submicron particle dispersions are of critical importance for the understanding of particulate systems in general, and colloidal systems in particular. Research in this area, has been focused on the development of on-line instrumentation and the interpretation of signals from concentrated particle dispersions (i.e., the multiple scattering problem). This paper reports on the development of on-line particle characterization methodology as an integrated sampling and dilution strategy combined with light scattering methods. A novel dilution system is presented. This dilution system is capable of reducing in few seconds, particle concentrations from 40% solids to concentrations in the range of milligrams/cm^3 where standard light scattering techniques can be applied (i.e., concentrations where single particle scattering approximations are valid). The potential of a combined dilution system with a multiwavelength turbidity detector is demonstrated using as case study the evolution of the particle size distribution in an emulsion polymerization reactor.

On-line methods for the characterization of micron and submicron particle dispersions are of critical importance for the understanding of particulate systems in general and colloidal systems in particular. Research in this area, has been focused on the development of on-line instrumentation and on the interpretation of measurement signals from concentrated particle dispersions. Commercial instrumentation, based on several detection principles, is now available in a variety of configurations for static and flow measurements, for both, laboratory and process applications (1,2). However, the development of new detection systems addresses only one part of the measurement and characterization problem of particle

[3]Corresponding author.

dispersions. Sampling, and the analysis of the fluid structure (particle-particle interactions) are also key elements in the development of any on-line characterization strategy. In fact, the type of sampling strategy determines, to a large extent, the type of detection system to be used. For example, with modern light scattering based technology, concentrated particle dispersions can be interrogated undiluted, directly within the process vessel, or from a slip or sample stream, by optically sampling through a window or through a fiber optic. Under these conditions, measurements at backscattering angles will have to be conducted, and in both instances, the angular position of the source and the detector(s) will determine the sampling volume from which the concentration can be established. The signal from the detector(s) will contain information on the particle size, the particle shape, and the fluid structure arising from the flow field and the colloidal forces particular to the system being analyzed. Clearly, understanding of the detector signal will enable both, the characterization of the particles and the characterization of the process stream in terms of structures like flocs, aggregates, etc. Understanding of the nature of the sampling strategy will then enable the identification of the relationship between the characterization at the detector and the process variables. The combined problem of sampling and measurement from concentrated dispersions is clearly a formidable one for which a general solution is still forthcoming. Nevertheless, for a large number of important industrial process involving stable colloidal dispersions (i.e., emulsion polymerizations), it is possible to sample the process without compromising the integrity of the sample relative to the process. Two approaches can be identified, grab-sample and continuous on-line sampling. The first approach is the most common and consists of capturing a sample, which is then generally taken to the laboratory for analysis with standard particle characterization techniques. Most of the laboratory particle characterization techniques require dilution of the sample to bring it within a concentration range for which the validity of the method can be ensured. The dilution step is particularly important for light scattering and particle counting techniques where single particle measurement conditions should be approached. It is evident that, if standard particle characterization techniques are to be used, the second sampling approach also requires a strategy for continuous dilution. This paper reports on the development of an on-line particle characterization methodology based on an integrated sampling and dilution strategy combined with light scattering methods. A novel dilution system is presented which is capable of reducing in a few seconds, particle concentrations from 40% solids to concentrations in the range of milligrams/cm^3 where standard light scattering techniques can be applied (ie, concentrations where single particle scattering approximations are valid). The potential of a combined dilution system with a multiwavelength turbidity detector is demonstrated using as case study the evolution of the particle size distribution in an emulsion polymerization reactor.

Dilution System

The main problems in the design of integrated on-line sampling dilution and measurement systems stem from, the difficulties in obtaining a representative sample, sample transportation delays, the residence time necessary in the dilution vessel to homogenize sample and diluent in the appropriate proportions, and the time required for the measurement. For stable homogeneous emulsions a representative sample of the process can be readily obtained. The transportation lags can be minimized, and accounted for, by bringing the dilution and detection system as close as possible to the process. However, given the high dilution ratios required for light scattering and other spectroscopy measurements (approx: 1:75000), large volumes of diluents, and therefore comparatively long residence times in the dilution vessels are

$$C_s(n) = \frac{C_s(n-1)\ q_s(n)}{q_s(n) + q_d(n)} \tag{1}$$

required (3-9). Several types of automatic dilution systems have been commercially available for general dilution purposes and new improvements and designs are continuously reported (3,4). Successful dilution systems for process applications have been reported (5-9). However, none of these systems allows for continuous sampling and dilution, and amount to processes in series in which the sampling, transportation, dilution and measurement times are additive. A continuous sampling and dilution scheme with considerable reduction in the dilution time and diluent consumption can be easily achieved by having the dilution steps in parallel (10-13), as shown in Figure 1. The number of parallel dilution steps depend on the desired dilution range needed. The first step consists of continuously delivering a sample from the reactor (C_s (1) g/ml) at a flow rate $q_s(1)$ ml/min . This stream is mixed with a first diluent stream with flow rate q_{d1} ml/min. The combined sample and diluent streams are allowed to mix uniformly in the tube. Uniform mixing can be achieved with the configuration shown in Figure 1 and/or with the use of in-line static mixers. Repetition of the sampling process ($C_s(2)$, $q_s(2)$) process with a second stream ($q_d(2)$) and successive dilution streams will achieved the desired dilution range within a very short time interval. This time interval solely depends on the efficiency of in-line mixing. The concentration at the nth dilution step is given by:

Minimization of the diluent consumption can be easily achieved through manipulation of the flow rates for sampling and dilution .Once the desired concentration level has been achieved, the sample stream leaving the dilution system can be connected to any of the flow through commercially available light scattering detectors.

Preliminary Results

The dilution in parallel system discussed in the previous section has been implemented using a Manostat multicassette peristaltic pump (10-13). To take advantage of the short dilution times achieved with this system, and on the basis of literature reports (14-16), in which it has been demonstrated that it is possible to recover the completed particle size distribution of latex dispersions form multiwavelength turbidity measurements, a Hewlett-Packard 8440 diode array spectrophotometer was selected as a detector. Diode array spectrophotometers are capable of recording the complete Uv-vis spectra in a fraction of a second and therefore enable adequate averaging and conditioning of the signal prior to interpretation. The complete reactor configuration including the sampling and dilution system is shown in Figure 2. Typical on-line continuous measurements of a polystyrene emulsion polymerization reactions are shown in Figures 3-4. Notice the presence of styrene monomer droplets at approximately 280 nm in Fig 3. Similarly, notice in Figures 3 and 4 the transformation of the spectra, from a the spectrum characteristic of a styrene monomer emulsion to that typical of a polymer latex.

Conclusions And Future Work

Preliminary results with the continuous sampling and dilution system demonstrate that the dilution ratios required for the use of standard flow-through light scattering techniques can be achieved. Furthermore, in every instance it was possible to maintain the concentration within the linear range of the spectrometer. The dilution times measured are in the order of seconds, thus suggesting that this approach is suitable for the monitoring of a large variety of processes containing particles. The relative magnitude of time constants for the spectrophotometer measurements and the dilution system, to the time constants typical of emulsion polymerization processes, strongly suggest that the configuration proposed herein is suitable for the monitoring and control of emulsion polymerization reactors. The complete characterization of the dilute reactor contents as functions of the reaction time will be reported separately.

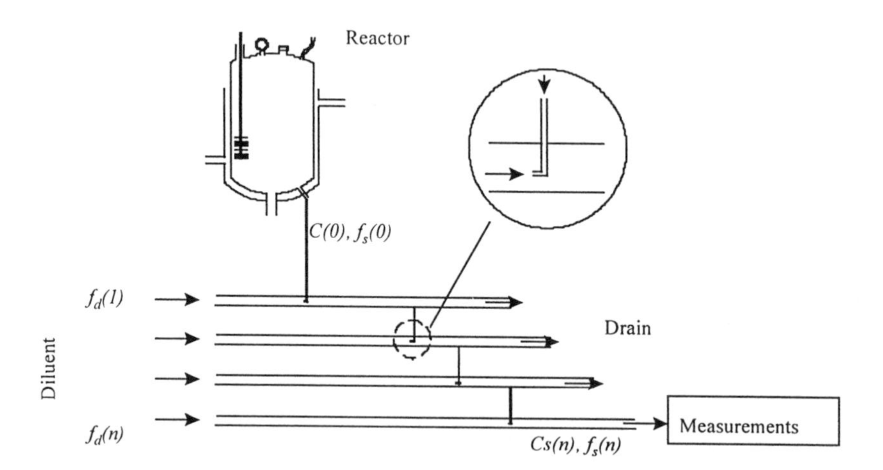

Figure 1. Parallel sampling and dilution configuration

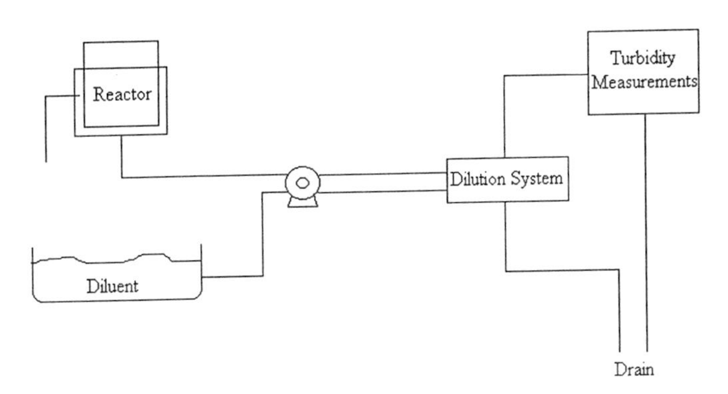

Figure 2. Schematic of the emulsion polymerization reactor layout.

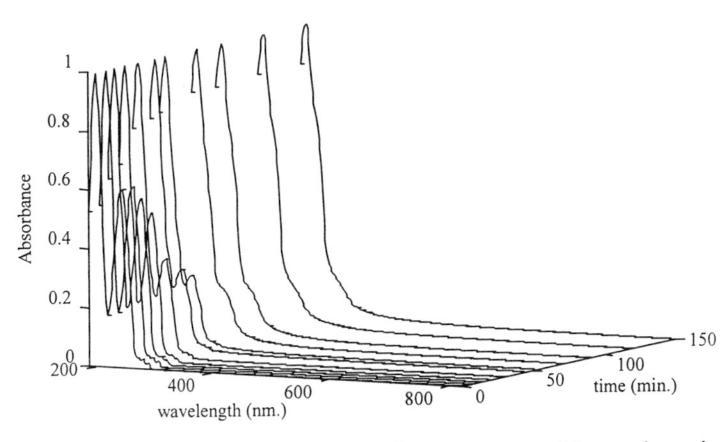

Figure 3. Real time turbidity spectra of a styrene emulsion polymerization reaction

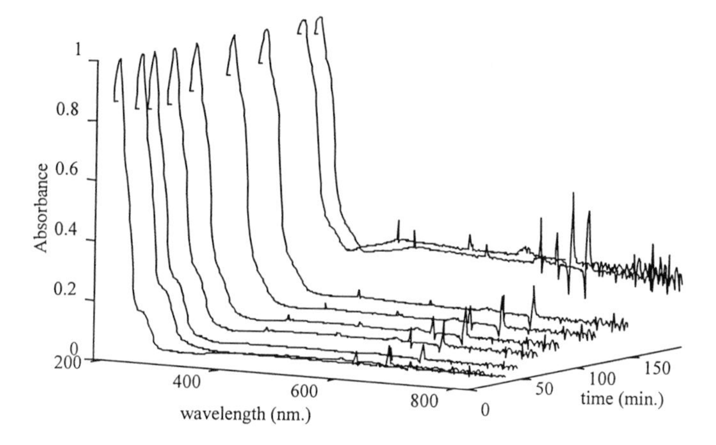

Figure 4. Real time turbidity spectra of a second styrene emulsion polymerization reaction.

Acknowledgments

This work has been supported in part by the University of South Florida Center for Ocean Technology (ONR Grant No N00014-94-1-871), the American Water Works Research Foundation Grant No196-94, BTG Inc., and the Engineering Research Center (ERC) for Particle Science and Technology at the University of Florida #ERC-94-02989.

Literature Cited

1. Kaye H. K. *Chem Eng,* **1995.**
2. Kerker, M. *The Scattering of Light and Other Electromagnetic Radiation*; Academic Press: New York, NY,1969;
3. Hashizume Y., Kariyone, A. and Ryuzo, H. U. S. Patent 5221521, 1993;
4. White, L. B., U. S. Patent 5297431, 1994;
5. Ponnuswamy, S., Shah, S. L. and Kiparissides, C. *J Pol Sci* **1986**, *32*, pp
6. Nicoli, D. F. and Elings, V. B. U. S. Patent 4794806, (1989)
7. Kourti, T. and MacGregor, J. F. *ACS Symposium Series*; ACS: **1991**, *vol.* 472.
8. Brandolin, A. and Garcia-Rubio, L. H. *ACS Symposium Series*, **1991**, *chapter* 4, pp. 472.
9. Leiza, J. R., de la Cal, J. C., Montes, M. and Asua, J. M., *Process Control and Quality*, **1992,** *vol.* 4, pp. 197-210.
10. Sacoto, P., Lanza, Fco., Suarez, H. and Garcia-Rubio L. H., "Automatic Dilution System for the Characterization of Suspensions", Internal Report, University of South Florida, Tampa, November, 1995.
11. Lanza, Fco., Undergraduate Research Progress Report, University of South Florida, submitted to the ERC on Particle Science and Technology, University of Florida, Tampa, December 4th, 1995.
12. Sacoto, P., Undergraduate Research Progress Report, University of South Florida, submitted to the ERC on Particle Science and Technology, University of Florida, April 10th, 1996.
13. Elicabe, G. and Garcia-Rubio, L. H., *J. Coll. and Interface Sci*, **1988,** ***vol.***129, pp 192.
14. Elicabe, G. and Garcia-Rubio, L. H., *Adv. Chem. Series*, **1990**, *chapter* 6, pp. 227.
15. Brandolin, A., Garcia-Rubio, L. H., Provder, T., Kohler, M. and Kuo, C., *ACS Symposium Series*, **1991,** *chapter* 2, pp. 472.
16. Chang S., Koumarioti Y. and Garcia-Rubio L. H., *J. Coll. and Interface Sci.*, **1996**, Submitted for Publication.

Chapter 4

Multiangle–Multiwavelength Detection for Particle Characterization

C. Bacon and Luis H. Garcia-Rubio[1]

Department of Chemical Engineering, University of South Florida, Tampa, FL 33620

Recent developments in the spectroscopy analysis of particle dispersions have demonstrated that complementary information on the joint particle property distribution is available from angular measurements of the combined absorption and scattering spectra. In this paper, new instrumentation capable of simultaneous absorption and scattering measurements at several angles and wavelengths is presented. Experimental results with a recently constructed multiangle-multiwavelength detection system demonstrate that this technology can be used for the characterization of the joint property distribution (size-shape-chemical composition) of dilute and concentrated dispersions, and for on-line particle characterization applications. Other potential uses for the multiangle-multiwavelength technology are theoretically explored using Mie and Rayleigh-Debye-Gans scattering models.

The use of light scattering methods for the determination of the particle size distribution of suspensions of micron and submicron particles is well established (1-5). Commercial instrumentation is now available in a variety of configurations for static and flow measurements for both laboratory and process applications. In recent years, there has been an emphasis on the development of on-line methods for the characterization of concentrated particle dispersions. The focus of the characterization efforts has been on the determination of the particle size distribution, although it is known that the scattering and flow properties of micron and submicron particle dispersions are functions not only of the size distribution, but of the joint particle property distribution (size-shape-chemical composition-charge) (6-8). Until now, elements of the joint property distribution have been measured separately under the assumption that the measurements respond to a single property. Recent developments in spectroscopy analysis of particle dispersions have demonstrated that

[1]Corresponding author.

complementary information on the joint particle property distribution is available from angular measurements of the combined absorption and scattering spectra (7-15). In this paper, new instrumentation capable of simultaneous absorption and scattering measurements at several angles and wavelengths is presented. Experimental results with a recently constructed multiangle-multiwavelength detection system demonstrate that this technology can be used for the characterization of the joint property distribution (size-shape-chemical composition) of dilute and concentrated dispersions, and for on-line particle characterization applications. Other potential uses for the multiangle-multiwavelength technology are theoretically explored using scattering models.

Theoretical Analysis

The information content, in terms of molecular parameters, of the UV/vis spectra of particles, recorded as functions of the angle of observation, can be readily investigated using as a basis the models derived from the Rayleigh-Gans-Debye theory (RGD). For this purpose, the discussion has been divided into two sections addressing the scattering dominated regime and simultaneous absorption and scattering.

Non-Absorbing Particles. For the conditions where there is negligible absorption and the Rayleigh-Debye-Gans approximations are valid (dilute dispersions, $d/\lambda << 1$ and $n_1/n_0 \cong 1$), the angular dependence of the scattered intensity is given by (1, 3),

$$\left.\frac{i_s}{I_0}\right)_\theta = \frac{2\pi^2 \left[n\,(dn/dc) \right]^2 C}{N_A\ R^2\ \lambda^4 \left(1/M + 2BC \right)} (1 + \cos^2\theta) P(\theta) \tag{1}$$

where

$$P(\theta) = \left(\frac{1}{N}\right)^2 \sum_i \sum_j \sin(\mu\, r_{ij}) / (\mu\, r_{ij}) \tag{2}$$

and $\mu=(4\pi/\lambda)\sin(\theta/2)$. Equation 1 is the fundamental relationship for the scattering of unpolarized light by non-absorbing monodisperse particles and macromolecules in suspension. In equation 1, the refractive index of the suspending medium and the refractive index of the particles are inversely proportional to the wavelength (3). The contrast for scattering measurements is given by the refractive index ratio n_1/n_0 (which is equivalent to n). This ratio generally increases the contrast, and therefore the sensitivity of light scattering measurements, as the wavelength is decreased. Furthermore, at a given angle, θ, the ratio $(1 + \cos^2\theta)/\lambda^4$ will also increase with decreasing wavelength. Thus, suggesting that measurements at several angles and wavelengths provide not only enhanced sensitivity but also desirable redundancy in the data for improved statistics (8).

Notice from equation 1 that, even at small angles, by recording the spectrum as function of the wavelength improved resolution for the particle size distribution can be obtained. Furthermore, if it is recognized that the specific refractive index increments for complex particles and macromolecules (ie. cells, coated particles, copolymers, proteins, etc.), can be approximated as a weighted sum of the refractive index increments of the moities present (3), and that the refractive indexes themselves are functions of the wavelength, then, it is clear that the conditions required for the characterization of the joint particle property distribution (3) will be met by recording the UV/vis spectrum as function of the viewing angle. This of course is better accomplished if absorption is present.

Equation 2 represents the form factor equation developed by Debye to account for the shape of the particles. There are many shape functions derived for different shapes (spheres, rods, random coils, cubes, etc.) as a function of angle and wavelength. Using these shape functions in conjunction with the Rayleigh scattering equation will allow for the extraction of shape from the angular measurements over a range of wavelengths. These equations show that it is essential to measure the scattering and absorption intensities over a range of wavelengths and angles in order to analyze the complete joint property distribution (shape-chemical composition-particle size distribution).

Absorbing Particles. For particles of arbitrary shape in the Rayleigh-Gans-Debye regime it has been shown that (1-8),

$$\left.\frac{i_s}{I_0}\right)_\theta = \left(\frac{9\pi^2 V_p}{2\lambda^4 R^2}\right)\left|\frac{m^2 - 1}{m^2 + 2}\right|^2 C\,(1+\cos^2\theta)P(\theta) \qquad (3)$$

where

$$m = \frac{n(\lambda) + i\,k(\lambda)}{n_0(\lambda)} \qquad (4)$$

and V_p represents the volume of the particle.

Equations 3 and 4 indicate that, under the assumption of additivity of chromphore absorption (13-15), the combination of angular measurements at several wavelengths may allow for the estimation of the chemical composition, size and shape of particles and macromolecules. The details of the calculations are given in reference (8). The instrumentation required to accomplish the measurements is described below.

Hardware Design and Development

The multiangle-multiwavelength spectrophotometer was designed and constructed utilizing Ocean Optics Inc. miniature fiber optic spectrophotometers. These miniaturized spectrophotometers measure a specified range of wavelengths simultaneously. Four spectrophotometer cards, allowing simultaneous measurements at four angles, were placed in a pentium computer for data acquisition and analysis. UV/vis transparent fibers are connected from the spectrophotometer cards to collimating lenses which are attached to the optical board (see Figure 1). These lenses provide a parallel beam of light to illuminate the center of the cylindrical scattering cell. The four detectors (lens/fiber combination), or five including the incident source/backscattering detector, are currently free standing to allow ease of movement, several angles of observation and the possibility of superimposing a field (flow, electrical, magnetic fields) to the sample. The fiber optics/lenses configurations have been optimally placed relative to the scattering cell to operate within the linear range of the detectors and to increase the sensitivity for the detection of the scattered light. The incident light source is comprised of both visible and UV sources to allow a complete wavelength range (190 to 900 nm) for the angular measurements.

Improvements have been made to the basic system as described above. An option for focusing the incident light has been added for increased sensitivity in the scattering system. The collection lenses will be focusing lenses which will essentially measure the image of the scattering volume in the cell. This will be especially important for dilute dispersions and low scattering samples. Immediately prior to the detector's collection lens is an adjustable iris which controls the acceptance angle of the detectors. Note the additional angular measurement, pure backscattering. This was accomplished using a bifurcated fiber which allows for the projection of incident light as well as the measurement of 180° backscattered light. This additional measurement will allow for the scattering measurements of concentrated dispersions which is important to industry. One of the additional problems that has been eliminated for certain angular measurements is the refractive index complications due to the curvature of the cylindrical cell. To eliminate this a octagonal quartz cuvette was designed where the corresponding angular measurements can be made perpendicular to the face of the cell.

Experimental Studies

The initial set of experiments for the multiangle-multiwavelength detection system has been designed to develop the protocols for calibrating the instrument and ensure reproducibility. For this purpose, well characterized polymer latexes have been used. The validating instruments used are the Hewlett Packard UV/vis spectrophotometer, Wyatt Techonology's goniometer system (now out of distribution), as well as their DAWN multiangle laser light scattering system. The standard deviation in all of the measurements and their replicates was less than 2.5%. In other words, the instrument provides reproducible results.

Figure 2 represents the multiangle-multiwavelength spectrophotometer's measurement of a 50 nm 3% polystyrene dispersion and a 500 nm 3% polystyrene dispersion. The intensity ratio has been normalized to the maximum intensity of the 60 degree measurement; thus, the magnitudes of the spectra are relevant. The patterns of the spectra are unique for polystyrene and will help in the characterization of particles. As would be predicted the magnitudes of the spectra for the differently sized particles are different. This measurement indicates that the particle size as well as the chemical composition may be obtained from the multiangle-multiwavelength measurements. It is already known that multiangle laser light measurements will lead to decent approximations of the particle size distribution.

Figure 3 is the transmission spectra from a UV/vis Hewlett Packard spectrophotometer of a dilute 1 micron polystyrene dispersion. The next figure, figure 4, is the multiangle-multiwavelength spectrophotometer's response to the same dispersion (at a slightly higher concentration). The wavelength range plotted is small due to the type of lamp that was used, a white light source that was very strong in the region of 400 to 800 nm. The new multiangle-multiwavelength configuration (focusing lenses and an octagonal cuvette) was used for this sample. This configuration also proved reproducible. The pure backscattering angle was also measured for this sample.

Another important aspect of the multiangle-multiwavelength spectrophotometer will be the ability to measure concentrated dispersions. Utilizing the backscattering angles, as well as pure backscattering (180 degrees), concentrated dispersions can be characterized in terms of their particle size. This can currently be done empirically, but with the aid of reflectance models, it can also be done theoretically and quantitatively. Figure 5 shows poly(methylmethacrylate) samples for 5 different size distributions. The dispersions were very concentrated at 30% solids. The size distributions were also very narrow. Again the patterns of the spectra are distinctive but there are differences in the slopes of the scattering spectra in the higher wavelengths. These type of patterns indicate that chemometric methods may also be used to analyze the multiangle-multiwavelength data.

Conclusions and Future Work

Simulation studies in the Rayleigh-Gans-Debye scattering regime for differently shaped particles have demonstrated that each particle shape its own characteristic scattering pattern as a function of the angle of observation and the incident wavelength of light. This strongly suggests that particle morphology can be extracted from multiangle-multiwavelength spectrophotometer measurements. In response to this information, a UV/vis multiangle-multiwavelength spectrophotometer system has been designed and constructed utilizing state-of-the-art miniature fiber optic spectrophotometer technology. The new instrument has been tested using dilute and concentrated dispersions of well characterized polymer

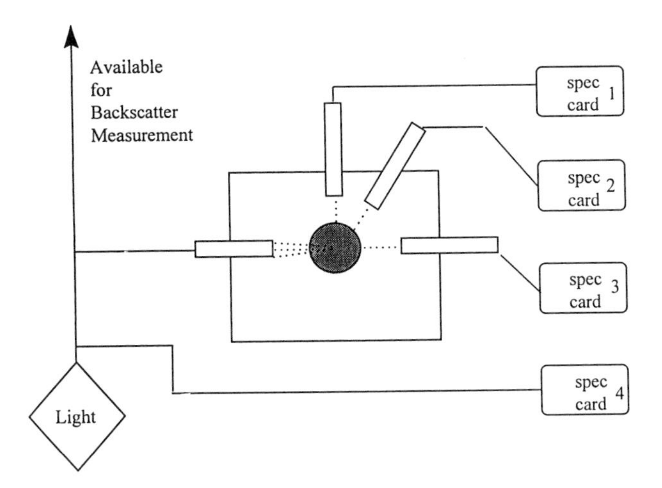

Figure 1. Schematic of the multiangle-multiwavelength detection system for particle characterization.

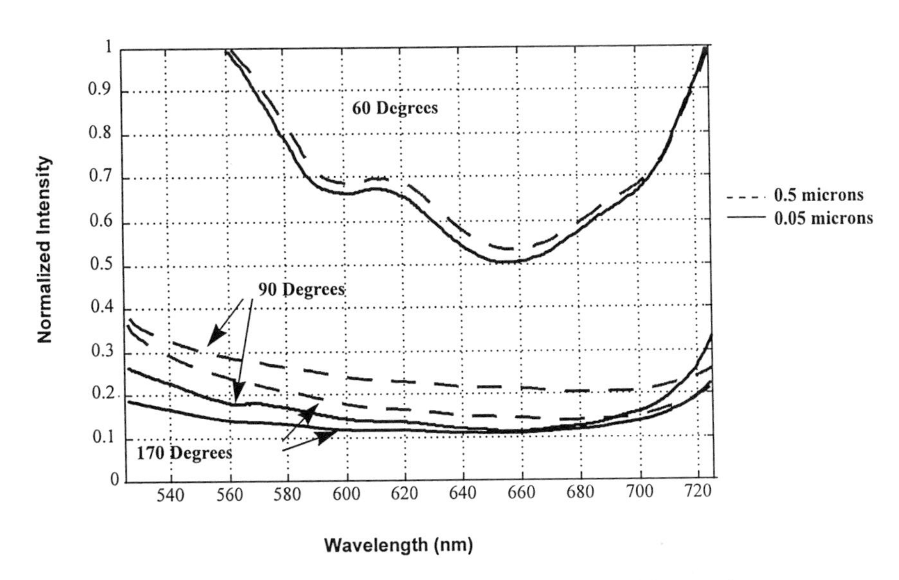

Figure 2. Angular scattering of 3% solids 50 nm and 500 nm polystyrene dispersions.

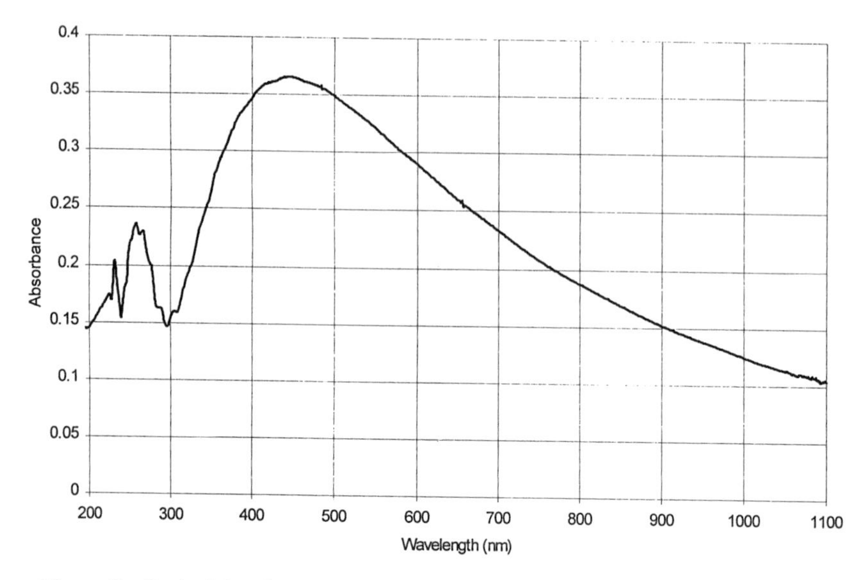

Figure 3. Optical density measurement of dilute 1 micron polystyrene standard.

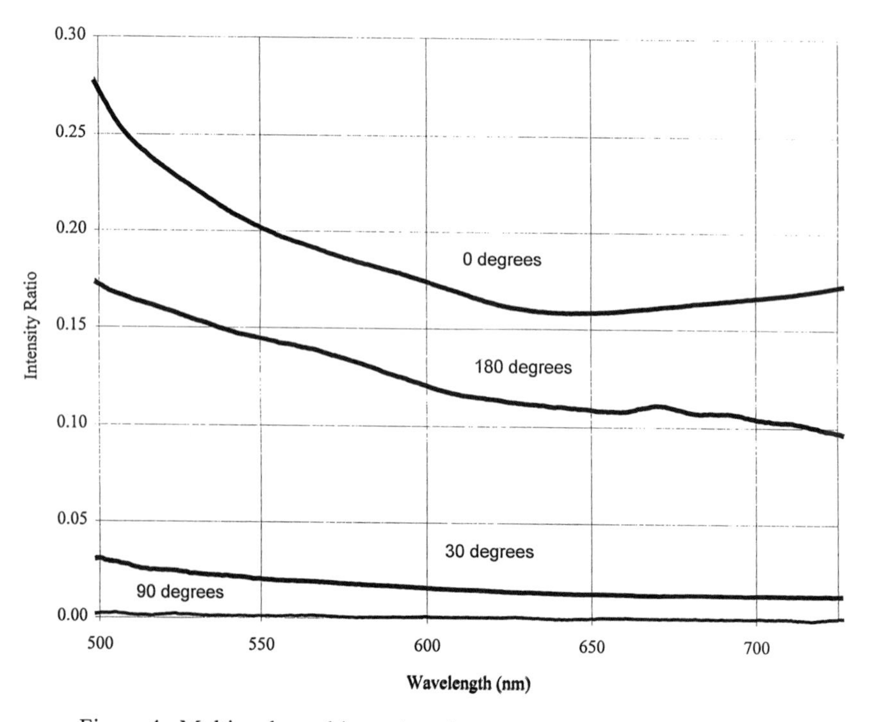

Figure 4. Multiangle-multiwavelength measurement of dilute 1 micron polystyrene standard.

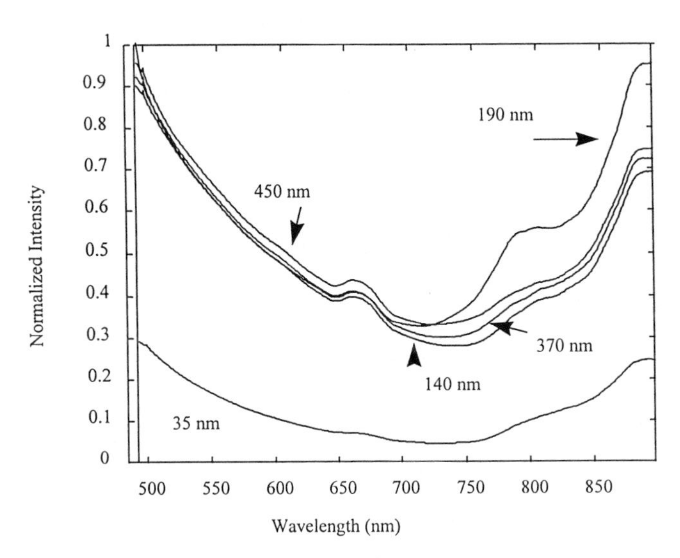

Figure 5. Multiangle-multiwavelength 170^0 backscattering for 30% poly(methyl methacrylate) standards ranging in size between 50-500 nm.

lattices. The results indicate that the multiangle-multiwavelength measurements are sensitive to size, concentration, and chemical composition.

Currently, the multiangle-multiwavelength spectrophotometer is being modified for easier calibration and added sensitivity to scattered light. Once this is accomplished, sample standards of various morphologies will be tested in the instrument. The scattering measurements will be compared with theoretical results obtained from the Rayleigh-Gans-Debye equations and the inversion of the joint particle size-chemical composition-shape distribution will be attempted.

Acknowledgments

This work has been supported in part by the University of South Florida Center for Ocean Technology (ONR Grant No N00014-94-1-871), and the Engineering Research Center (ERC) for Particle Science and Technology at the University of Florida #ERC-94-02989.

Literature Cited

1. van de Hulst, H. C. *Light Scattering by Small Particles*; Wiley: New York, 1957;
2. Kerker, M. *The Scattering of Light and Other Electromagnetic Radiation*; Academic Press: New York, 1969;
3. *Light Scattering from Polymer Solutions*; Huglin, M. B., Editor; Academic Press Inc: London, 1972;
4. Schmitz, K. S. *An Introduction to Dynamic Light Scattering by Macromolecules*; Academic Press: San Diego, 1990;
5. Wyatt, P. H.; *Appl. Opt*. **1980**, *7*, pp. 1879
6. Bohren, C.F. and S.B. Singham, *J. Geoph. Res.* **1991**, *96*, pp 5269
7. Al-Chlabi, S.A.M. and A.R. Jones, *J. Phy.s D: Appl. Phys.* **1995**, *28*, pp 1304
8. Bacon C. P., Honors Thesis, University of South Florida, Tampa, FL, 33620, 1994;
9. Elicabe, G. and Garcia-Rubio, L. H., *J. Coll. and Interface Sci.* **1988**, *129*, pp 192
10. Elicabe, G. and Garcia-Rubio, L. H., *Adv. Chem. Series*; Chapter 6, 1990; Vol. 227.
11. Brandolin, A., Garcia-Rubio, L. H., Provder, T., Kohler, M. and Kuo, C., *ACS Symposium Series*; Chapter 2, 1991; Vol. 472.
12. Chang S., Koumarioti Y. and Garcia-Rubio L. H., *J. Coll. and Interface Sci.* **1996**, Submitted for Publication.
13. Garcia-Rubio, L.H. and N. Ro, *Can. J. Chem.* **1985**, *63*, pp 253
14. Garcia-Rubio, L.H, *Macromolecules* **1987**, *20*, pp 3070
15. Garcia-Rubio, L.H., *Chem. Eng. Comm.* **1989**, *80*, pp 193

Chapter 5

Analysis of Macromolecules Using Low- and Right-Angle Laser Light Scattering and Photon Correlation Spectroscopy

Norman Ford, Trevor Havard, and Peter Wallace

Precision Detectors, 10 Forge Park, Franklin, MA 02038

The objective of this paper is to introduce a new concept for a light scattering detector design which will combine the measurement of the diffusion constant of macromolecules and particles in a flowing stream with the measurement of the light scattering intensity at two angles.
This new detector design will therefore provide the hydrodynamic radius of macromolecules eluting from any separation device and provide size information of a particle or molecular weight distribution from a chromatographic separation without any prior knowledge of sample concentration or chemical composition. This measurement operates using the principle of Dynamic Light Scattering (DLS) (*1,2*) also known as Quasi Elastic Light Scattering (QELS) and Photon Correlation Spectroscopy (PCS).
The instrument is designed in such a way that several other measurements can also be made simultaneously in the same light scattering cell. These measurements include a 90° static (Raleigh) light scattering measurement as well as a 15° static (Raleigh) light scattering measurement. There is also a provision to make a 90° static depolarized light measurement in the future. The design concept of this device is detailed by Norman Ford in 1994 (*3*). This light scattering detector optics bench is commercially known as the PD2000.

Light Scattering Apparatus.

The typical laser light source to be used by the detector is provided by a laser diode of 100 mW of power at a wavelength of 790 nm. The optical bench is an axial design (*4,5,6*) using the solid state laser which is focused into the flow cell using a mushroom shaped condensing lens (Fig 1,2). The flow cell is 10 µl in volume with an optical volume of less than 0.01 µl for the Raleigh static type detectors.
The intention of using small optical volumes provides two major benefits: the ability to discriminate against stray particles in a flowing system by reducing the probability of detection, the duration of the particles in the detected area.
A small-polarized beam is focused into a very small optical volume providing a unique advantage in that the need for re-calibration is reduced even when changing solvents (*4,6*).

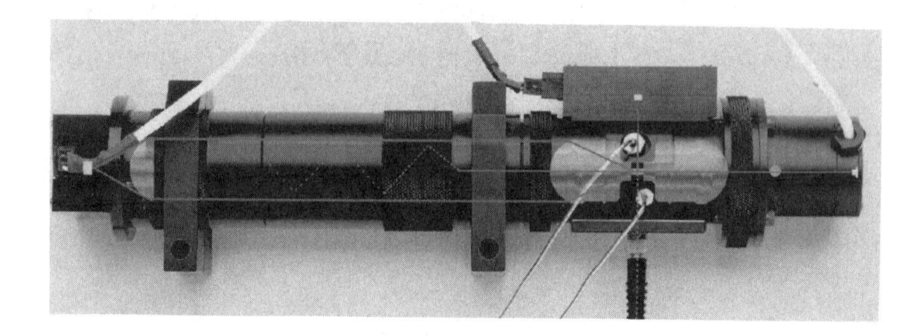

Fig. 1 Picture of the PD2020-DLS Light Scattering System with Photon Correlation Spectroscopy and Static Light Scattering.

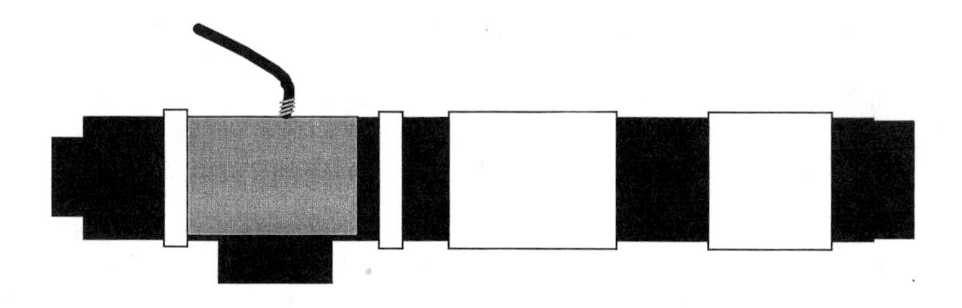

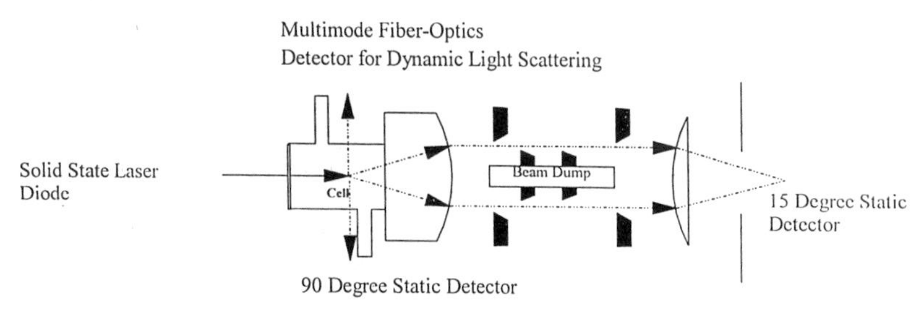

Fig. 2 Schematic of PD2020/PCS Design for the in Flow Measurement of Macromolecules and Particles using Static and Dynamic Light Scattering.

Experimental System.

The system comprised of a Waters 590 pumping system operating at flow rates between 0.1 and 1 ml/min; a Waters 712 auto-sampler operating between 50 and 100 μl injection volumes; Tetrahydrofuran (THF) pre-filtered through a 0.22 membrane filter; a Jordi Associates mixed bed column (10 mm. ID x 250 mm.) and a Waters 410 refractometer.

Static Light Scattering Theory (Raleigh).

The use of low angle and multi-angle light scattering detectors for the characterization of macromolecules to obtain molecular weights and the root mean square radius or radius of gyration using chromatographic separations have been cited in the literature since the 1970's. The mathematics that describes the relationship between molecular weight and scattered light has been well-established (*7.8.9*). The relationship using polarized single wavelength laser light in the PD2000 to determine molecular weight is described in equations (1) and (2). Light scattering intensities of the 90° and 15° angle produce the following equations:

$$Ls_{(90°)} = K_{(90°)} M_W \, c \, (dn/dc)^2 \, P\theta_{(90°)} \qquad \text{Equation (1)}$$

$$Ls_{(15°)} = K_{(15°)} M_W \, c \, n \, (dn/dc)^2 \, P\theta_{(15°)} \qquad \text{Equation (2)}$$

where:

Ls = the excess Raleigh Light Scattering Signal.
K = the optical constant for the detector.
(dn/dc) = the change in refractive index as the concentration changes.
Pθ = the ratio of scattered intensities at angle θ to that of 0°.
c = concentration.
n = solvent refractive index.

Using the combination of a Refractometer and PD2000

In the case of the 90° detector the intensity of light is independent of the refractive index of the solution. This is an advantage when using the PD2000/RI detector design, because it provides an opportunity to change from one solvent system to another, while maintaining constant detector parameters. The incorporation of the light scattering detector into a temperature-controlled oven with a refractometer improves the accuracy and precision of the measurements in the following ways:-

1. Minimized and constant inter-detector volume.
2. Sensitivity and stability in a temperature controlled oven.
3. Portability of the PD2000 detector combination across SEC/FFF/HPLC/Ion Exchange systems without total re-calibration.

In order to successfully use the light scattering detector, the refractometer is used to calculate dn/dc and concentration. The equation that describes the use of a refractometer for dn/dc and concentration slice calculations is as follows:

$$RI_{(sig)} = K_{(RI)} \, c \, (dn/dc) \qquad \text{Equation (3)}$$

By dividing equation (1) by (3), a new relationship for the dual detector can be derived which enables the detector with the accompanying software algorithms to become a true absolute detector independent of the SEC system.

$$\frac{Ls_{(90°)}}{RI_{(sig)}} = \frac{K_{(90°)} M_w \,(dn/dc)\, P\theta_{(90°)}}{K_{(RI)}} \qquad \text{Equation (4)}$$

A single well characterized standard with a known dn/dc can be used to calibrate the optical constant $K_{(90°)} / K_{(RI)}$.

$$K(High) = \frac{K_{(90°)}}{K(RI)} \qquad \text{Equation (5)}$$

Macromolecules with molecular weights below 12 nm. produce little angular dissymmetry between the 15° and 90° detectors. This can be seen in fig (3) for angular measurements predicted by the Debye function for random coils.
In the case where there is large angular dissymmetry at 50 nm. fig (4) it becomes obvious for the need for measurements close to the 0° angle for accurate measurements.

Any monodisperse standard can be used to calibrate the 90° detector .

$$\frac{Ls_{(15°)}}{RI_{(sig)}} = \frac{K_{(15°)} M_w \, n \,(dn/dc) P\theta_{(15°)}}{K_{(RI)}} \qquad \text{Equation (6)}$$

A similar equation can be derived for the 15° detector that includes the refractive index (n) of the solvent. Therefore, we now have three constants and the inter-detector volume to consider to fully calibrate the PD2000/RI detector system :

The polydispersity of low molecular weight standards between 100,000 and 10,000 routinely approach 1.04 +/- .01. These values are determined by light scattering and sedimentation ultra centrifugation for the molecular weight averages and by vapor pressure osmometry, membrane osmometry and end group analysis to determine the molecular number average. There is always experimental error between the two physical and colligative absolute molecular weight measurements. This error usually determines the second decimal place in polydispersity for well - characterized standards supplied from the manufacturer.

Determination of Rg using two angles.

Large molecules scatter less light at high values of (θ) than at low angles because of interference effects caused by the fact that light scattered from one part of the molecule

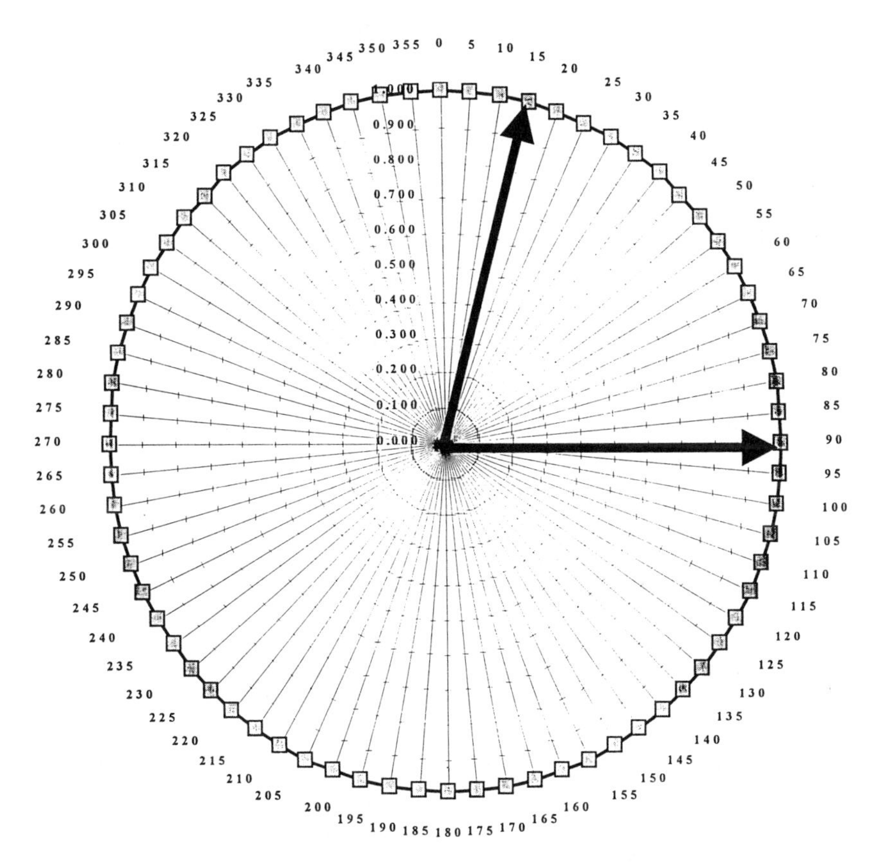

Fig. 3 There is a Minimal Effect on Angular Dissymmetry at an Rg less than 10 nm (wavelength = 790 nm.)

travels a different distance from and therefore is not exactly in phase with light scattered from another part of the molecule. This phenomenon is quantified by defining the light scattering form-factor,

$P(\theta)$ = scattered intensity at angle θ / scattered intensity at angle 0°

Calculations show that P (θ) can be written as a series

$$P(\theta) = 1 - 1/3(q^2 Rg^2) + \dots\dots\dots \qquad \text{Equation (7)}$$

where:

$q = 4\, n \sin(\theta/2) / \lambda_0$

Rg is the radius of gyration of the molecules
n is the index of refraction of the fluid.
λ_0 is the wavelength of light in a vacuum.

Using measurements taken at scattering angles of θ = 15° and 90°
So from equation 7.

$$P(\theta) = 1 - 26.3(Rg.n/\lambda_0)^2 \qquad (\text{for } \theta = 90°)$$

$$P(\theta) = 1 - 0.897(Rg.n/\lambda_0)^2 \qquad (\text{for } \theta = 15°)$$

It is now possible to solve $P(\theta)$ at either angle and derive the value Rg. The current calculations are carried out using the Debye function for gaussian coils and all errors and a full explanation of the calculations can be found in reference (*3*).

Light Scattering Theory (Photon Correlation Spectroscopy PCS).
Molecules undergoing Brownian Motion cause intensity fluctuations in the scattered light intensity due to constructive and destructive interference. This fluctuation in scattered light has a time scale that is related to the diffusion speed of the movement of the molecules, and hence, their size.
The residence time, is the time that the particles are illuminated by the laser. At 1 ml/min we have determined the residence time to be approximately 10,000 microseconds. It takes about 20 to 100 microseconds to obtain correlation's due to particle diffusion effects, and using an equivalent autocorrelator of 1024 channels it will take 2000 microseconds to obtain a correlation and at least 5 correlation's will be obtained during the residence time of the particle. Over a 1-second period 500 correlations will be obtained. As we increase the number of correlations obtained the auto-correlation function's precision and accuracy will become acceptable in order to report a hydrodynamic radius.

The motion of molecules in solution, governed by Brownian mechanics, is directly related to the translational diffusion coefficient of the molecules. When monochromatic, polarized laser light illuminates a solute the intensity of the light scattered from the molecules fluctuates. The time required for these fluctuations to transpire conveys information about the dynamic properties of the molecules. PCS is a technique, which calls upon the relationship between the time dependence of these intensity fluctuations and the translational diffusion coefficient of the scatterers. The hydrodynamic radii of the solute molecules can be calculated by employing the Stokes-Einstein equation, which relates the molecular diffusion coefficient to the radii.

The PD2000 PCS instrument has four major components: a light-scattering spectrometer, a photon detector box, a correlator box, and a laboratory computer.

The avalanche photodiode resides in the small photon detector box, which is connected to the light-scattering spectrometer by a fiber-optic cable, and to the correlator box by its power cord.

The correlator box is the heart of this operation. It contains the correlator board along with its DSP (Digital Signal Processor) chip that calculates the autocorrelation functions and diffusion constants of the samples. The correlator box collects information from the photon detector box via a coaxial cable, and transmits processed information to the computer via an RS-232 serial cable.

System Operation

Laser light enters the flow cell along the optical axis of the instrument. A fiber optic cable is positioned over the cell at 90° to the incident beam. This fiber transmits the scattered light to the photon detector box. The avalanche photodiode produces an electrical pulse for each detected photon; these pulses are then sent to the correlator board. The correlator performs calculations to determine the correlation functions associated with the detected intensity fluctuations. These correlation functions are then used to compute the translational diffusion coefficient and ultimately the hydrodynamic radius of the sample. This information is fed via an RS-232 cable to a computer running the autocorrelation software.

The collection of photons into time windows or channels and the subsequent analysis (autocorrelation) of this data yields the Transitional Diffusion Coefficient (D). Once (D) has been determined, the Hydrodynamic Radius (Rh) can be calculated using the Stokes-Einstein equation.

$$D = \frac{k_B T}{6\pi\eta Rh} \qquad \text{Equation (8)}$$

where k_B is Boltzmann's constant, T is the absolute temperature in Kelvin, and η is the solvent viscosity. Thus, a large molecule will have a diffusion constant smaller than that of a small molecule. Therefore, the fluctuations will take place more slowly but due to the Raleigh equation the scattered light intensity measurements of large molecules will produce more photons and higher accuracy.

In this manner, measurements can be taken in 5 -100 seconds. The flow rate through the cell can be adjusted to collect sufficient results to measure the diffusion constant across a particle size distribution, which is produced by the size exclusion separation. Thus, it is a fast, accurate, and non-invasive method of obtaining macromolecular size and distribution information.

Biological and Synthetic macromolecules are ideal candidates for measurement using Photon Correlation Spectroscopy (PCS). Macromolecules scatter enough light for measurement at relatively low concentrations. Biological and synthetic macromolecules are often synthesized in very well defined molecular weights which allows for very monodisperse solutions. This means that one can study size and conformational changes of proteins, polysaccharides, and other supermolecular assemblies, as well as measure the size of synthetic structures like dendrimers that have hydrodynamic radius of less than 12 nm. This is important, as the measurement of the radius of gyration at below 12 nm is very difficult; it can produce large inaccuracies and is practically impossible with most proteins less than 100,000 daltons. Furthermore, Photon Correlation Spectroscopy (PCS) allows the study of aggregation phenomena and conformational changes as a function of temperature and physiological and non-physiological solutions. Photon Correlation Spectroscopy (PCS) does not perturb the sample and is able to detect very small changes in size and conformation of macromolecules.

Results

Sets of polystyrene standards were run at 1ml/min as well as Dow 1673 to test if the this new system could obtain Rg and Rh values simultaneously in flow mode. Using the combination of the (PCS)/Raleigh light/RI system, the results (Table 1) obtained for the Rg agree well with the literature values fig (5). The exception was the 8.5 million polystyrene standard which gave a lower than expected Rh value. This was due to the amount of time that the sample was left to dissolve which was probably insufficient for complete dissolution.

Each injection was at 200 ul except for the 117,000 and 17,200 molecular weight sample, a more recent test from December 1996, which was run at .2 ml/min with a 50 sec collection interval.

The exponent calculated for the hydrodynamic radius versus log molecular weight matches perfectly the values reported in Dynamic Light Scattering edited by Robert Pecora - page (195).

If we calculate the product of the molecular weight and the concentration (CM), we get a realistic idea of how this prototype will respond to concentrations at various molecular weights. For example, the lowest calculated value was for the 44,000 daltons sample which had a CM of 396,000. The 1,090,000 dalton sample has a CM of 1308,000. We can now predict that the concentration of the 1,090,000 sample could be reduced by a factor of 3.3 (1308/396) times and the (PCS) detector will still give an accurate determination of the Rh for this sample. In figures (6,7,8,), it is possible to see the determination of Rh on the fly.

For each calculated Rh value, it is also possible to determine the precision of the measurement from the log of the correlation function fig. (8) Table II. The ratio of Rg to Rh provides very useful information about the geometry of the molecules. Polystyrene gave a range for Rg/Rh from 1.30 to 1.48. This is because the Rh value is related to the

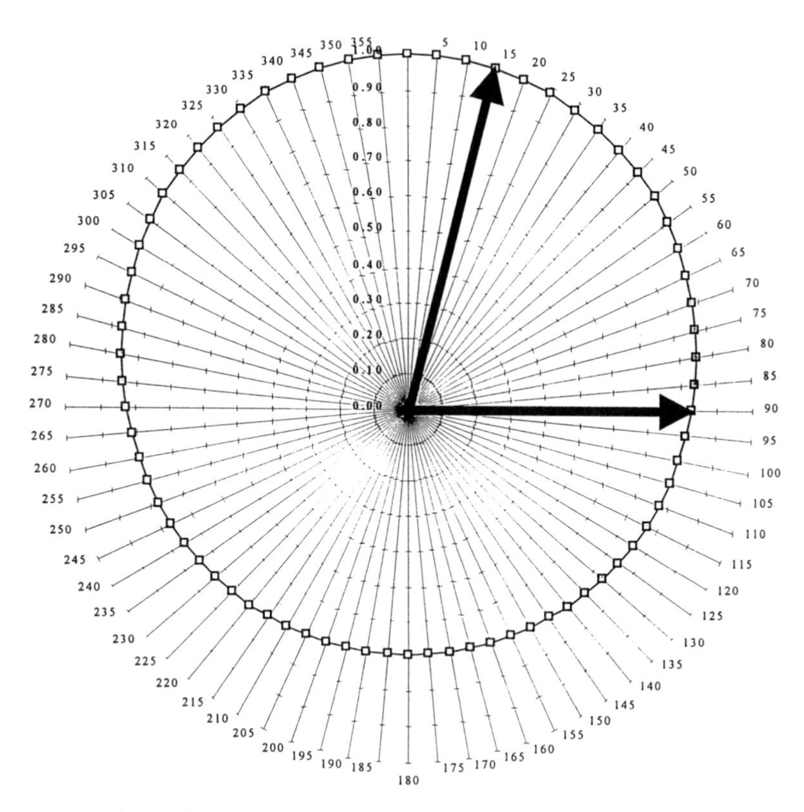

Fig. 4 The Effect of Angular Dissymmetry of a Random Coil of Rg = 50nm. (wavelength = 790 nm.)

Table I Table of Rh and Rg values for Polystyrene Standards run in flow mode using SEC.

Molecular Weight M_w	Conc. (mg/mL)	Conc X M_w	Rh (nm.)	Rg (nm.)	Ratio Rg/Rh
8,500,000	0.35	2,975,000	57.10	124	2.17
1,090,000	1.2	1,308,000	35.00		1.38
761,000	1.3	989,300	28.50		1.44
348,000	3.5	1,218,000	18.26		1.31
186,000	3.4	632,400	12.85		1.48
100,000	10 (25uL)	500,000	10.35		
44,000	9	396,000	5.69		
Exponent			0.569	0.569	
Lit Value Exp.			0.56	0.58 +/- .01	

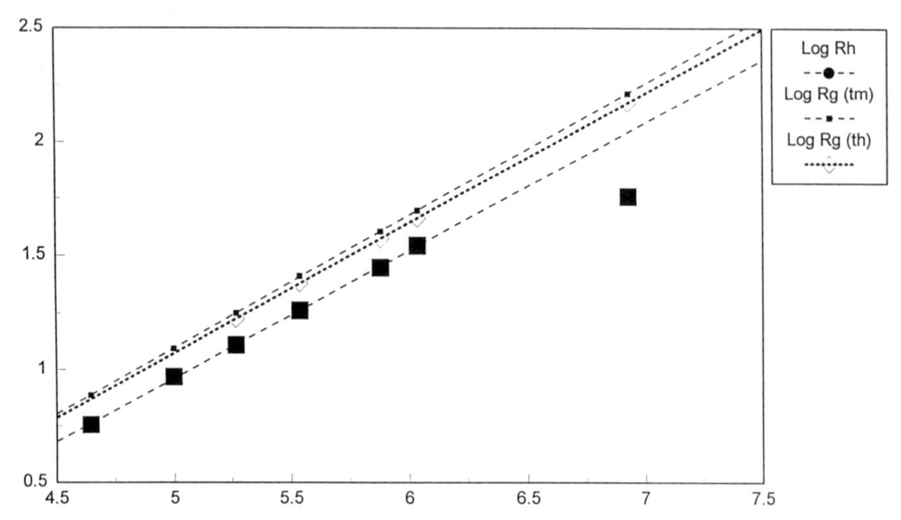

Fig. 5 Comparison of Rg and Rh using the PD2020 in Flow mode for Polystyrene Standards Compared to Results of (tm) Mourey et al. ref. *5*.

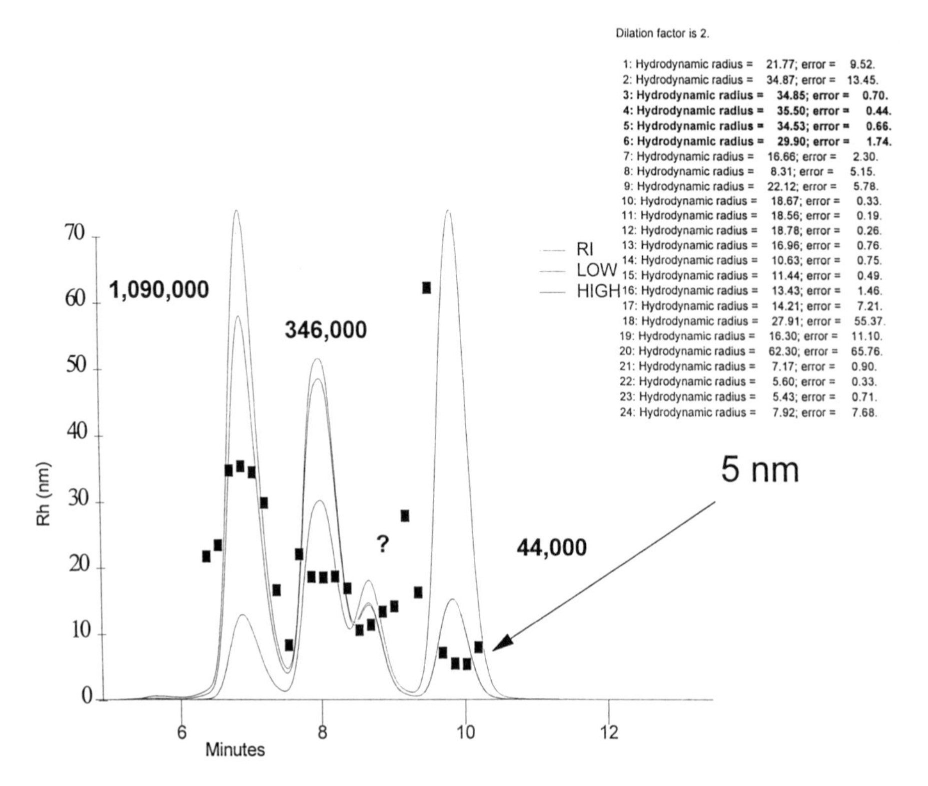

Fig. 6 Polystyrene Standards in Flow Mode Measuring Dynamic and Static Light Scattering

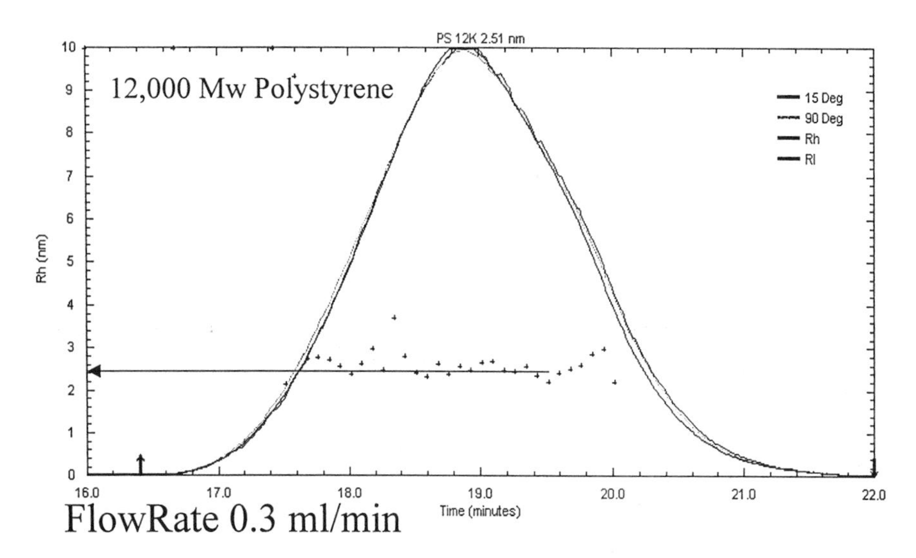

Fig. 7 Low Molecular Weight Measurements for 12,000 Mw PS Standard with a Size Measurement of Polystyrene Standard (Rh =2.5nm)

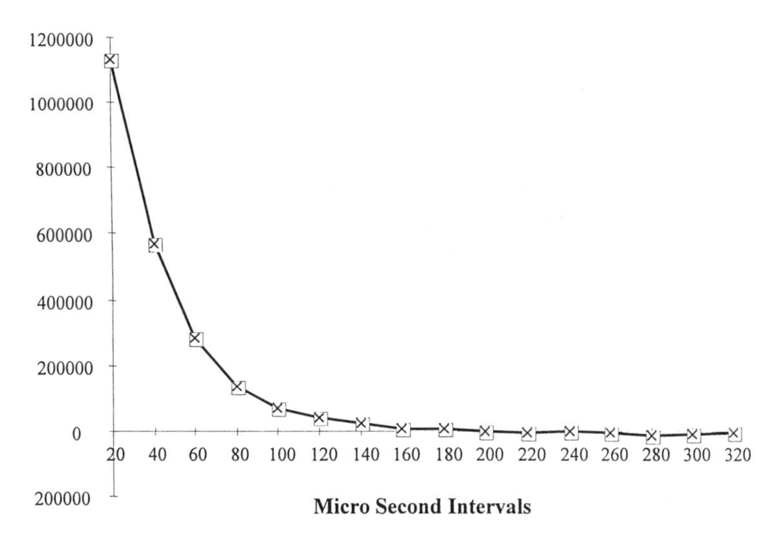

Fig. 8 PCS Correlation Function for QELS in Flow Mode for 100,000 MW Polystyrene in THF at 0.1 ml/min.

Table II The Slice Numbers for the Polystyrene Standard Demonstrate the Rh Calculation as well as the Error of the Linear Fit for the Auto-Correlation Function.

1: Hydrodynamic radius = 1.71; error = -NAN.
2: Hydrodynamic radius = 10.28; error = 0.82.
3: Hydrodynamic radius = 18.31; error = 3.87.
4: Hydrodynamic radius = 8.95; error = 0.58.
5: Hydrodynamic radius = 8.26; error = 0.17.
6: Hydrodynamic radius = 10.36; error = 0.06.
7: Hydrodynamic radius = 9.33; error = 0.07.
8: Hydrodynamic radius = 9.03; error = 0.20.
9: Hydrodynamic radius = 8.87; error = 0.44.
10: Hydrodynamic radius = 9.26; error = 0.35.
11: Hydrodynamic radius = 8.26; error = 1.01.
12: Hydrodynamic radius = 8.57; error = 0.15.
13: Hydrodynamic radius = 9.29; error = 0.35.
14: Hydrodynamic radius = 11.13; error = 2.00.
15: Hydrodynamic radius = 9.09; error = 0.19.
16: Hydrodynamic radius = 9.31; error = 0.14.
17: Hydrodynamic radius = 9.56; error = 0.31.

frictional coefficient of the molecule and Rg is a mass averaged radius. In the case of polystyrene in THF, which has the conformation of a random coil in a good solvent, Rg is greater than Rh. If the polystyrene standards were branched the Rg /Rh ratio will tend to 1. For highly branched macromolecule the Rg/Rh value is less than 1. This kind of study is material for future publications.

Conclusion

This paper has demonstrated the first commercial application of a light scattering detector to measure hydrodynamic radius through a flowing system. The advantage of this system is that it can measure the molecular weight, radius of gyration and hydrodynamic radius simultaneously where all the light scattering measurements are made in a single cell. Being the first commercial instrument to measure a hydrodynamic radius in a flowing system, it presents a whole new set of experimental possibilities: The ability to mix and adjust pH of macromolecules systems and monitor the aggregation effect simultaneously. The option of monitoring the effect of thermal stability of macromolecular assemblies. These methods can utilize conventional HPLC systems to introduce the samples into the cell manually or automatically.

Other applications include sizing of particles eluting from a process chromatography as well as FFFC or Hydrodynamic separations.

References.

1. Robert Pecora Ed. *Dynamic Light Scattering of Macromolecules in Solution.* ISBN 0_306-42790-1
2. B. E. Dahneke Ed. *Measurement of Suspended Particles by Quasi Light Scattering Light,* ISBN 0-471-87289-X
3. Ford N.C. *U.S.A.Patent No. 5305073,* **(1994)**
4. Ford N. C.; Frank L.; Frank. R. *Hyphenated techniques in Polymer Characterization,* T. Provder, Ed. ACS Advances in Chemistry Series Vol. 247.
5. T. Moury and H. Coll, *J Applied Polymer Sci.* **(1995)** Vol 56, pp 65-72
6. T. Havard Inter. *GPC Symposium Proceedings* **(1994)** ; Waters Corporation, Milford, MA, **(1994)**, pp 99.
7. B.H. Zimm , P. Doty, R. S. Stein. *Polym. Bull* **(1945)** 1(415), 90-119
8. P. Debye. *J. Applied Phys.* **(1944)** 15, 338
9. C.Tanford. *Physical Chemistry of Macromolecules* **(1961)** ISBN 0 471 84447 0

Chapter 6

High-Resolution Particle Size Analysis of Mostly Submicrometer Dispersions and Emulsions by Simultaneous Combination of Dynamic Light Scattering and Single-Particle Optical Sensing

D. F. Nicoli[1], K. Hasapidis[2], P. O'Hagan[2], D. C. McKenzie[1], J. S. Wu[1], Y. J. Chang[1], and B. E. H. Schade[3]

[1]Particle Sizing Systems, 75 Aero Camino, Suite B, Santa Barbara, CA 93117
[2]Particle Sizing Systems, 668 Woodbourne Road, Suite 104, Langhorne, PA 19047
[3]Particle Sizing Systems, Annapaulownastraat 1, 3314JK Dordrecht, Holland

Dispersions and emulsions containing particles/droplets predominantly in the submicron size range are ideal candidates for particle size analysis using the combined techniques of dynamic light scattering (DLS) and single-particle optical sensing (SPOS). We have developed an instrument which combines results from both of these techniques, yielding particle size distribution (PSD) results over a wide size range with unprecedented resolution. The DLS subsystem offers a "broad-brush" picture of the overall PSD, the parameters of which can be used to identify the "end point" of a production process. The complementary SPOS subsystem provides a detailed picture of the upper portion of the PSD (> 0.5 or 1.3 μm, depending on the sensor principle utilized), with much higher resolution and accuracy than are provided by ensemble techniques, notably laser diffraction. In the case of emulsions, the "tail" of largest droplets/aggregates often strongly influences final product quality. An Autodilution system and specialized software permit the PSD results from the two subsystems to be coupled quantitatively.

Accurate characterization of the particle size distribution (PSD) of particulate-based suspensions and dispersions has proven to be of increasing importance for ensuring the success of a wide variety of process materials and final products[1,2]. Knowledge of the detailed shape of the PSD, including such parameters as the volume fraction of large "outlying" particles (aggregates or over-size primaries) or the percentage of fine particles in the population distribution, is often crucial for optimizing the physical properties of a wide variety of particulate systems. Implementation of statistical process control (SPC) and "zero-defect" manufacturing methods requires the development of more precise and reliable methods of particle size analysis.

As anyone who is even casually acquainted with particle size analysis can attest, this field contains a bewildering array of techniques and instruments, each claiming to hold the key to reliable PSD information for most applications. The physical techniques which underlie the various instruments vary greatly in principle of operation, as well as reliance on operator expertise and knowledge of sample parameters. In general, most instruments are claimed to be effective over large size ranges. However, upon closer examination using "real" samples, these instruments display large differences in absolute accuracy, precision and resolution. They produce different PSD results, which are often conflicting and confusing. Finally, in many cases, the best choice of a method and instrument for a research setting often conflicts with the needs of QC analysis in a production environment.

A good example in point concerns the popular technology usually referred to as laser, diffraction. In reality, this well-known approach to particle size analysis consists of a combination of two distinct physical techniques. These are Mie scattering (MS), which is most effective for particle diameters smaller than a few microns, and Fraunhofer diffraction (FD), which is applicable for sizes above approximately 1.5 micron (μm). Based on theory alone, a combination of the MS and FD techniques, which forms the basis of every modern "laser diffraction" instrument, should provide the optimal method for analyzing most samples over a wide size range. In practice, however, this approach often suffers from several significant shortcomings. In the present context, it is particularly poorly suited to characterizing PSDs which straddle the region lying above and below one micron. This size region includes a wide variety of colloidal dispersions.

In particular, production of oil-in-water emulsions requires particle size results of highest resolution and absolute accuracy, in order to arrive at final preparations and products which are stable and possess the desired physical properties. In most cases the emulsion droplets are predominantly below one micron in diameter. These "mostly-submicron" emulsions typically have volume-weighted mean diameters in the range of 0.1 to 0.5 μm. It is difficult to characterize these mostly-submicron emulsions in such a way that the PSD thus obtained is a useful predictive parameter. The goal, ultimately, is zero-defect manufacturing of both intermediate and final products. This translates into assurance of long shelf stability: i.e. avoidance of coalescence of oil droplets. It also means having the ability to affect desired final properties. These depend not only on the mean diameter of the PSD but also on its size range, or "polydispersity".

The success of emulsion preparation often hinges on limiting the amount of the "dispersed phase" which resides in the large-diameter "tail" of the PSD -- i.e. the fraction of droplet volume located above 1 um in the PSD. While these "off-scale" particles are typically located several standard deviations away from the mean diameter and constitute a seemingly insignificant fraction of the total amount of dispersed phase, they nevertheless are often critical in determining the final properties of the overall product. For example, their presence can signal that an injectable pharmaceutical emulsion is unsafe for intravenous administration. Alternatively, in the case of a food emulsion, this fraction of "outlyers" may produce an unappealing taste or appearance, or warn of impending phase separation.

Limitations of Laser Diffraction: Need for a Fresh Approach

There are at least three main reasons why the combination of Mie scattering and

Fraunhofer diffraction is poorly suited to measuring typical emulsions. First, the results of the analysis from each method must be successfully merged, with proper normalization, in order to produce a single, accurate, overall PSD. Unfortunately, this is a nontrivial task, given the fact that each of the constituent methods of analysis is itself an "ensemble" technique, where the underlying signal is derived from the simultaneous influence of a large number of particles of all sizes. Hence, by definition an inversion calculation is required to characterize even in an approximate way the underlying distribution of particle diameters. Only relative PSD information is thereby obtained, without any "absolutes" in particle number or concentration to permit unambiguous connection of the two disparate pieces of the PSD.

A second problem concerns the approximations which are required for analysis of the Mie scattering (MS) data. First, both the real and imaginary parts of the refractive index of the particles must be relatively well known (in addition, of course, to the refractive index of the suspending solvent). These affect significantly the variation of scattered intensity with angle for each particle size, upon which the entire MS analysis is based. Also, even if these parameters are well known, there is the fundamental limitation which arises from the ensemble nature of the measurement. It is difficult to invert the angular intensity data and obtain a stable and reliable value for both the mean diameter and the width, or standard deviation, of the distribution, even in the relatively simple case of a unimodal size distribution.

Finally, with the Fraunhofer diffraction (FD) technique it is virtually impossible to obtain an accurate estimate of the amount of particle volume, or mass, which resides in the outermost "tail" of the PSD -- i.e. for particles/droplets larger than about one micron. This shortcoming is particularly important in the case of oil-in-water emulsions, as well as other mostly-submicron colloidal systems. The amount of the total dispersed particle volume, or mass, which resides at the largest particle sizes, relatively far from the mean diameter (e.g. above one micron), is typically less than 0.1% of the entire volume of the dispersed oil phase, in the case of an emulsion. The difficulty for both the MS and FD measurements, which provide the basis for the final "laser diffraction" result, is that the amount of material contained in the large-diameter, outlying "tail" represents too small a fraction of the total particle/droplet mass to have a measurable, reproducible effect on either the scattered or diffracted intensities. Thus, for "mostly-submicron" applications of typical interest the laser diffraction instrument usually lacks the necessary dynamic range in detectable signal, independent of the fact that in theory it possesses an extremely large dynamic range in size -- an interesting, but largely irrelevant (and misleading), specification.

An Alternative to MS+FD (Laser Diffraction): DLS+SPOS. It is primarily for these reasons that we have focussed on a fresh approach for the measurement of colloidal suspensions and dispersions having PSDs which are predominantly in the submicron size range. Like the combination of Mie scattering and Fraunhofer diffraction which comprises "laser diffraction", it also combines two different physical methods of particle size analysis. Recent advances in techniques for particle size analysis[1,2] have yielded instruments which provide some of these needed capabilities. The purpose of this article is to describe a particularly promising approach, offering significant improvements in resolution and accuracy over existing, commercially available methods. We shall review the technical advantages of this new instrumentation, based on a combination of two proven techniques. The first is dynamic light scattering[1-6] (DLS), or photon correlation spectroscopy (PCS). Like the two methods discussed above, DLS is also an ensemble technique. It is effective when the majority of particles are smaller than approximately one micron. The second technique is that of single-particle optical sensing[7,8] (SPOS). It is most effective and easy to use when the majority of particles are larger than this size. Unlike DLS or the previously mentioned methods (MS and FD), the SPOS technique detects and measures particles one at a time. These two techniques, while radically different in principle and type of output, will be seen to be naturally complementary. Indeed, the advantages of one substantially compensate for the disadvantages of the other, thereby making for a favorable relationship.

Review of the DLS Technique

For PSDs which are predominantly submicron, the DLS technique has a number of advantages over competitive methods. In particular, it enjoys a key advantage over classical Mie scattering (which describes the variation in scattering intensity with particle size as a function of angle), in that it is an absolute technique. That is, for relatively "monodisperse" (i.e. narrow) size distributions, the PSD obtained from a DLS measurement is independent of particle composition, provided the concentration is sufficiently dilute that multiple scattering and interparticle interactions are effectively eliminated. A brief review of the principles underlying the DLS technique[3-6] may be useful.

Particles in a stationary suspension illuminated by a focussed laser beam produce scattered waves, which add together "coherently" everywhere in space. Brownian motion, or random-walk diffusion, of the particles causes the phases of the individual waves to fluctuate, thereby resulting in a fluctuation of the net scattered intensity, measured at a given angle by a PMT detector. The diffusivity, D, of the particles is inversely proportional to the mean (spherical) particle radius, R_H, as described by the Stokes-Einstein relation,

$$D = kT/6\pi\eta R_H \tag{1}$$

where k is Boltzmann's constant, T the temperature (K), and η the shear viscosity of the solvent (usually water).

Quantitative information about the distribution of particle sizes can be obtained from the random fluctuations in the scattered intensity signal, I(t), by use of a simple mathematical operation known as autocorrelation. We refer to the scattered intensity autocorrelation function (ACF) as $C(\Delta t)$. Typically obtained using fast digital electronics, $C(\Delta t)$ is defined as the running sum of products of the intensity I(t), measured at time t, and the intensity $I(t-\Delta t)$, measured previously for many different values of Δt.

For the simplest kind of size distribution, consisting of uniform particles, $C(\Delta t)$ is simple: a single, decaying exponential function, with decay time τ inversely proportional to D, (i.e. proportional to the particle diameter),

$$C(\Delta t) = \langle I(t) \bullet I(t-\Delta t) \rangle = A \exp(-2DQ^2 \Delta t) + B \quad (2)$$

where A is an instrument constant and B is the baseline of the ACF, approximately equal to the square of the average scattering intensity. The brackets $\langle\ \rangle$ refer to a running sum of products, over time t. Constant Q, the "scattering wavevector", connects the time scale of the intensity fluctuations with the particle diffusivity, D,

$$Q = (4\pi n/\lambda) \sin \theta/2 \quad (3)$$

where n is the refractive index of the suspending solvent, λ the laser wavelength, and θ the scattering angle (usually 90°). Equations 1-3 are unaffected by the composition of the particles -- hence, the absolute nature of the DLS technique.

Particle systems of interest, of course, are not uniform; rather, they usually contain a significant range of particle diameters. The resulting autocorrelation function is described by,

$$C(\Delta t) = A\left\{\sum_i f_i \exp(-D_i Q^2 \Delta t)\right\}^2 + B \quad (4)$$

where each particle size has a corresponding diffusivity D_i (given by equation 1), and its presence is expressed by the appropriate decaying exponential function. The weighting of the i-th particle size in the ACF is represented by coefficient f_i. This, in turn, is related to the number N_i and volume V_i of particles of radius R_i by

$$f_i = N_i V_i^2 G_i^2 \quad (5)$$

Quantity G_i^2 is the scattering intensity "form factor", which accounts for <u>intra-</u>particle interference, described by Mie theory. For particle diameters smaller than about 100 nm (= 0.1 μm), defined as the "Rayleigh region", there is no angular dependence; G_i^2 is essentially unity and can be ignored. In this case the weighting coefficient f_i varies as the square of the particle volume, or the <u>6th</u> power of the diameter. For substantially larger sizes, G_i^2 varies with scattering angle and particle size, and the weighting f_i falls to a lower effective power of particle diameter.

The challenge for a DLS particle size analyzer is to determine as reliably as possible the set of weighting coefficients f_i in equation 4, given the measured ACF, $C(\Delta t)$. This set f_i constitutes the intensity-weighted PSD. Manipulation of this distribution, using equation 5, and elimination of G_i^2 then yields the corresponding volume-weighted PSD ($\bullet\ N_iV_i$) and number-weighted PSD ($\bullet\ N_i$), provided the index of refraction of the particles is known. The important limitation of the DLS technique is not accuracy <u>per se</u>, in the sense of calibration, because of the absolute nature of the technique. Rather, the key issue is its effectiveness in characterizing polydisperse PSDs. This depends on the details of the algorithm used to invert the ACF. Fortunately, there are many important applications, including mostly-submicron emulsions, which have simple PSDs, that can be approximated by single-peak ("unimodal") distributions having moderate standard deviations.

Examples of Typical Results Obtained by DLS

For simple PSDs which can be approximated by a single-peak (unimodal) distribution, the method of cumulants[9] often yields a reliable result. This "Gaussian" analysis, consisting of a least-squares fit of a quadratic in Δt to $\log_e[C(\Delta t)-B]$, yields two useful parameters: an intensity-weighted mean diameter, d_{int}, and standard deviation (approx. half width), SD. Figure 1 shows a typical volume-weighted unimodal PSD obtained by DLS for an oil-in-water emulsion made by homogenization (used for pharmaceutical applications). In this case the dispersed oil phase represents 10% (wt.) of the total emulsion. The volume-wt mean diameter, d_{vol}, estimated from d_{int} using the "rules" of classical light scattering, outlined above, is 152 nm, and the SD parameter is 36 nm, or 24 % of d_{vol}. The low value (0.3) of the goodness-of-fit parameter, χ^2, indicates that, statistically speaking, this Gaussian result should be reliable.

As a second example, Figure 2 shows the vol-wt PSD obtained by DLS with the same fitting procedure for a similar oil-in-water emulsion, but one formulated with 20% oil. Here, we find a considerably larger vol-wt mean diameter, d_{vol} = 225 nm, with an SD parameter of 65 nm, equal to 29% of d_{vol}. Again, we find a low value (0.35) for χ^2, indicating a reliable analysis and a realistic shape for the true PSD.

For more complex distributions, such as those which can be approximated by a bimodal, a more sophisticated analysis algorithm is required. We have frequently obtained PSD results with superior resolution and reliability by utilizing an inverse Laplace transform (ILT) procedure[10-13], including proprietary features, to analyze $C(\Delta t)$. The resolution and overall effectiveness of this ILT procedure, compared to alternative approaches, can be objectively judged by measuring clean bimodal mixtures of narrow, unaggregated latex standards. For example, our approach is able to characterize accurately such mixtures as a 3:1 ratio (volume), respectively, of 91 and 261 nm latex standards. More closely-spaced bimodals, or even certain classes of trimodal distributions, can often be analyzed successfully.

Characterizing Protein Aggregation by DLS. A challenging and useful application of this multimodal analysis capability is the characterization of the extent of aggregation of a relatively narrow, "primary" distribution of particles. Figure 3A shows the volume-weighted PSD obtained from our multimodal Nicomp analysis for a globular protein (0.7 mg/ml, MW approx. 50K) stored at elevated temperature (37°C) for several days, to promote aggregation. The main peak, centered at 6.4 nm (vol-wt PSD), represents the "native" (unaggregated) protein, accounting for almost 95% of the total particle mass. In addition, there is a second peak at 50 nm, accounting for 4.6% of the mass, and a third, very minor peak at 180 nm, representing the remaining protein mass.

The result shown in Figure 3A represents a simplification of the true PSD. The secondary peaks provide a semi-quantitative approximation of the actual expected distribution of particle aggregates, which should resemble a continuous, decaying "tail", where the % volume, or mass, decreases smoothly with increasing diameter. The more extensive the aggregation, the higher should be the relative intensities of the secondary peaks, and the larger their mean diameters.

Figure 3B shows the original intensity-weighted PSD obtained from inversion of the ACF data, from which the vol-wt result in Figure 3A was obtained. Figure 3B

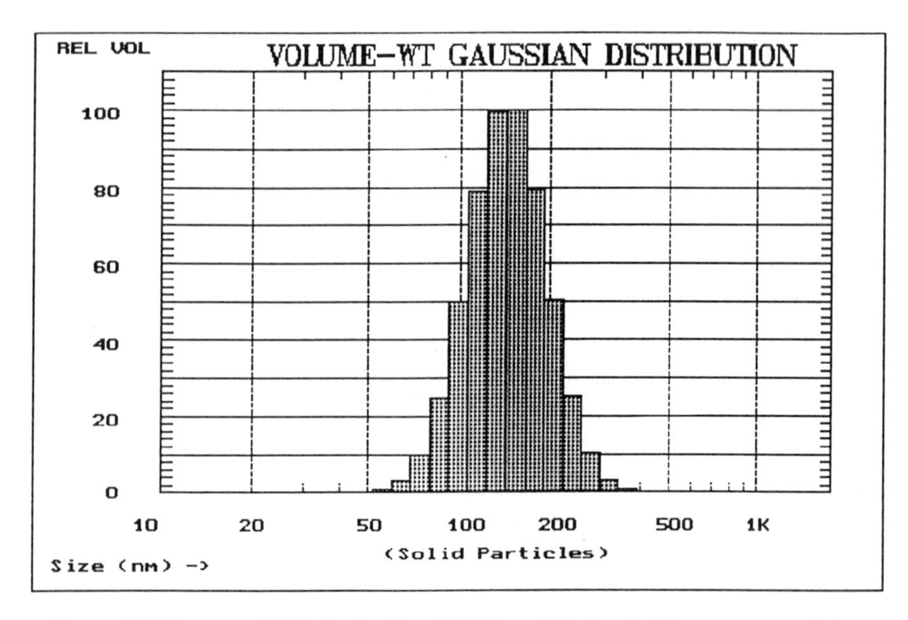

Figure 1. Volume-wt PSD obtained by DLS for a 10% (wt.) oil-in-water emulsion, using Gaussian (cumulants) analysis.

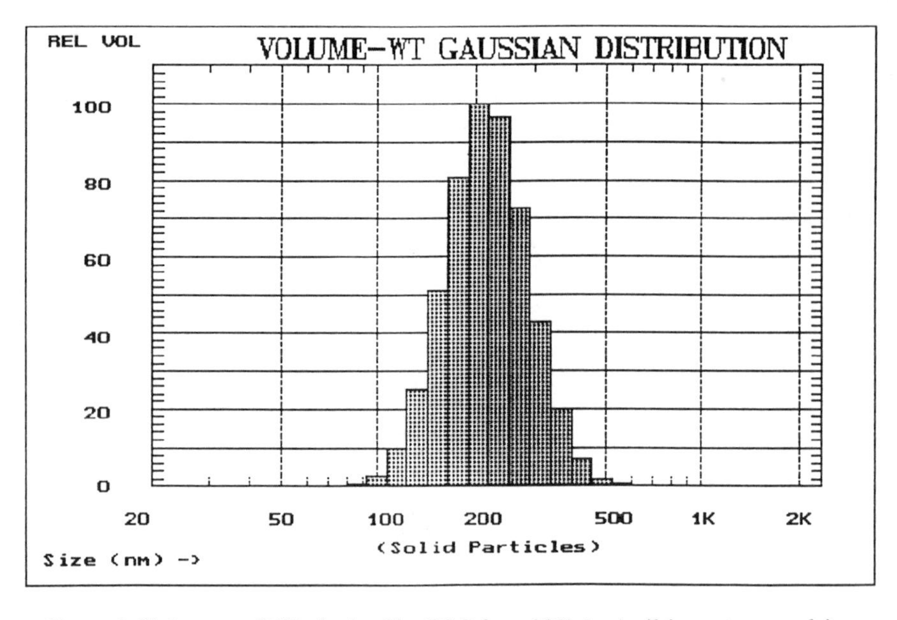

Figure 2. Volume-wt PSD obtained by DLS for a 20% (wt.) oil-in-water emulsion, using Gaussian (cumulants) analysis.

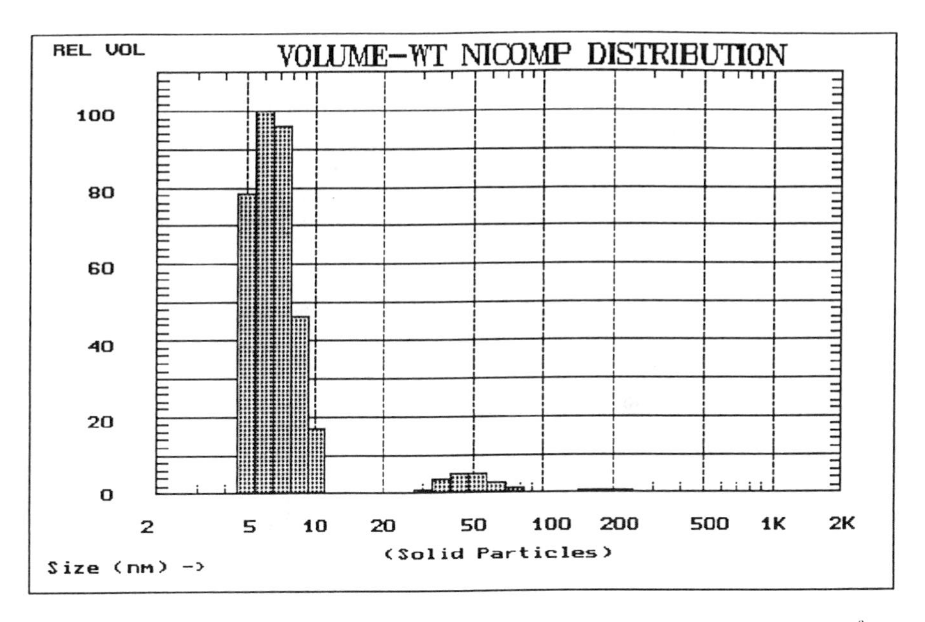

Figure 3A. Volume-wt PSD obtained by DLS for a protein (MW • 50K) held at 37°C for 5 days to promote aggregation, using the Nicomp multimodal (ILT) analysis.

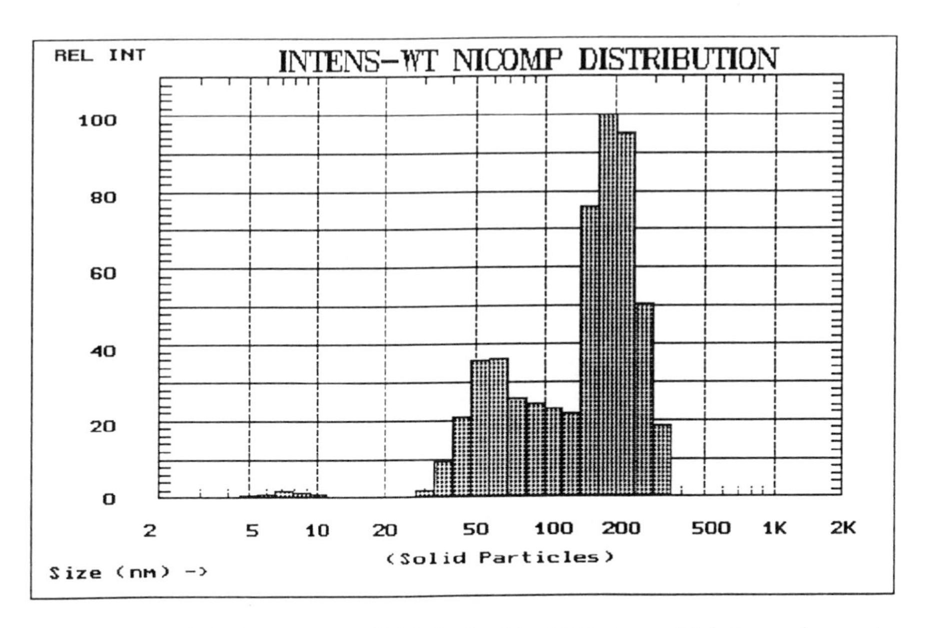

Figure 3B. Original intensity-wt PSD obtained by DLS, from which the volume-wt PSD in Figure 3A was calculated.

provides a graphic illustration of the strong influence that larger particles, such as aggregates, have on the scattered intensity. They have a disproportionate effect on the ACF and, therefore, on the resulting computed PSD. This behavior is ideal for detecting small amounts of aggregation.

Measurement of the same protein stored at 4°C, in order to minimize aggregation, is summarized in Figures 4A and 4B. The major peak, representing the native, unaggregated protein, is located at the same size, 6.4 nm, as that found above for the "stressed" sample (Figure 4A). The remaining minor peaks, representing an aggregate "tail", are also present, but they are now located at substantially smaller diameters and have much lower relative intensities (Figure 4B), indicating much less extensive aggregation. Results like these promise to enhance the reputation of the DLS technique for characterizing "difficult" systems, for which the traditional cumulants approach usually yields overly simplified, even misleading, PSD results.

In summary, a DLS particle sizing instrument should be judged by its ability to characterize reliably a relatively complex, polydisperse PSD, like that described by Figures 3A,B and 4A,B and to distinguish it unambiguously from a broad, smooth Gaussian-type distribution, like that seen in Figures 1 and 2. The χ^2 parameter usually indicates which analysis result is preferred -- Gaussian or ILT (multimodal).

Review of the SPOS Technique

Particles suspended in liquid or gas are made to flow across a small "photozone" -- a narrow, slab-like region of uniform illumination, typically produced by a laser diode. The flowing particle suspension is sufficiently dilute that the particles pass through the photozone one at a time, thereby avoiding coincidences. When a particle passes through the sensing zone, its presence gives rise to a signal pulse in an optical detector, the magnitude of which depends on the mean particle diameter and physical principle of detection -- either light scattering (LS) or light extinction (LE). The optics and detection system are designed so that the pulse height increases smoothly and monotonically with particle diameter. The true PSD is constructed in real time (typ. rate $\leq$ 10,000 particles/sec) using fast digital signal processor (DSP) technology, by comparing the measured pulse heights with a standard sensor response, or calibration, curve. This technique was originally developed for the field of contamination analysis and has been used extensively for applications as diverse as pharmaceutical parenteral solutions[14], semiconductor process liquids[15] and hydraulic fluids[16].

For particles smaller than approximately 1.5 μm, the method of light scattering provides the high sensitivity needed for detecting and sizing such ultra-fine particles[17-20]. The practical lower limit is approximately 0.2 to 0.4 μm, depending on the type of particles being analyzed, the power of the laser light source, the details of the sensor detection system and the cleanliness of the solvent used to dilute the starting concentrated sample. For particles larger than about 1.3 μm, the method of light extinction[7,8,14,16] is preferred.

When a particle enters the photozone in an LE-type SPOS sensor, it acts like a tiny lens, deflecting light away from the detector, located opposite the source of illumination, across the flow channel, as shown schematically in Figure 5. A small fraction of the photozone area is therefore effectively blocked, by an amount equal to the cross-sectional area of the particle. This produces a negative-going detected pulse, with a

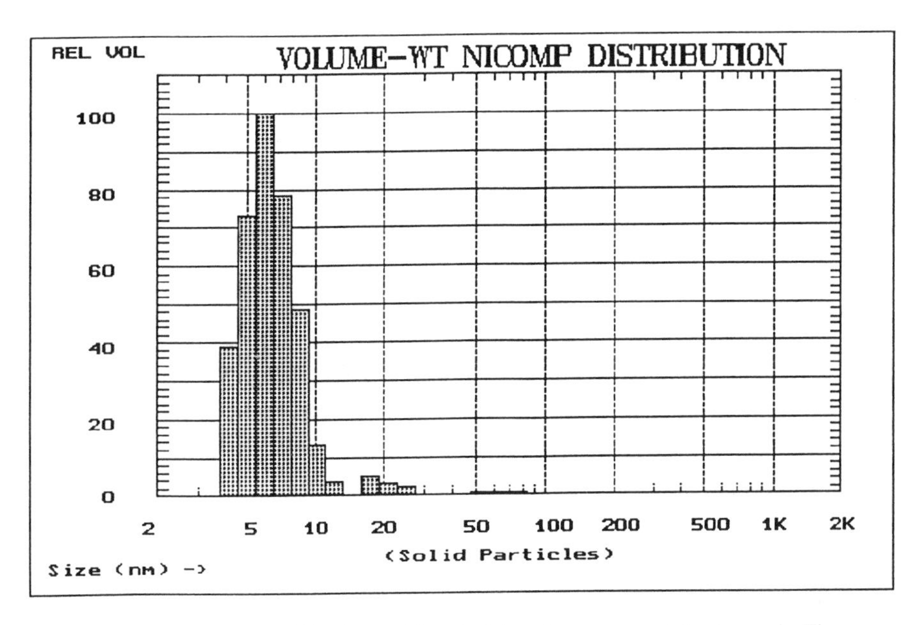

Figure 4A. Volume-wt PSD obtained by DLS for the same protein shown in Figures 3A and 3B, held at 4°C for 5 days to minimize aggregation, using the Nicomp multimodal (ILT) analysis.

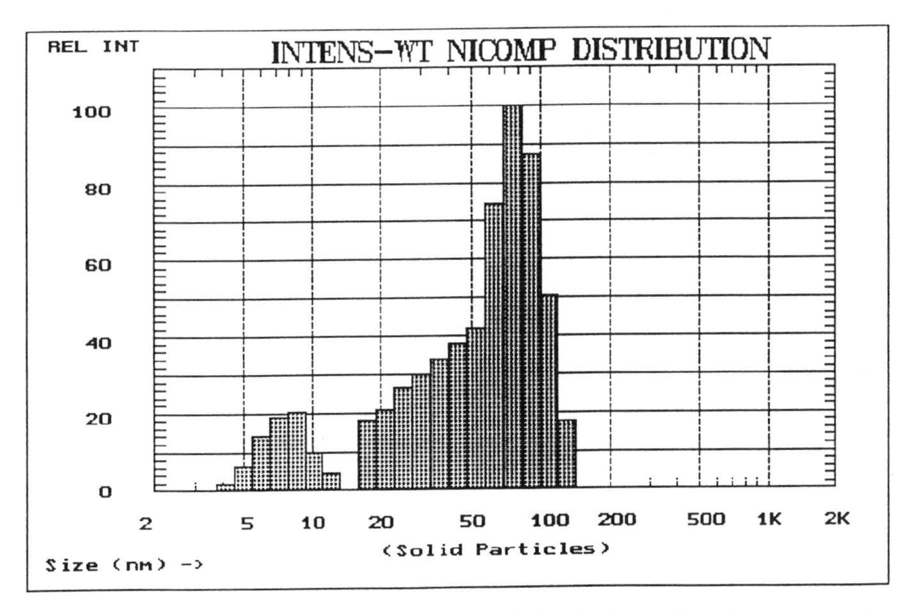

Figure 4B. Original intensity-wt PSD obtained by DLS, from which the volume-wt PSD in Figure 4A was calculated.

height proportional to the square of the particle diameter for particles smaller than the photozone width (typ. 30-50 μm) and linear with diameter for those substantially larger.

We have recently developed a new type of hybrid SPOS sensor,[21] which incorporates both the light scattering and light extinction responses. By combining the LS and LE signals appropriately, one obtains a smooth, monotonic response, or calibration, curve over a relatively large particle size range. This hybrid approach (Patents pending[21]) has the advantage of improved sensitivity -- i.e. lower minimum detectable diameter, typically 0.5 μm or lower -- conferred by the LS response, while still maintaining the large maximum allowable particle diameter (i.e. large dynamic size range), which is characteristic of the LE method. A more detailed discussion of this new approach to particle size analysis by SPOS will be the subject of a future report.

For the purpose of this article, we shall focus most of our attention on results obtained using the simple LE-type SPOS method. Extinction-type sensors are generally more practical for use in a combined DLS/SPOS instrument. They permit the size range above approx. 1.3 μm to be measured with a much higher degree of resolution and accuracy than can be achieved with virtually any other type of technique, especially Fraunhofer[22,23]. The SPOS method is able to reveal not only a "broad brush" profile of the PSD, but also its detailed shape, at both the fine and coarse ends of the distribution. This exceptional quality of analysis allows the SPOS method either to stand on its own, for PSDs which are mostly above one micron in size, or, as we shall see, to be used to complement the DLS technique, for PSDs which are mainly submicron.

Distinguishing Property of SPOS: High Resolution. The SPOS technique has several characteristics that make it ideal for analyzing a wide variety of particulate-based systems. The most obvious, of course, is its resolution: one can't do better than to count and size particles one at a time. The only other technique which offers comparable resolution is the classical electrozone method[22,23]. This well-known technique, which also sizes particles individually measures the small change (pulse) in resistance between two chambers of partially conducting fluids separated by a fine orifice, which is caused by a particle passing momentarily through the orifice. The SPOS technique is the optical analog of the electrozone method. However, there is a critical difference between the two approaches: the flow channel for the former is much larger than the orifice required by the latter. Hence, clogging of the flow path, while a persistent problem for electrozone-based instruments, rarely occurs with SPOS-type analyzers (assumining that a sensor of the appropriate size/diameter range has been chosen to match the application in question). Additional advantages of SPOS over electrozone technology include compatibility with any solvent (i.e. electrolyte not required) and much wider dynamic size range (typ. at least 300:1 in diameter). Also, SPOS sensors are able to measure particles at much higher count rates (20:1 to 50:1, depending on the size range). This property results in much better particle counting statistics, which translates into more accurate PSDs.

Advantages of SPOS over Ensemble Techniques. There are two important attributes of the SPOS method which make its PSD results generally superior to those produced by ensemble techniques. First, by its very principle of operation, SPOS yields a true population, or number-wt, PSD, from which the volume-wt PSD can be computed directly and unambiguously. By contrast, the laser diffraction technique is mass-sensitive, yielding a volume-wt PSD as a starting point. Hence, it is relatively blind to the smallest particles in the distribution. Also, a laser diffraction measurement yields a highly simplified, or "stylized", PSD, often missing important details.

A second important attribute of the SPOS method is that it is highly robust in an analytical sense. The PSD which it generates is stable and reproducible, devoid of major artifacts. This is because the raw data of particle counts vs diameter require no mathematical manipulation. The addition of new particles to an existing distribution cannot alter the shape of the PSD that was obtained up to that point. The opposite description applies to the laser diffraction method, for which particles of all sizes contribute simultaneously to the raw diffracted intensity data. Hence, a mathematical algorithm must be used to extract the approximate PSD. Due to the "ill-conditioned" nature of such procedures, the presence of particles of certain sizes may alter the response of the instrument to particles of other sizes, often resulting in serious, undesireable artifacts in the computed PSD.

Examples of Typical Results Obtained by SPOS

First, we show a typical PSD result obtained by SPOS for a sample in which the majority of particles lie within the range of the light-extinction (LE) sensor -- i.e. larger than approx. 1.3 μm. Figure 6A shows the population PSD obtained using our AccuSizer 770 Optical Particle Sizer with Autodilution for moderately-fine zeolite particles (used for catalyst applications), suspended in water. This population PSD constitutes the "raw data" produced by the SPOS measurement. Approximately 86,000 particles were counted and sized during the 90-second measurement period.

As seen from Figure 6A, most of the particles in this sample fall within a single, smooth, nearly-symmetric peak, extending from roughly 10 to 200 μm and centered at about 30 μm. In addition, there is a secondary population of ultra-fine particles located between the starting threshold of 1.3 μm (and likely below that size, as well) and 2 μm, but essentially no particles between that point and the onset of the main population of zeolite particles at 10 μm.

Using this PSD as a starting point, it is straightforward to obtain precise corresponding area-wt and volume-wt PSDs, simply by multiplying the particle counts in each sample by the square or cube, respectively, of the mean diameter for that channel. The volume-wt PSD is frequently the desired result, in order to ascertain where the largest fraction of the total particle mass is located. The volume-wt PSD corresponding to Figure 6A is shown in Figure 6B. It is a smooth, nearly-symmetric peak centered at approximately 80 μm, with most of the particle mass located between 20 and 200 μm. Of course, the ultra-fine particles seen in Figure 6A have essentially disappeared from the volume-wt PSD, given their enormous relative underweighting by volume. It is also worth noting that the smooth shape of the main peaks in both Figures 6A and 6B is not the result of mathematical smoothing. Rather, it is the consequence of the high level of statistical accuracy achieved, given the large number of particles counted in most channels.

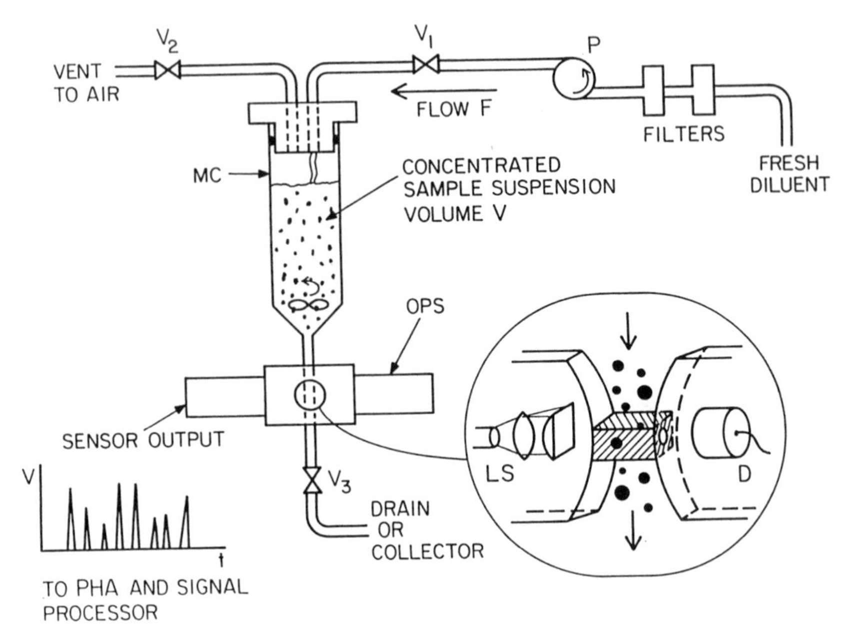

Figure 5. Schematic diagram of the AccuSizer SPOS system, including light-extinction (LE) particle sensor and Autodilutor.

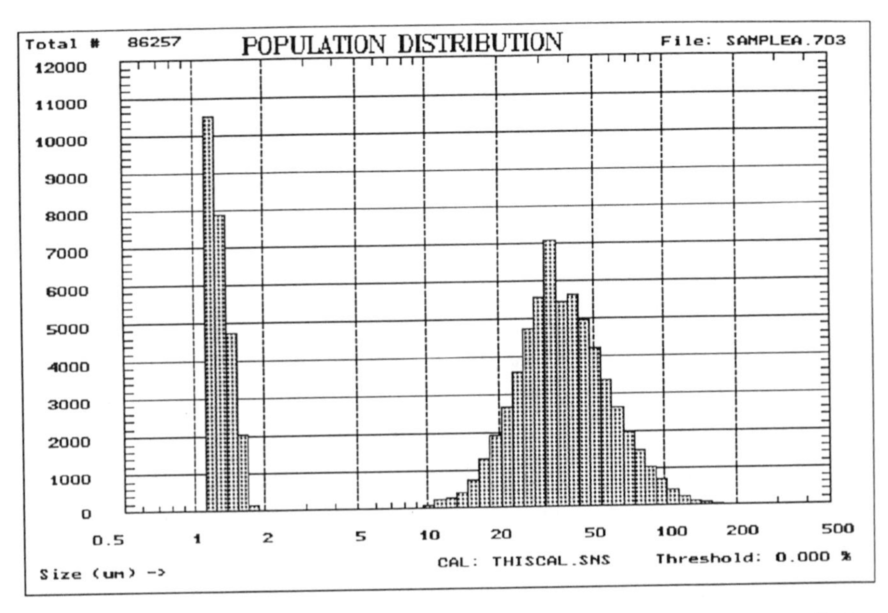

Figure 6A. Population PSD obtained by SPOS for one sample of catalyst (zeolite) powder, dispersed in water.

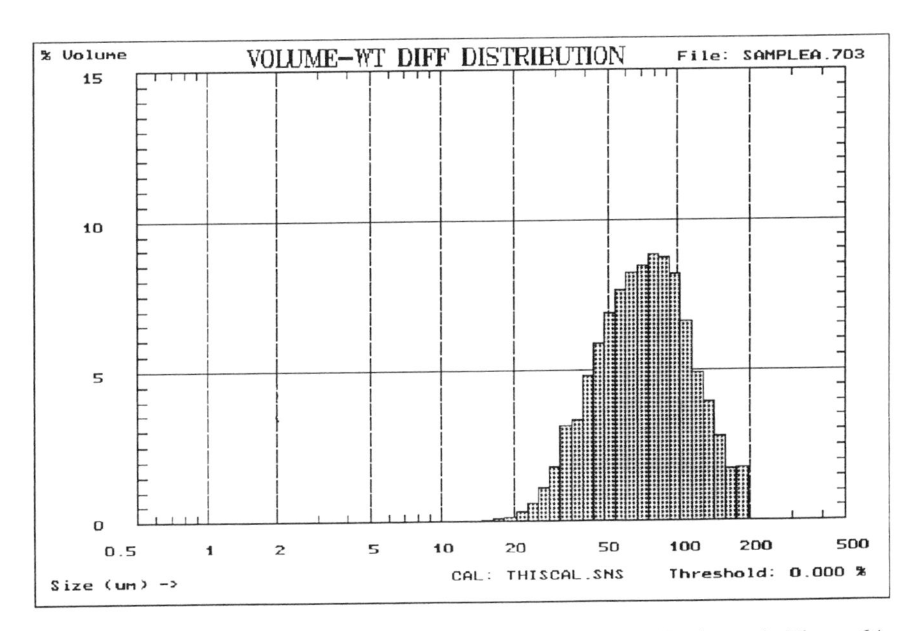

Figure 6B. Volume-wt PSD corresponding to the population PSD shown in Figure 6A.

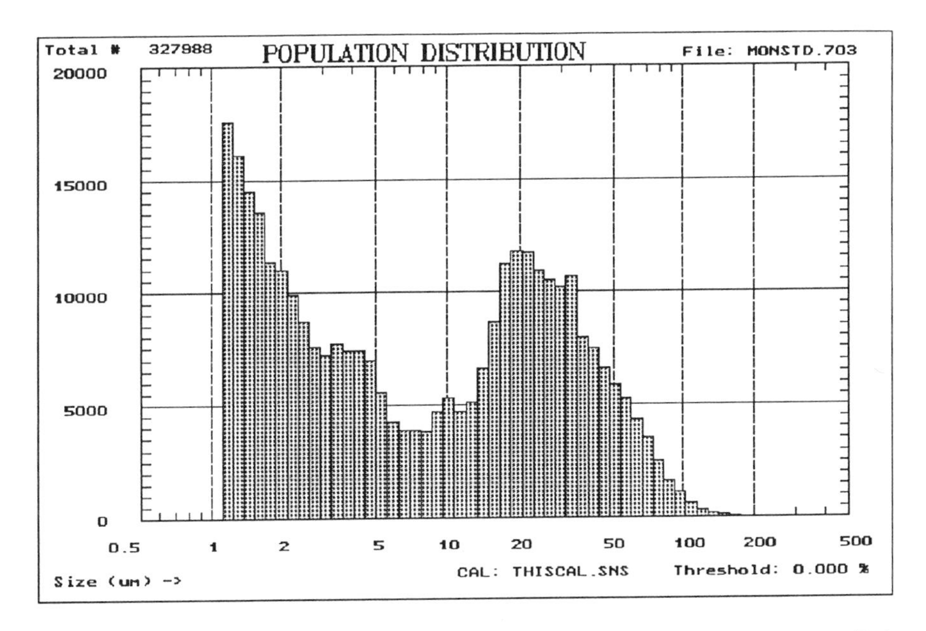

Figure 6C. Population PSD obtained by SPOS for a second sample of catalyst (zeolite) powder, dispersed in water.

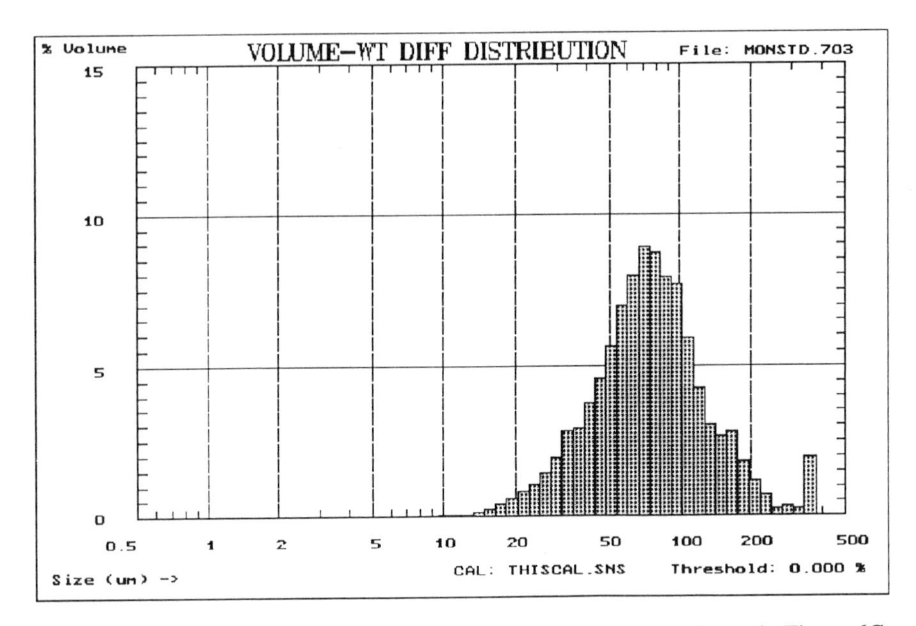

Figure 6D. Volume-wt PSD corresponding to the population PSD shown in Figure 6C.

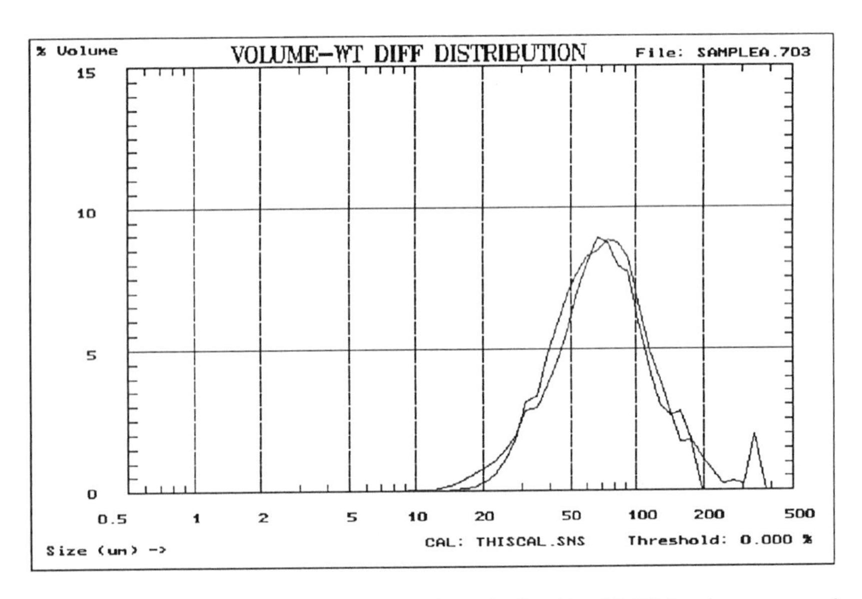

Figure 6E. Comparison of the volume-wt PSDs obtained by SPOS for the two samples of catalyst (zeolite) powder.

A Key Attribute of SPOS: Sensitivity to Fine Particles. There are two important attributes of the SPOS method which cause its results to be generally superior to those produced by ensemble-type techniques. First, as noted previously, SPOS yields a true population, or number-wt PSD, from which the volume-wt PSD is computed directly. By contrast, for example, the ensemble technique of laser diffraction is sensitive to particle mass, yielding a volume-wt PSD as a starting point. Hence, it is relatively "blind" to the smallest particles in the distribution. Of potentially equal importance is the fact that laser diffraction yields a highly simplified, or "stylized", PSD, which often will not yield a reliable number-wt PSD.

The importance of the dependence of the SPOS technique on particle number, rather than mass (or volume), can be illustrated by analyzing a second sample of zeolite particles, produced under different process conditions. Figure 6C shows the population PSD obtained for it. The shape of the distribution differs substantially, in one major respect, from that observed for the first sample. We now observe a significant population of fine particles which fill the gap between 1.3-2 μm and 10 μm. Given the applications for which this catalyst material was intended, the presence of so many fine particles lying below the desired distribution would likely identify this sample as "inferior" and result in modification of variables in the production process.

The volume-wt PSD for this second sample is shown in Figure 6D. It is seen to be nearly identical to the volume-wt distribution found for the first sample, shown in Figure 6B. The comparison is easier to make by looking at an overlay of the two volume-wt PSDs, shown in Figure 6E. Based on this comparison alone, one would be hard pressed to conclude that there is any significant difference between the two samples. The conclusion is inescapable: If this kind of powder were characterized only by a volume-wt PSD, it would essentially be impossible to judge its quality, from the point of view of the existence or absence of overly-fine particles.

Second Key Attribute of SPOS: Absence of Artifacts. The second essential characteristic of the SPOS method is the fact that it is highly "robust", in an analytical/mathematical sense. Its results are highly stable and reproducible, void of major artifacts. The reason for this, of course, is that the raw data produced by an SPOS measurement consist simply of particle counts vs diameter. Generation of the PSD, whether simply by population or weighted by diameter, requires no obscuring mathematical manipulations or assumptions regarding the shape of the underlying distribution. From the very definition of SPOS, the arrival of new particles in the sensor, of whatever size and shape, cannot alter the shape of the PSD which has already been obtained from previous particles. In mathematical terms, the SPOS response is said to be "additive". This characteristic can be tested by "spiking" a sample with one or more latex "standards", of well-defined size. The nature of the PSD which results (discussed below) will serve to highlight the uniqueness of the SPOS response relative to that typically produced by an ensemble technique.

Figure 7A shows the number-wt PSD obtained for a fine (1000 Mesh) silicon carbide powder, used for abrasive applications, suspended in water with a nonionic surfactant (Triton X-100) and ultrasonication. More than 370,000 particles were counted and sized during a two-minute measurement period. In Figure 7A we see that the great majority of particles are larger than the 1.3-μm measurement threshold, while relatively few are larger than approximately 15 μm. The number-wt PSD is peaked at about 5 μm.

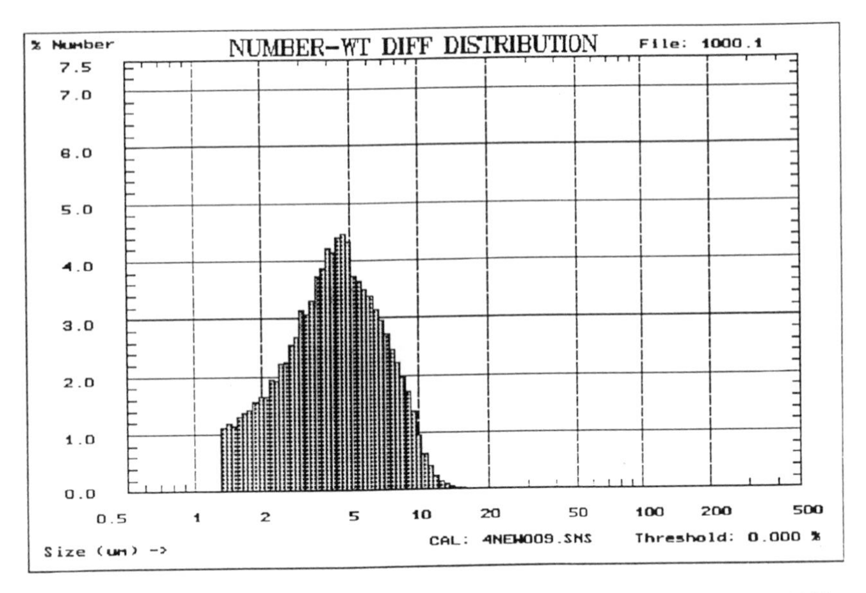

Figure 7A. Number-wt PSD obtained by SPOS for a silicon carbide powder (1000-Mesh) dispersed in water and surfactant (Triton X-100).

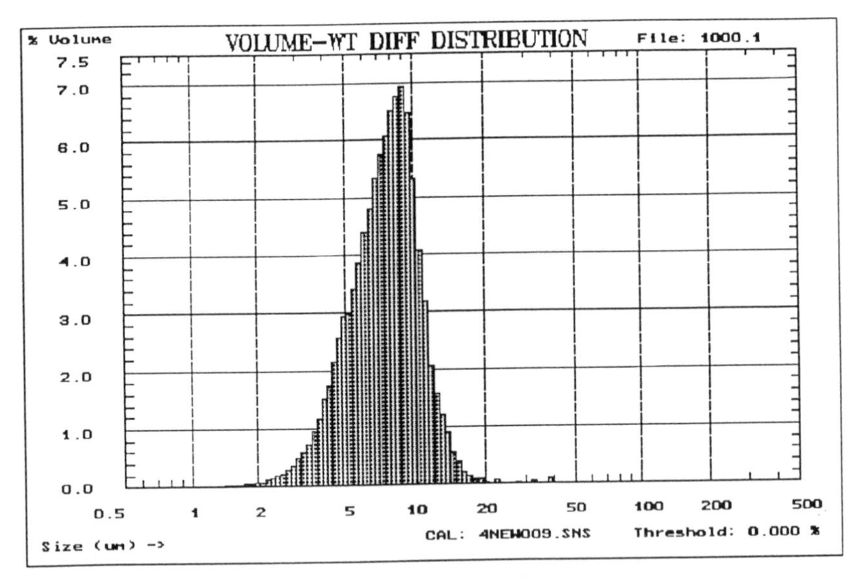

Figure 7B. Volume-wt PSD corresponding to the number-wt PSD shown in Figure 7A.

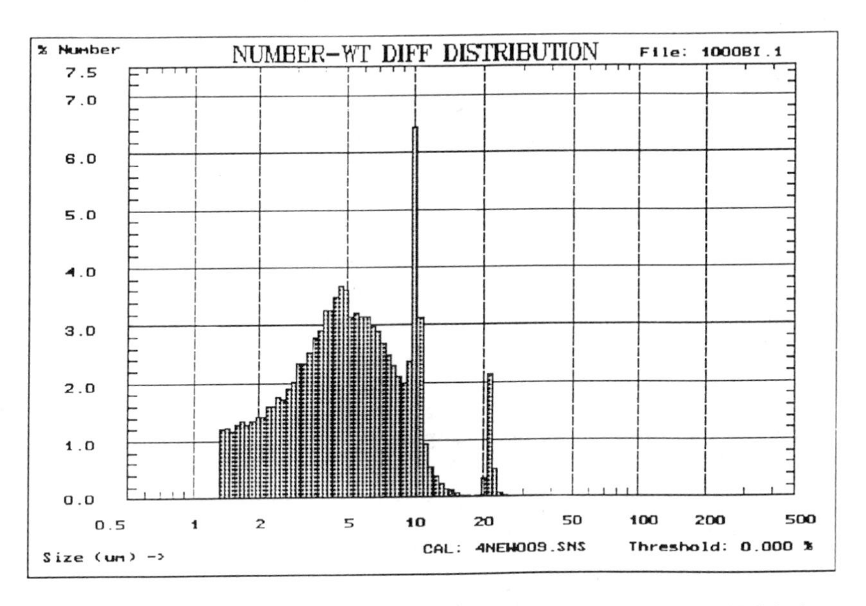

Figure 7C. Number-wt PSD obtained by SPOS for the same sample of 1000-Mesh silicon carbide sample, to which were added a few drops each of two polystyrene latex standards (9.7 and 20.5 μm).

Figure 7B shows the volume-wt PSD corresponding to Figure 7A. The most obvious consequence of weighting the particle counts by the cube of the particle diameter is to narrow significantly the width of the distribution peak. We also observe a moderate asymmetry in the shape of the PSD, a consequence of the finer particles in the population PSD. Finally, the volume-wt PSD cuts off abruptly at approximately 20 μm, with evidence of just a few "outlyers" above this diameter.

Figure 7C shows the number-wt PSD obtained for the same sample of 1000-Mesh silicon carbide powder, suspended in water/Triton X-100, to which was added a few drops each of two concentrated polystyrene latex standards, with mean diameters of 9.7 and 20.5 μm. As can be seen by comparing Figures 7C and 7B, the resulting number-wt PSD for the "spiked" sample is essentially the original PSD plus two narrow peaks due to addition of the narrow latex standards. Clearly (and obviously), addition of the latter did not appreciably change that portion of the PSD ascribed to the original, unspiked sample.

If this same analysis were repeated using the laser diffraction method, an entirely different conclusion would likely result. Because particles of all sizes contribute simultaneously to the raw diffracted intensity data, a mathematical inversion algorithm must be used to extract the approximate PSD. Owing to the "ill-conditioned" nature of such procedures[22], the presence of particles of certain sizes may, in effect, alter the response of the instrument to all the remaining particles, regardless of size. Unfortunately, serious, undesirable artifacts in the PSD can easily occur, such as the disappearance of peaks which should be present and/or the appearance of others which, in reality, don't exist.

Autodilution Simplifies SPOS Measurements. The development of a simple system for automatic dilution of concentrated samples has greatly increased the attractiveness of the SPOS technique. This Autodilution system (Pat.), described elsewhere[24], maximizes the particle concentration while avoiding coincidences, thereby minimizing the run time and maximizing the statistical accuracy of the PSD. Typically, a small aliquot of concentrated sample is added to the mixing chamber of the Autodiluter, usually having a volume, V, of 50 to 100 cc. Filtered diluent is pumped at a constant flow rate, F, into the chamber, the fluid contents of which empties into the particle sensor at the same flow rate. The output particle concentration therefore decreases exponentially in time, with a decay time constant, τ, (also called the residence time) equal to V/F -- typically 1-2 min. Particle sizing begins as soon as the particle concentration falls to a preset level, typically about 10,000 particles/cc.

Given the exponential dilution method, the starting concentration of particles can be inferred from the measured population distribution, allowing the absolute PSD to be determined. This attribute contrasts sharply with most ensemble-type sizing techniques, which can produce only relative PSDs. This feature is extremely useful for quality-control applications, where hard quantitative information is desired. An example, based on an oil-in-water emulsion, will be discussed below.

This same continuous dilution method is also ideally suited for a DLS-type instrument. However, in this case, the requirements are less stringent, because the DLS measurement is independent of the absolute particle concentration, provided it is sufficiently dilute. Therefore, the Autodiluter simply must adjust the particle concentration within an optimal range: high enough to ensure a good signal/noise level (relative to contaminant particles in the suspending solvent), but low enough to avoid changes in the ACF due to multiple scattering or interparticle interactions (typically Coulombic repulsions).

SPOS Results Obtained from Typical "Submicron" Emulsions

Figure 8 shows the population PSD which was obtained from the 10% emulsion of Figure 1, using the SPOS technique (AccuSizer 770) and continuous Autodilution. The plot shows the precise population PSD for the outermost "tail" of the emulsion, for particle diameters larger than 1.3 μm. A significant number of particles (69,233) extend out to 10 μm and beyond, even though the vast majority of oil droplets are in the submicron range, essentially invisible to the LE-type sensor. The decaying curve shown in Figure 8 has not been mathematically smoothed. Rather, the number of oil droplets counted in each size channel is sufficiently large that the statistical fluctuation (given by the square root of the number) in most of the channels is negligible.

A powerful result emerges from the SPOS measurement of Figure 8. One can determine with high accuracy the absolute droplet volume, or mass, which resides in the measured PSD, because of the ability to actually count and size each particle in that size range. Then, knowing the amount of concentrated emulsion which was injected into the mixing chamber of the Autodiluter, the system can easily compute the percentage of original total oil volume contained in the measured PSD. In the case of the 10% oil emulsion shown in Figure 8, we find that the oil droplets occupying the tail region of the PSD between 1.2 and 20 μm represent 0.0017% of the total oil volume of the original emulsion.

Figure 9 shows a similar result which was obtained for the 20% injectable oil emulsion shown in Figure 2. The shape of the population curve is qualitatively very similar to that seen in Figure 8. However, the sample was diluted to a much greater extent -- only 50 μl was injected into 85 ml of water in the Autodilution system, compared to 1 ml of the 10% emulsion in Figure 8. The number of oil droplets sized was much larger -- 406,812. Therefore, one isn't surprised to find a much larger percentage of the total oil volume residing in the tail of the PSD -- 0.094%. In absolute percentage terms, this still represents a relatively small amount of oil in the tail of the distribution.

Once again, the absolute nature of the SPOS technique can be appreciated by "spiking" the previous 20% emulsion sample with a small amount (50 μl) of a narrow polystyrene latex standard, 5 μm in mean diameter. Figure 10 shows the resulting PSD obtained for this composite sample. As can be seen by comparing Figures 9 and 10, the resulting population PSD for the spiked sample has the same shape as that found for the pure emulsion sample, except for the additional narrow peak representing the added latex beads. Clearly, addition of the latter has in no way changed the shape of the PSD for the emulsion component alone. Additional proof of this can be seen in Figure 11, which shows the PSD obtained by subtracting, channel by channel, the PSD of Figure 9 from that of Figure 10. To first approximation, all that remains is a narrow peak at 5 μm, representing the added latex spike. Such a test would surely fail using existing laser diffraction technology.

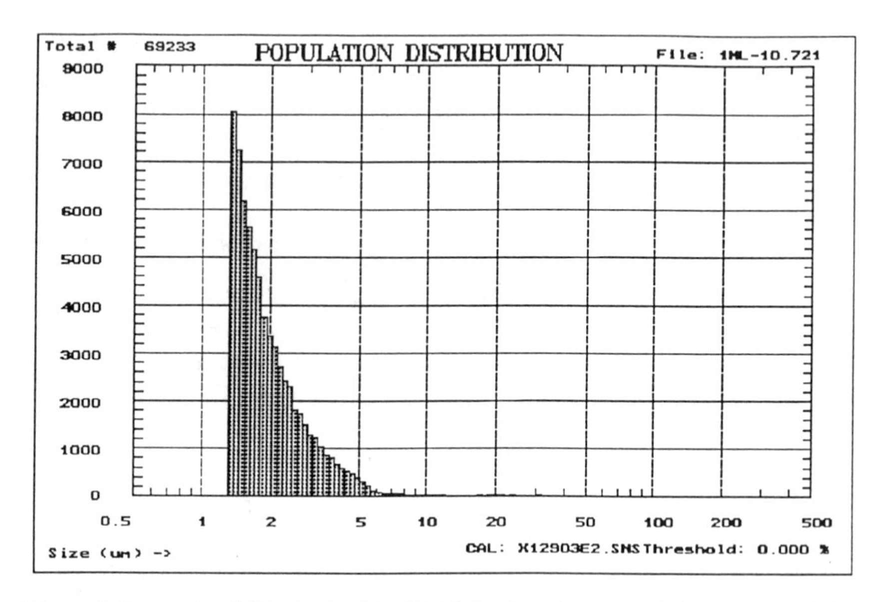

Figure 8. Population PSD obtained by SPOS for the 10% (wt.) oil-in-water emulsion of Figure 1.

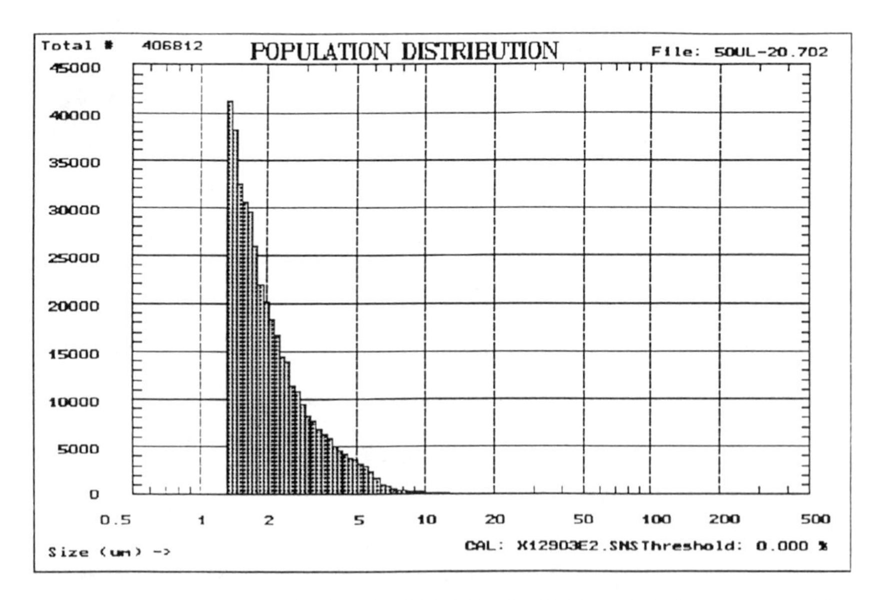

Figure 9. Population PSD obtained by SPOS for the 20% (wt.) oil-in-water emulsion of Figure 2.

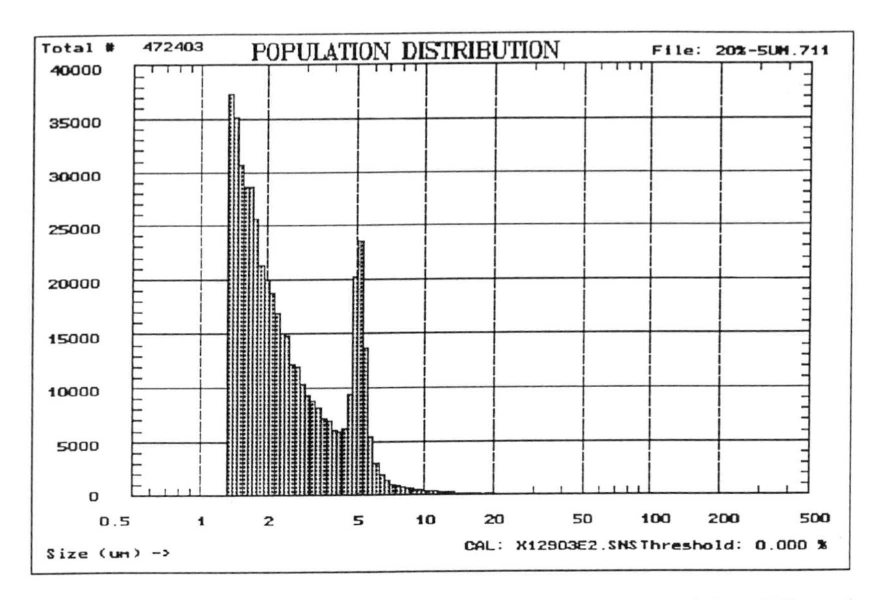

Figure 10. Population PSD obtained by SPOS for the 20% (wt.) emulsion of Figure 9, to which was added a "spike" of latex standard (5 μm).

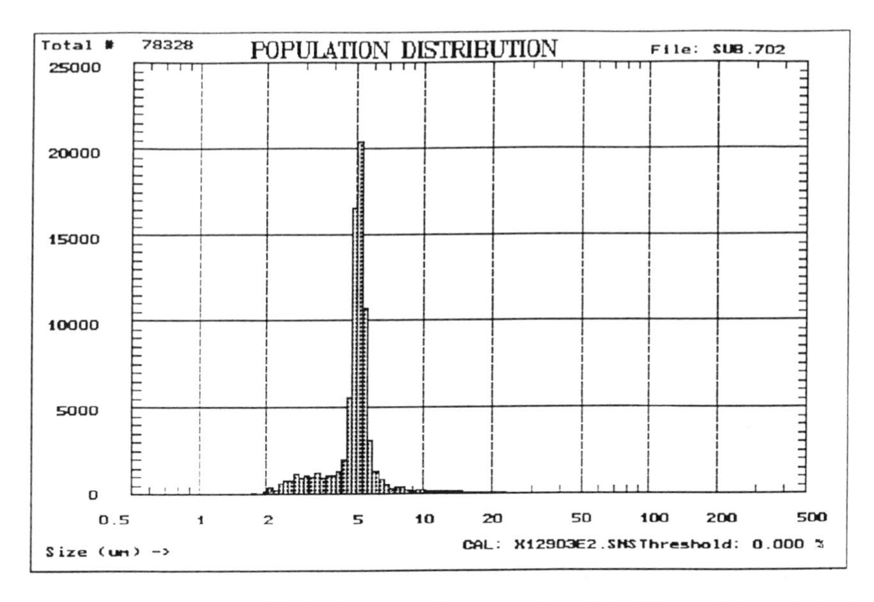

Figure 11. Population PSD obtained by subtracting the PSD of Figure 9 from that of Figure 10, channel by channel.

A Combination Analysis System: DLS + SPOS

It is useful to review the strengths and weaknesses of these two complementary techniques. First, DLS can provide reproducible and accurate results for simple PSDs that can be approximated by a smooth, Gaussian-like shape (narrow or broad), or by a bimodal of sufficient separation. The only requirement is that the majority of particles be smaller than approx. 1 μm. The principle advantage of the DLS approach, apart from its ability to measure extremely small particles ($\leq$ 5 nm), is that the PSD results are independent of the composition of the particles. For a wide variety of colloidal dispersions DLS provides a decent "broad-brush" picture of the actual PSD, using only the parameters of mean diameter and width. However, because of mathematical limitations, it is unable to characterize small "details" of the PSD.

By contrast, SPOS is ideally suited to measuring the detailed structure of the PSD. This includes quantitative determination of the fraction of fine particles, located well below the mean diameter, and the population of oversize primary particles or aggregates, located well above the mean diameter. The only significant disadvantage of SPOS is that it ceases to be practical or effective below approx. 0.5 μm, depending on the specific application in question.

The particles which are sized by SPOS are also counted. Hence, the absolute particle volume contained within a given size range can be determined easily and with high accuracy. From the total amount of sample mass, or volume, injected at the start, together with the dilution factor, the instrument can easily determine the fraction of the original sample mass, or volume, which lies within that size range.

The DLS portion of a combined DLS/SPOS instrument is ideal for monitoring a wide variety of manufacturing processes involving mostly- submicron PSDs. These include oil-in-water emulsions made by homogenization or Microfluidization and polymer latices made by emulsion polymerization. DLS provides a fast, reproducible means for identifying the "end point" for such processes. The SPOS portion of the combined system, by contrast, is ideal for evaluating the quality of a dispersion, after the nominal end point has been reached. Typically, this involves the quantitative determination of the population of large, "off-spec" particles, which can seriously affect the quality of a final product -- coating, adhesive, ceramic casting, etc.

Figure 12 shows a display/printout obtained from our combined AccuSizer-Nicomp 377 system, for the same 20% oil-in-water emulsion discussed earlier. The instrument performs the SPOS and DLS analyses automatically, in rapid succession. The two PSDs produced by the respective technologies are displayed together on the same screen. In Figure 12 we see the population PSD for the "tail" of the emulsion, for particle diameters > 1.3 μm. In the upper-right corner of the display we see the overall vol-wt PSD obtained from DLS. These two results can be interchanged on the screen, if one wishes to focus more on the DLS part of the analysis. Also, the weightings of each PSD can be changed.

SPOS is able to give quantitative significance to the decaying tail in Figure 12. The emulsion particles which are larger than 1.3 μm represent 0.034% of the total particle volume in the PSD. This result illustrates the real advantage of the SPOS capability; accurate determination of the volume percentage of the largest emulsion droplets is invaluable for assessing the overall quality and stability of the emulsion.

In conclusion, it should be apparent that the DLS and SPOS techniques have properties which make them naturally complementary. When used in combination, as

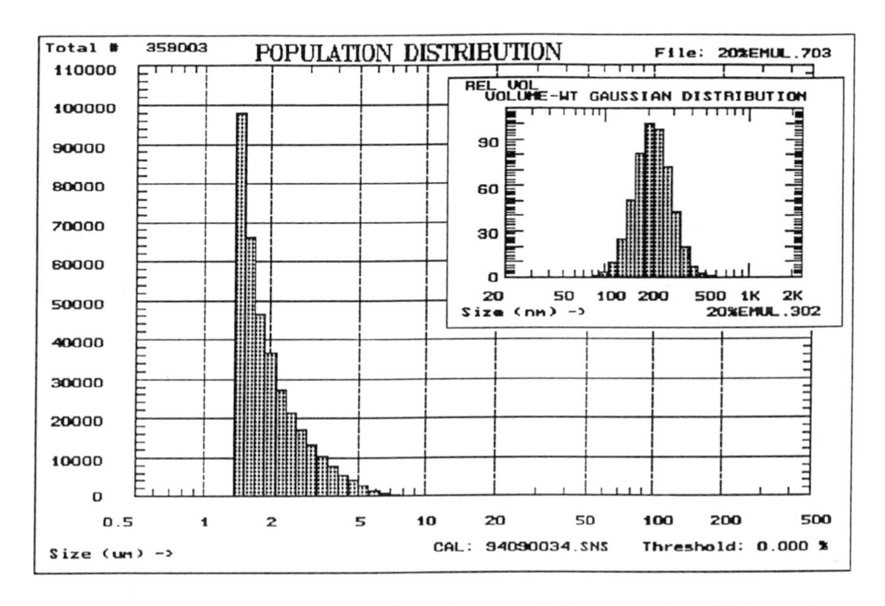

Figure 12. Simultaneous display of the volume-wt PSD obtained by DLS and the population PSD obtained by SPOS, for the 20% oil-in-water emulsion of Figures 2 and 9, from the combined DLS-SPOS instrument.

an integrated system, they can provide particle size results of exceptional accuracy and resolution over a wide dynamic size range.

Literature Cited

1. Particle Size Distribution: Assessment and Characterization; Provder, Th., Ed.; ACS Symp. Series 332; Washington, D.C., 1987.
2. Particle Size Distribution II: Assessment and Characterization; Provder, Th., Ed.; ACS Symp. Series 472; Washington, D.C., 1991.
3. Chu, B. Laser Light Scattering; Academic Press: New York, NY, 1974.
4. Pecora, R. Dynamic Light Scattering, Applications of Photon Correlation Spectroscopy; Plenum Press: New York, NY, 1985.
5. Measurement of Suspended Particles by Quasi-Elastic Light Scattering; Dahneke, B.E., Ed.; Wiley-Interscience: New York, NY, 1983.
6. Pike, E.R. In Scattering Techniques Applied to Supramolecular and Non-Equilibrium Systems; Chen, S.H.; Nossal, R.; Chu, B., Eds.; Plenum Press: New York, NY, 1981.
7. Knapp, J.Z.; Deluca, P.P. Supplement to J. Paren. Sci. & Tech. 1988, 42, 1.
8. Knapp, J.Z.; Abramson, L.R. In Proc. Int'l Conf. on Particle Detection, Metrology and Control; Arlington, VA, 1990.
9. Koppel, D.E. J. Chem. Phys. 1972, 57, 4814.
10. Provencher, S.W. Computer Phys. Commun. 1982, 27, 213, 229.
11. Ostrowsky, N.; Pike, E.R. Opt. Acta 1981, 28, 1059.
12. Morrison, I.D.; Grabowski, E.F.; Herb, C.A. Langmuir 1985, 1, 496.
13. Bott, S.E. In Particle Size Distribution: Assessment and Characterization; Provder, Th., Ed.; ACS Symp. Series 332; Washington, D.C., 1987.
14. Barber, T.A. In Proc. Int'l Conf. on Liquid Borne Particle Inspection and Metrology; Arlington, VA, 1987.
15. Lieberman, A. In Semiconductor Processing; Gupta, D.C., Ed.; ASTM STP 850; Amer. Soc. for Testing and Mat'ls: 1984.
16. Verdgan, B.M.; Stinson, J.A.; Thibadeau, L. In Proc. 43rd Nat'l Conf. of Fluid Power; Amer. Soc. Fluid Power: Chicago, IL, 1988.
17. Kerker, M. The Scattering of Light and other Electromagnetic Radiation; Academic Press: New York, NY, 1969.
18. Van de Hulst, H.C. Light Scattering by Small Particles; General Publishing: Toronto, ON, 1981; also Dover: New York, NY, 1981.
19. Lieberman, A. In Proc. Int'l Conf. on Particle Detection, Metrology and Control; Arlington, VA, 1990.
20. Knollenberg, R.G.; Gallant, R.C. In Proc. Int'l Conf. on Particle Detection, Metrology and Control; Arlington, VA, 1990.
21. Wells, D.; Nicoli, D.F. "Single-Particle Optical Sensor with Improved Sensitivity and Dynamic Size Range"; 1997, Patent pending.
22. Allen, T. Particle Size Measurement; 3rd Ed.; Chapman and Hall: New York, NY, 1981.
23. Modern Methods of Particle Size Analysis; Barth, H.G., Ed.; Wiley-Interscience: New York, NY, 1984.
24. Nicoli, D.F.; Elings, V.B.; "Automatic Dilution System"; 1988, U.S. Patent # 4,794,806.

Chapter 7

Developing Fiber Optic Probes for Noninvasive Particle Size Measurements in Concentrated Suspensions Using Dynamic Light Scattering

B. B. Weiner, Walther W. Tscharnuter, and A. Banerjee

[1]Brookhaven Instruments Corporation, 750 Blue Point Road, Holtsville, NY 11742

Particle size measurements as a function of concentration for several common colloidal suspensions are presented. The results, one up to 25% wt/vol, shows the advantage of using single mode fiber optics as a practical tool. A comparison is made between a previous probe design and a new one. Several of the limiting features of such measurements are reviewed.

Dynamic light scattering (DLS) is a well-known technique for submicron particle sizing. DLS is one type of quasielastic light scattering (QELS). Many reviews exist to explain the theory as well as the limitation.[1,2] For accurate particle sizing DLS applies normally in dilute suspensions where multiple scattering and particle-particle interactions are insignificant. In such cases, the autocorrelation function (ACF) formed from the scattered laser light may be analyzed to yield intensity-weighted particle size distribution information.

Normal concentrations at which DLS measurements are made vary from volume fractions of 10^{-6} to 10^{-3}. Yet, measurements at higher concentrations are sometimes useful. In some liquid-liquid emulsion systems there is the concern that dilution will change the size distribution. In some dispersions it may be difficult to maintain the same concentration of dispersants in the diluent, because their identity and concentration are unknown to the analyst wishing to determine particle size. In some oxide systems, notably those made from alumina, it is claimed that maintaining constant conditions at the particle surface is nearly impossible upon dilution due to the large number of ionic equilibria. Finally, as a matter of pure convenience, it is advantageous to make measurements without the need for a dilution step.

What is high concentration? The answer depends on the system of interest. For paint systems over 50% w/v is high. For many proteins 2% w/v is considered high concentration. And for organic pigments used in inkjet printers, 0.5% w/v is considered high.

Van Keuren, Wiese, and Horn[3] review other methods for measurements in concentrates. In addition, they demonstrate that their method (also called FOQELS) is capable of measurements at various angles and up to generally higher concentrations than the results reported here. Their method consists of dipping fibers directly into the liquid, making sure the distance between the fixed launching and the movable receiving fiber is only a few tens of microns to avoid multiple scattering. Although their results are impressive, and help in identifying the different influences that arise in concentrates, we do not find their approach viable for use in an industrial environment.

Recently Pusey detailed a series of elegant measurements in concentrates using two-color DLS to suppress multiple scattering. (Pusey, P.N.; paper presented at 71st Colloid & Surface Science Symposium, Univ. of Delaware, July 1997). Measurements up to 50% wt/vol in latex suspensions were made. Multiple scattering was further surpressed by partial index matching. While such measurements go along way in explaining the fundamentals of particle-particle interactions and structure in complex liquids, they are not very practical for routine measurements.

We have instead adopted the technique pioneered by Dhadwal et. al.[4] Here the fibers are sealed in a probe that is close , but external, to the cell wall.

Theory

Dilute Systems. In a standard setup, the fluctuations in the scattered light intensity are analyzed using a digital autocorrelator. The ACF is formed as the ensemble average over the product of a number of photons collected during a sampling time Δt and another number delayed by a time τ. This is done as a function of the delay time τ. Since the scattered photons are correlated, this has given rise to an alternate name for the technique: Photon Correlation Spectroscopy (PCS). As long as Δt is small compared to any relaxation time of interest in the sample, the ACF may be analyzed to yield the time dependence of the scattered light.

Figure 1 shows a typical ACF, an exponential decay to a baseline B from an initial intercept I. The ratio I/B varies from 1 to 2. Values close to 1 are obtained in a poorly designed optical system, or in the case of strong multiple scattering, or when a strong local oscillator is present.

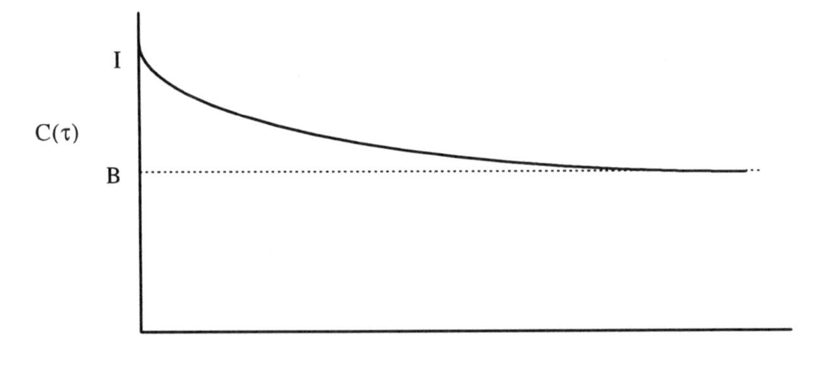

Figure 1. Typical autocorrelation function showing intercept (I) and baseline (B).

A local oscillator consists of a small amount of unscattered light, from the same light source, mixing with the scattered light on the detector. By monitoring the ratio I/B for a particular optical design, one can spot either the existence of unwanted stray light acting as a local oscillator or the onset of multiple scattering.

For a single exponential decay, corresponding to a monodisperse sample, in the absence of a local oscillator or multiple scattering, the data are fit to the following function:

$$C(\tau) = B \cdot [1 + A \cdot \exp(-2\Gamma\tau)]$$

where,

$$\Gamma = D_o q^2$$

and,

$$D_o = \frac{k_B T}{3\pi\eta d_H} \quad ; \quad q = |\vec{q}| = \frac{4\pi n_o}{\lambda_o} \cdot \sin\left(\frac{\theta}{2}\right)$$

Here D_o is the translational diffusion coefficient at infinite dilution (also called the free diffusion coefficient), q is the magnitude of the scattering wave vector, and Γ is called the linewidth. D_o is related to the particle size through the Stokes-Einstein equation, where k_B is Boltzman's constant, T is the absolute temperature, and η is the viscosity of the continuous medium. The hydrodynamic particle diameter, d_H, is obtained assuming a sphere. The scattering wave vector q is related to the refractive index n_o of the continuous medium, λ_o the wavelength of light in vacuo, and θ the scattering angle. The parameter A is related to the optical design and to multiple scattering. The value of A approaches unity in a properly designed system, *when no* local oscillator is present *and no* multiple scattering occurs.

Moderately Concentrated Systems. Pusey and Jones[5] have reviewed the more general case allowing for both particle-particle interactions and structure in the liquid. Here the effective diffusion coefficient, D_{eff}, obtained from the first cumulant, $\overline{\Gamma}$, is given by:

$$D_{eff}(q, \phi_v) = \frac{\overline{\Gamma}}{q^2} = D_o \frac{H(q, \phi_v)}{S(q, \phi_v)}$$

Cumulants are obtained by fitting the logarithm of the ACF to a power series in τ. The first cumulant, $\overline{\Gamma}$, is the coefficient of the linear term in the power series.

Here ϕ_v is the volume fraction, S is the static structure factor of the colloidal suspension, and H represents the hydrodynamic interactions between particles. S and H are functions of both the scattering angle (through q) and the volume fraction. As $\phi_v \rightarrow 0$ (low concentration limit) both S and H $\rightarrow$ 1, and $D_{eff} \rightarrow D_o$.

The effective diffusion coefficient may arise from mutual or from self diffusion or a combination Which one is actually measured depends on relative length scales. One of these length scales, $2\pi q^{-1}$, is determined by the experimental setup. In the instrument described below, the FOQELS, with a scattering angle in water of 155° and a wavelength of 670 nm,

$$2pq^{-1} = 258 \text{ nm}$$

The other length scale depends on the average distance between particles, r_m, and may be estimated, for monodisperse samples, from the number density, ρ_N. by:

$$r_m = \rho_N^{-1/3} = \left(\frac{6\phi_v}{\pi d^3}\right)^{-1/3}$$

If $2\pi q^{-1} \leq r_m$, then $D_{eff} = D_{self}$, the short-time self diffusion coefficient. However, if $2\pi q^{-1} >> r_m$, then $D_{eff} = D_{mutual}$, the mutual diffusion coefficient. Theory[4] shows that, initially, D_{self} decreases with increasing concentration. D_{mutual} initially increases with increasing concentration provided repulsive, particle-particle interactions dominate. This is expected to be the case for hard sphere interactions and with electrostatically or sterically stabilized suspensions. D_{mutual} initially decreases with increasing concentration provided attractive, particle-particle interactions dominate. This is expected to be the case with oil/water and water/oil emulsions where Van der Waals attractions play a dominate role. Both D_{mutual} and D_{self} approach D_o at low concentrations.

Multiple Scattering Effects. The effect of multiple scattering, at least in the initial stages of multipli-scattered light, is a more rapid decrease in the ACF.[6] The apparent diffusion coefficient is larger than for singly scattered light, and the calculated particle size is smaller. When the detected light is the result of many, many scattering events, one must treat the situation using the theory of Diffusing-Wave Spectroscopy.[7]

Local Oscillator. A local oscillator is a fraction of the primary beam that is mixed with the scattered light. Sometimes this is intentional; often it is not. When it is not intentional, the local oscillator arises either from stray light or strong scattering from dust particles or gel networks. In either of the last two cases the intensity of the local oscillator is not constant, which makes it very difficult to establish a baseline and to fit the data to a theoretical model. When the stray light is constant, and many times larger than the scattered light, the ACF becomes:

$$C(t) = B \cdot [1 + A \cdot \exp(-\Gamma t)]$$

The major difference is that the decay is twice as slow, falling as $\exp(-\Gamma\tau)$ rather than $\exp(-2\Gamma\tau)$. This has the effect of doubling the calculated particle size, if one is not aware of the presence of the local oscillator. In addition, A becomes much smaller, typically ≤ 0.08 .

Experimental setup of FOQELS

A schematic of the FOQELS is shown in Figure 2. The name FOQELS stands for ***F***iber ***O***ptic ***Q***uasi-***E***lastic ***L***ight ***S***cattering. Measurements described here were made with a 10 mW, 670 nm solid state laser. A special version of the FOQELS uses a 100 mW, 780 nm laser to measure ~3 nm proteins in dilute as well as concentrated solutions. The enclosure for the electronic and optical components is quite small, about the size of a large dictionary.

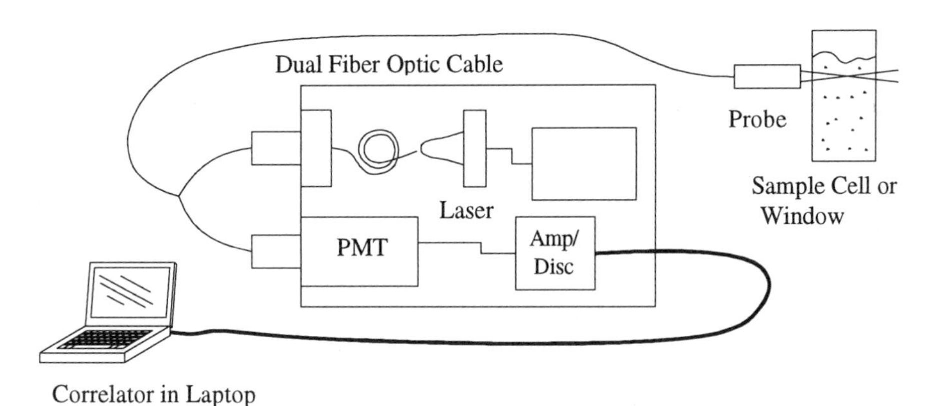

Figure 2. Schematic of FOQELS

A BI-9000AT digital autocorrelator board (Brookhaven Instruments Corporation) was used to accumulate the ACF. Since this correlator is capable of simultaneously correlating two inputs, it is possible to couple two optical setups to a single correlator board, allowing the simultaneous monitoring of either two positions in a sample or two completely different samples. The Windows-based software allows the accumulation of up to 16 separate inputs.

A close up of the probe is shown in Figure 3. Using standard FC/PC connectors the length of the cable may be changed easily; the standard is one metre, but cables up to 100 m have been used successfully. The probe itself is made of stainless steel, about 10 cm long and 5 mm in diameter. Normally, the tip of the probe is just touching the cell or window wall; however, the probe can be moved backwards a few millimetres in order to decrease the path scattered light has to travel, decreasing multiple scattering. If moved back too far, however, scattering from the cell wall may occur —a prime source of stray light. This is easy to detect as the count rate increases, the I/B ratio decreases, and the calculated particle size increases (local oscillator effect).

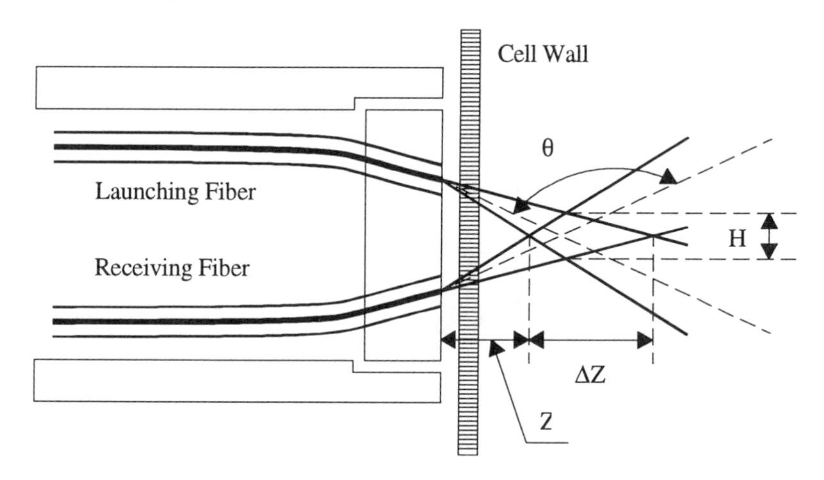

Figure 3. Close up of FOQELS probe.

In the current design the cell or window wall is 1 mm; however, special probes allowing up to 5 mm thick walls have been made. There is a change in angle due to the refractive index difference between the air and the liquid; however, for simplicity, it is not shown in this diagram.

The scattering volume is the cross-hatched area. It can be calculated by adding the volumes of two, right angle, circular cones that share a base of diameter H with total length of ΔZ. Two probes were used in this study, Type I and II. The volume of the Type I probe is 6.7×10^{-4} cm^3 and that of the Type II probe is 1.0×10^{-5} cm^3. For Type I $Z = 1.8$ mm and $\Delta Z = 1.5$ mm, and for Type II probe $Z=3.0$ mm and $\Delta Z = 4$ mm. By reducing the distance light must travel inside the cell by a factor of 3 or 4, measurements with a Type II probe should allow for higher concentrations before multiple scattering becomes a problem.

Compared to a standard setup (lenses, pinholes), either probe allows the measurement of higher concentrations because the scattering volume is much closer to the edge of the cell or window wall. Thus, multiple scattering from particles between the cell/window wall and the center of a standard cell is much reduced.

Experimental Results

All the results reported here were for samples in water. All the sizes were calculated assuming the viscosity is the same as the pure liquid. In all the measurements reported here a Type II probe was used, except for the comparison on TiO_2.

Figure 4 shows the results for TiO_2. The Effective Diameter is plotted as a function of concentration. Eff. Dia. is obtained from the first cumulant as explained above. With the Type II probe, concentrations as high as 1.2% w/v can be reached before a significant decrease in apparent size occurs. The Eff. Dia. obtained with a Type I probe shows a significant decrease in Eff. Dia. at concentrations above 0.3% w/v There is an increase in Eff. Dia. below 0.1% w/v using the Type I probe. In the plateau region the Eff. Dia. is approximately 275 nm.

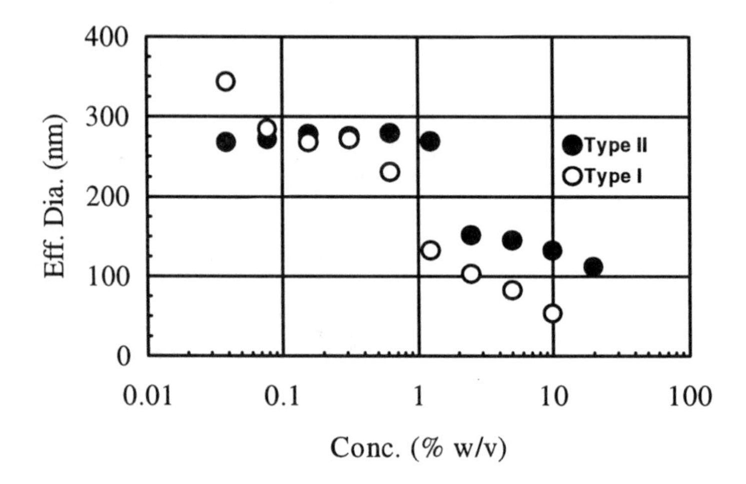

Figure 4. Size versus concentration for TiO_2 using two different probe types.

Figure 5 shows the results for Al_2O_3. In the plateau region the Eff. Dia. is approximately 155 nm until approximately 0.1% w/v. The curve has a mild negative slope until roughly 1% w/v, at which point there is a precipitous decrease in apparent Eff. Dia. In Figure 6 the I/B ratio, normalized to the lowest concentration, is plotted. Notice the decrease in I/B just above 0.1% w/v, mirroring the onset of the decrease in Eff. Dia. at about the same concentration.

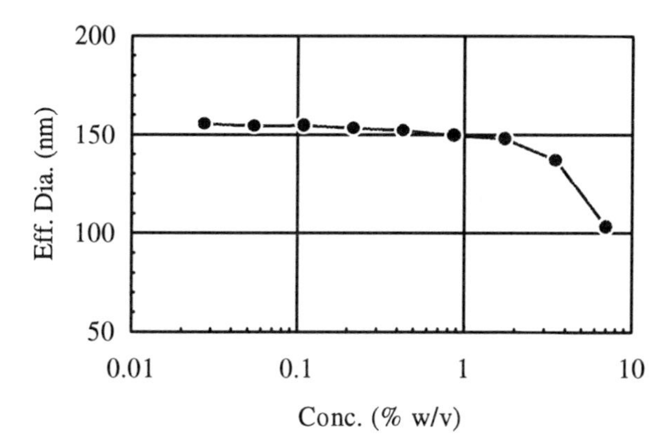

Figure 5. Size versus concentration for alumina.

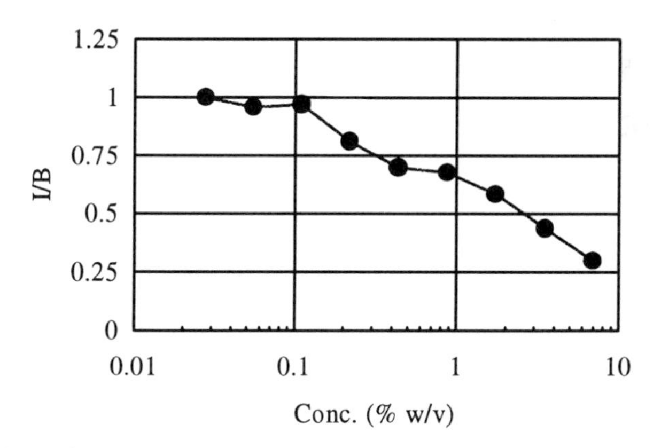

Figure 6. Intercept-to-baseline ratio versus concentration for Al_2O_3.

Figure 7 shows the results for an acrylic latex, one with a very low polydispersity. In the plateau region the Eff. Dia. is 130 nm. There is a slight increase around 1% w/v, followed by a more substantial decrease at about 3%, and finally a dramatic increase around 25%. This is the only sample that shows an oscillation before the onset of either a significant increase or decrease in Eff. Dia.

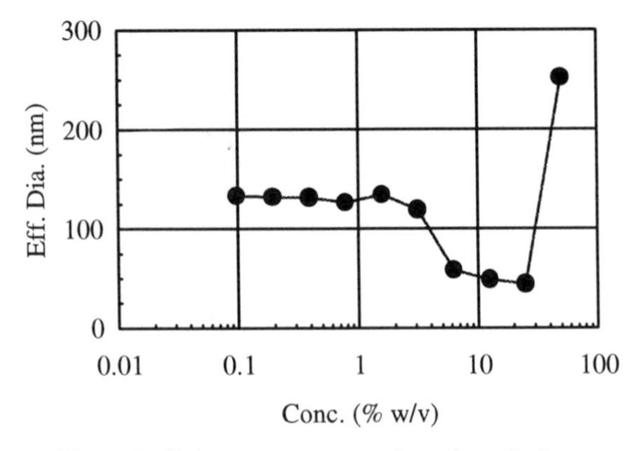

Figure 7. Size versus concentration of acrylic latex.

Figure 8 shows the results for a carbon black . In the plateau region the Eff. Dia. is 220 nm. There is a slight increase with concentration until approximately 0.25% w/v, above which the apparent Eff. Dia. rises dramatically.

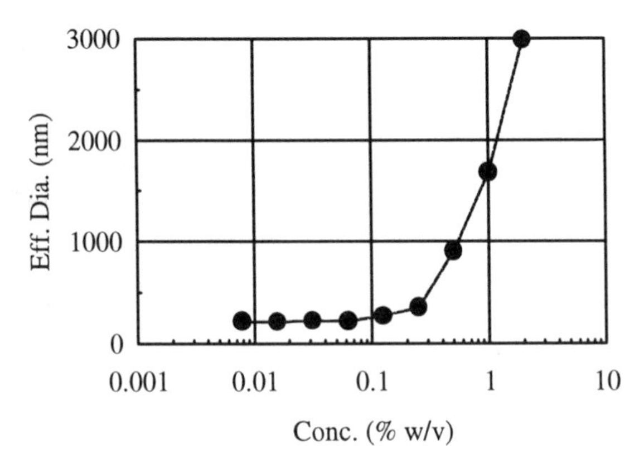

Figure 8. Size versus concentration for carbon black.

Figure 9 shows the results for Ludox, a well-known form of silica. In the plateau region the Eff. Dia. is 35 nm. There is a dramatic rise in the apparent Eff. Dia. starting at approximately 25% w/v. This is the highest concentration reached for any of the samples where measurements may still be used to determine a particle size.

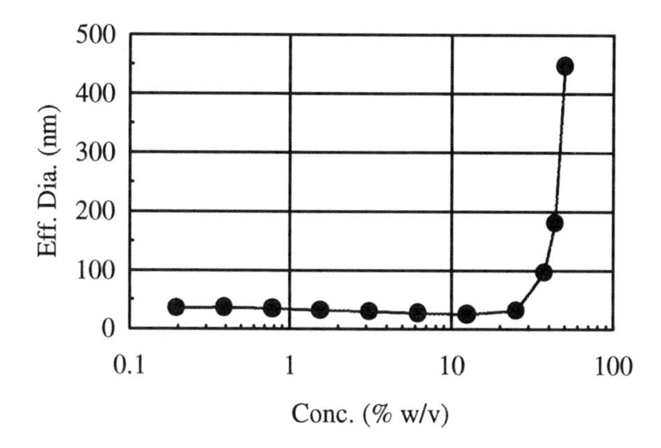

Figure 9. Size versus concentration for Ludox (silica).

Discussion of Results

The advantage of using the Type II probe is evident from Fig. 4, where the increase in the maximum concentration, over which the Eff. Dia. is constant, has increased by roughly a factor of 4, which reflects the decrease in path length of the beam in the cell by about the same factor.

In the discussions below the increase in Eff. Dia. with increasing concentration will not be ascribed to the local oscillator effect. The local oscillator effect occurs when any stray light is much larger than the scattered light. This is expected to occur at the lowest particle concentrations, and, when fully developed, it will double the particle size. This effect is only evident in the TiO_2 sample (Fig. 4) below 0.1% w/v, and only when the Type I probe was used which allowed the possibility of receiving flare light from the wall. The local oscillator effect cannot explain the large increase in Eff. Dia. at higher concentrations for any of the other samples using the Type II probe.

It is tempting to use the curves as is, without any further discussion, to estimate over which concentration range the FOQELS is useful for particle sizing. The maximum concentration, C_{max}, at which the apparent Eff. Dia. is still useful for estimating particle size, depends on the user's definition of acceptable differences. In the Table below we have made such an estimate for each curve. Though somewhat arbitrary, for the subsequent calculations the exact value of C_{max} is not critical.

Table: Calculating Diffusion Coefficient Type

	d_o	C_{max}	ρ_p	ϕ_v	r_m	D_{eff}
Silica	35 nm	0.25 g/mL	2.2 g/cm^3	0.102	60 nm	Mutual
Acrylic	130	0.03	1.2	0.024	360	Mixed
Al_2O_3	155	0.01	3.95	0.0025	920	Self
Carbon Blk	220	0.0025	1.9	0.0013	1,600	Self
TiO_2	275	0.012	4.2	0.0029	1,600	Self

Comparing r_m to the length scale determined by $2\pi q^{-1}$, allows one to assign D_{eff} to either self, mutual, or a mixture of these types of diffusion. In this way some of the changes in slope of the various curves may be explained.

The results for both Al_2O_3 and TiO_2 are easiest to understand. Both should display the effects of self diffusion, that is a modest rise in Eff. Dia. with increasing concentration. Instead both display a decrease of 30-40%, suggesting that multiple scattering dominates the results at higher concentration. Both are strong single scatterers on account of their size and refractive index, and multiple scattering is expected.

Comparing Figures 5 and 6 for Al_2O_3 one can argue that the Eff. Dia. does not decrease as rapidly as the I/B ratio. Finsy also noted this when working with polystyrene latex and a classical experimental setup that was not designed to probe higher concentrations.[8] He concluded that a little multiple scattering can be tolerated. Alternatively, at least in the case of Al_2O_3 , the rise expected from mutual diffusion may be partially compensated by multiple scattering in the middle concentration region. But eventually multiple scattering dominates.

For silica, which should display mutual diffusion effects, and assuming the repulsive forces dominate the particle interactions, one expects a decrease in particle size with increasing concentration. Instead there is a marked increase, more than an order of magnitude, which cannot be the result of a local oscillator. While it is possible that individual particles are aggregating at these higher concentrations, it is also known that silica forms reversible networks at these higher concentrations. Thus, the rise in the curve is probably due to structure.

It is hard to imagine that the dramatic rise in the carbon black curve, again more than an order of magnitude, can be solely explained by self diffusion. Viscoelastic measurements on similar samples also indicate that a structure is present in these samples at the higher concentrations shown here. (Kornbrekke, R., Lubrizol Corporation, Wickcliffe, Ohio, personal communication, 1996.) The sample is a weak scatterer, due to strong absorption in the visible, so it is no surprise that multiple scattering is not dominating the results.

The curve for acrylic latex, Figure 7, is perhaps the most interesting. It is the only curve that shows reversals in slope. This may be due to the fact that it is the only sample that is expected to show both mutual and self diffusion effects, the Eff. Dia. increasing and then decreasing with concentration. The rise by a factor of two at the highest concentration might be explained by the presence of colloidal crystals: an effect expected with charged systems such as this and ones that are very monodisperse as this sample was.

Based on the results for the acrylic latex, carbon black, and silica, we suggest that at some point in the concentration curve we are no longer measuring particle size; rather we are measuring the onset of various types of structures as suggested by Pusey and others[5]. If true, then Eff. Dia. is not the variable to plot. One should plot the relaxation time, $\tau_r = 1/\Gamma$. For the FOQELS, as employed in these measurements in water at 25 ^{o}C, each 100 nm in Eff. Dia. corresponds to 312 μs. Thus, the highest point on the carbon black curve of approximately 3,000 nm corresponds to a relaxation time of 9.4 ms.

Traditionally, in DLS the bulk viscosity of the liquid was used in the calculation of particle size, a value that was not compensated for by the presence of the particles.

As long as the sample was dilute, and the particles were not hindered in their ability to diffuse, the resulting particle sizes were consistent with the results from other techniques. Clearly, viscosity increases with increasing particle concentration, and one is tempted to use the bulk viscosity of the suspension to calculate particle size. However, as the viscosity increases with particle concentration, the calculated size is expected to decrease. While this effect would allow the size measurements to be extended over a higher concentration range for the latex, carbon black, and silica samples, it would not change the basic observation: the FOQELS technique as employed here is yielding information on structural changes in these samples.

Normally viscoelastic measurements are used to probe structure in complex colloidal suspensions. However, all such measurements require the application of a stress or shear. There is no such perturbation using light scattering. In this sense, one could think of the FOQELS in these particular applications as a zero-shear rheometer.

Summary

Using the FOQELS one can make useful particle sizing measurements at much higher concentrations than is currently possible with classic DLS instrumentation. The maximum concentration at which such measurements are useful depends on the strength of the single particle scattering and on particle-particle interactions. In particular, the appearance of either multiple scattering or structure limit the usefulness of the technique for particle sizing. However, the apparent onset of structure in complex fluids is itself an area where the FOQELS could be used.

Future measurements will include a probe where the path length in the cell is decreased further by the use of lenses, and measurements under a variety of salt and pH conditions, where applicable, would be useful in exploring particle-particle interactions. It would also be interesting to measure the bulk viscosity under the same conditions.

Acknowledgment

The authors want to thank Dr. Ian Morrison of the Xerox Corporation for pointing out that measurements of structure using light scattering are obtained at zero shear, a goal unattainable by standard rheometers.

Literature Cited

1. Finsy, R.; *Adv. Colloid Interface. Sci.* **1994**, *52*, 79-143.
2. Weiner, B. in *Particle Size Analysis*; Stanley-Wood, N.G. & Lines, R.G. Eds.; Royal Society of Chemistry, Cambridge, **1992**; pp. 173-185.
3. Van Keuren, E.R.; Wiese, H.; Horn, D.; *Langmuir* **1993**, *52*, 2883-2887.
4. Dhadwal, H.S.; Wilson, W.W.; Ansari, R.R.; Meyer, W.V.; *Proc. of Static and Dynamic Light Scattering in Medicine and Biology*; SPIE: New York, **1993**; Vol. 1884.
5. Pusey, P.N.; Jones, R.B.; *Annu. Rev. Phys. Chem.* **1991**, *42*, 137.
6. Dhont, J.K.G.; De Kruif, C.G.; Vrij, A.J.; *J. Colloid Interface Sci.* **1985**, *105*, 539.
7. Weitz, D.A.; Pine, D.J. in *Dynamic Light Scattering, The Method and Some Applications*; Brown, W. ed.; Oxford University Press, **1993**.
8. Finsy, R.; Ref. 1 above, pp. 91-94.

Chapter 8

Improvements in Accuracy and Speed Using the Time-of-Transition Method and Dynamic Image Analysis for Particle Sizing

Some Real-World Examples

B. B. Weiner[1], Walther W. Tscharnuter[1], and N. Karasikov[2]

[1]Brookhaven Instruments Corporation, 750 Blue Point Road, Holtsville, NY 11742
[2]Galai Productions Ltd., Migdal Haemek, 10500 Israel

The theory behind the particle sizing technique known as Time-of-Transition is discussed in detail. Specifically, sources of error are enumerated and accounted for quantitatively. Improvements in the accuracy of measuring large, airborne particles is shown to be a result of increasing the speed of rotation and the deflection angle of the wedge prism as well as increasing the data acquisition rate. Several practical example of the technique are demonstrated as well as one example involving image analysis.

There are many techniques for measuring particle size, especially above one micron: sieves, optical microscopy, sedimentation, laser diffraction, electrical and optical zone counters. Each technique has its own advantages and disadvantages. The zone counters yield high resolution, but cannot be used with airborne particles or particles on a microscope slide. Diffraction instruments yield results quickly, simply and with good repeatability, but they suffer from low resolution. Instruments based on sedimentation yield good resolution and are accurate, but the measurement time is long with broad distributions. Microscopy yields important information about shape, but is relatively slow, especially for broad distributions. Sieves are very inexpensive, but yield low resolution size distribution information and require skill in their use and long-term maintenance.

A technique known as time-of-transition (TOT) was introduced a few years ago (*1,2*). It offers an interesting combination of advantages. Measurements are made on single particles, so the resolution is relatively high. Unlike the zone-counters, the size is not determined by the pulse height. Instead, the size is determined by the pulse width. Therefore, calibration is not required, and, to first order, the results are not dependent on the optical properties of the particle.

Early implementations of the TOT technique were relatively slow, for reasons explained in the theory section of this paper. As a result, the TOT technique did not find as many useful applications as it does today. The purpose of this paper is to show the limitations as they exist today, discuss the improvements, and present some interesting data taken in the past few years.

Theory

Fundamental Principle of TOT. Figure 1 shows the basic optics. A collimated HeNe laser beam, wavelength λ = 632.8 nm, passes through a wedge prism (WP) which causes the beam to deviate from the optical axis by the deflection angle θ_d. The WP is made to rotate at an angular frequency $\omega = 2\pi\nu$. A lens (LA) focuses the beam down to a spot size ($1/e^2$ diameter) of 1.2 microns using a lens of focal length F. The rotating, focused, deflected beam describes a circle of diameter D in space, where, for simplicity, it is assumed, for the moment, that $D >> d_p$, the particle diameter. Particles are presented in a variety of ways to the beam: entrained in flowing or stirred liquid; on a raster-scanned microscope slide; entrained in flowing air; or simply sedimenting in air or liquid. A photodiode is placed directly behind the particles, perpendicular to the optical axis.

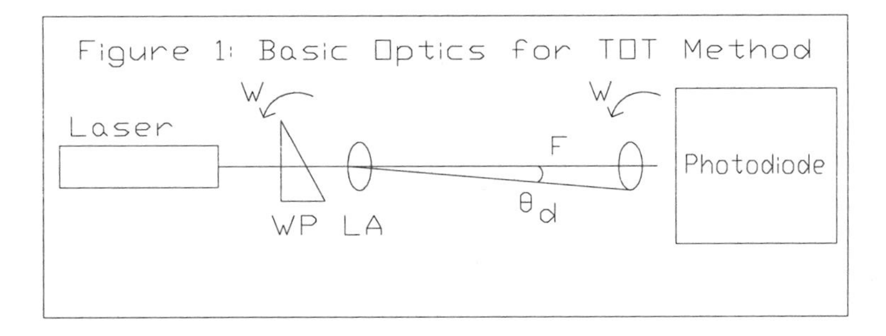

The signal on the photodiode is lower during the time the beam is crossing the particle. The ideal case is shown in Figure 2. By measuring the pulse width $\Delta\tau$ and multiplying by the tangential velocity, $V_T = \omega F \tan\theta_d$, one obtains the distance the beam traveled across the particle. Equating this distance with the particle size makes the TOT technique one of the simplest available for calculating particle size from raw data.

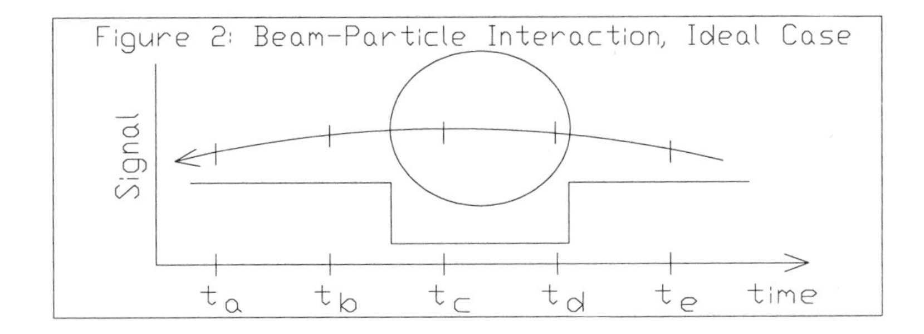

The technique is easily adaptable to different size ranges by changing the focal length, the deflection angle, and the rotational frequency. For the results described in this paper, two focal lengths F were used: 1.6 cm and 3.15 cm, corresponding to spot sizes of 1.2 μ and 2.4 μ, respectively. One version of the instrument, the Galai CIS-1, uses a rotational frequency of 80 Hz; a faster version, the Galai CIS-100, uses 200 Hz. Three deflection angles have been used: 1° or 2° with the CIS-1 and 4° with the CIS-100 . The configurations and resulting V_T are summarized in Table I.

Table I: Tangential Velocities

	CIS-1 (ν=80 Hz)		CIS-100 (ν=200 Hz)	
θ_d	F = 1.6 cm	F=3.15 cm	F = 1.6 cm	F=3.15 cm
1°	0.140 μ/μs	0.276 μ/μs	------------	------------
2°	0.281 μ/μs	0.553 μ/μs	------------	------------
4°	------------	------------	1.41 μ/μs	2.77 μ/μs

The lower rotational frequency and smaller deflection angles of the CIS-1 make it difficult to measure accurately larger particles, especially ones sedimenting under gravity in air. They either move too fast or are not small compared to the circling diameter D. By using a 4° instead of a 1° or 2° deflection angle, by increasing the rotational frequency from 80 Hz to 200 Hz, and by increasing the data rate from 1 MHz to 10 MHz one can now make measurements that are an order of magnitude faster. Thus, large particles from dry powders can now be measured in air, although for large and dense particles, one is still limited to either suspending the particles in a flowing liquid or presentation on a moving microscope slide covered with a sparse, random layer of the dry particles.

In practice, there are potential errors in applying this technique. Each is described below along with the method used to correct or compensate for the errors involved.

Off-Center Interactions: Chords. Figure 3 shows a typical off-center interaction between the circling beam and the particle: the beam transits across a chord. If such a pulse were allowed, the resulting size distribution would be broadened and shifted to smaller average sizes.

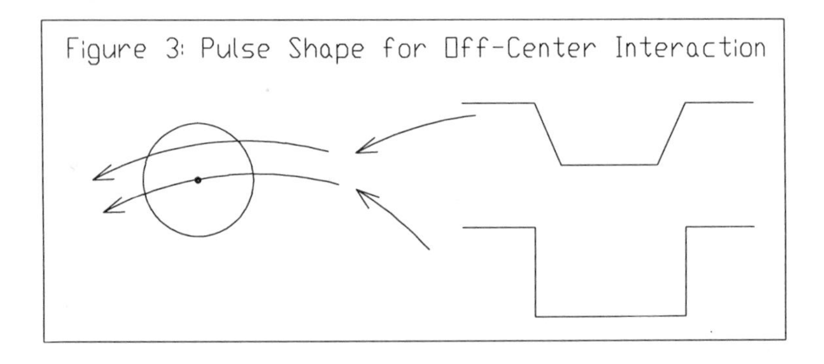
Figure 3: Pulse Shape for Off-Center Interaction

The pulse shapes for a transit across a diameter and across a chord are also shown in Figure 3. Note the faster rise and fall times (higher absolute slopes) for the transit across the diameter. In the ideal case the minimum rise/fall time is set by the spot size and V_T. Pulses with rise/fall times greater than a few times the minimum are rejected and do not effect the resulting size distribution.

If the criterion is too strict, many transit crossings are rejected and the statistical error is high; however, the distribution width is more accurate, allowing good resolution of multiple peak distributions. If the criterion is too lenient, the statistical error is low; however, the distribution is artificially broadened, lowering the resolution.

By default the criterion is set such that the standard deviation of the distribution of a monodisperse sample is about ± 17%. An alternate setting yields ± 10%, although the experiment duration increases. These values place a practical limit on the resolution of the instrument. Electro- and optical zone counters as well as field-flow fractionation devices offer higher resolution, some capable of resolving peaks just a few percent apart. However, compared to Fraunhofer diffraction devices, the resolution offered by the TOT method is much higher.

To test the limits of the theory, the criterion for rejecting chords was set very tight, resulting in a measured standard deviation from a monodisperse sample of ± 2.5%. This shows the ultimate capability for resolving peaks is quite high, though not necessary in many practical cases of interest.

Out of Focus Particles. Figure 4 shows an expanded view of the focused beam. The radius w_o of the beam at the waist is 0.6 μ (when F = 1.6 cm), and the radius w(z) a distance z away is given (*3*) by:

$$w(z) = w_o \cdot \sqrt{\left(1 + \left(\frac{\lambda \cdot z}{\pi \cdot w_o^2}\right)^2\right)}$$

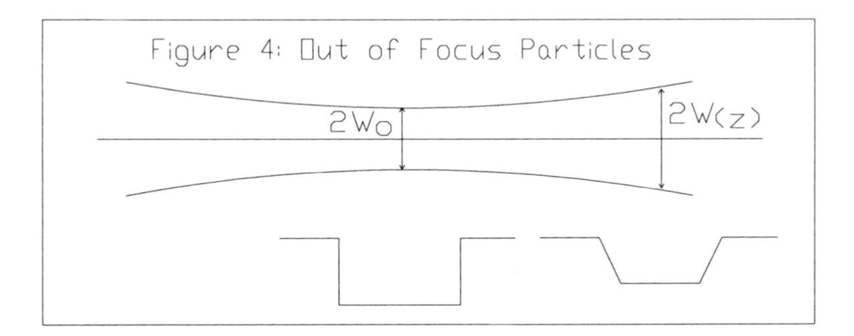

When a particle is out of focus it appears larger with a blurred boundary. In addition, the intensity per unit area of the beam is less than it is at the waist. The resulting pulses, one due to a transit at the waist and one due to a transit at w(z), are also shown. The pulse associated with a transit at w(z) has a smaller amplitude, a greater width, and a longer rise/fall time than for a pulse associated with a transit at the waist. If the rise/fall times are not equal and sufficiently short, the pulse is rejected as coming from a chord. In addition, by comparing the pulse width with the normalized pulse amplitude, one gets an additional rejection criterion. Specifically, an otherwise acceptable pulse is rejected unless it passes the normalized amplitude criterion. In this way it is possible to reject out of focus particles.

Beam Diameter and Small Particle Limitation. When the focused beam diameter (1.2 μ, F = 1.6 cm) is no longer small compared to the particle size, the pulse shape is no longer ideal: the leading and trailing edges are rounded. Since the beam has a Gaussian profile (TEM_{00}), it is not difficult to deconvolute the measured pulse to produce a pulse corresponding to the particle size only. From this deconvoluted pulse the size is determined. In this way particles as small as 0.5 μ have been determined with good agreement. For presentation purposes only, the results may be extrapolated to 0.1 μ. By decreasing the wavelength and spot size, one could, in principle, make measurements down to approximately 0.3 μ. The lower practical limit is set by the divergence of the beam, and for the current configurations, using a HeNe laser, this is 0.5 μ

The accuracy of the size obtained from the deconvolution, especially for particles less than about 10 μ, depends on the accuracy with which the focused beam diameter is known. For this reason, the optics must be kept clean; otherwise, the spot size increases and the resulting particle sizes are overestimated. In critical cases, in this lower size range, use of a good quality quartz cell reduces the spot size to its theoretical limit compared to a slightly larger spot induced by the optical quality of some inexpensive, common, plastic cells. The repeatability and resolution, however, are not affected. If they are the primary concern, then a disposable plastic cell is preferred.

Circling Beam Diameter and Large Particles. The diameter of the circling beam is given by $D = 2F \tan\theta_d$. With $\theta_d = 4°$ and F = 3.15 cm, D is 4.4 mm. As long as the particle diameter d_p is much smaller than D, the curvature of the circle as it transits across the particle can be ignored. Figure 5 shows the true path of a beam when D is not very much greater than d_p.

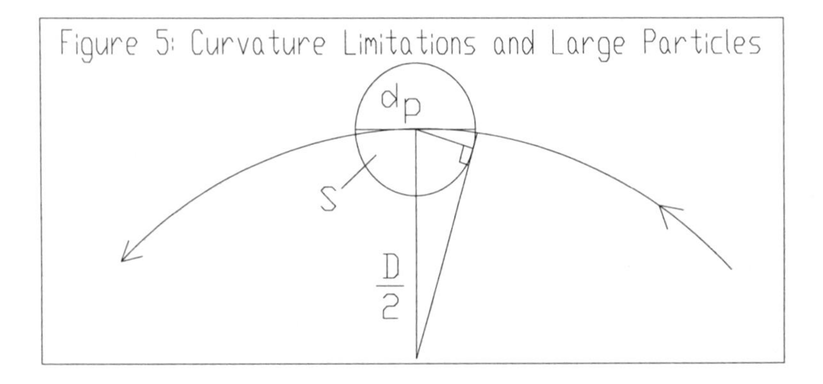

Figure 5: Curvature Limitations and Large Particles

The arc S is always greater than the particle diameter d_p . The pulse width yields the arc length, not the particle diameter. The relationship between the two is

$$d_p / D = \sin(S / D)$$

This equation is used to correct for the difference. In the worst case ($\theta_d = 4°$, F = 3.15 cm, d_p = 3,600 μ), the arc S is 17% larger than d_p. For all other configurations the difference is smaller and in all cases easy to compensate. In the most common configurations, the correction amounts to less than a few percent.

Particle Motion. Figure 6 shows four positions on the circling beam at which a particle moving with a velocity $V_\perp$ perpendicular to the optical axis might interact. The measured pulse width in all four cases corresponds to some distance x that is related to the actual particle diameter d_p .

The apparent speed of the beam, V_R, is obtained from the vector sum of the tangential velocity and $V_\perp$. It is given by:

$$V_R = V_T \cdot \sqrt{\left(1 + \frac{1}{R^2} - \frac{2\cos\theta}{R}\right)}$$

where $R \equiv V_T / V_\perp$ and θ is the angle from the x-axis to the center of the particle. The apparent size of the particle $x = \Delta\tau V_R$; whereas, the particle diameter is $d_p = \Delta\tau V_T$. Assuming the beam and particle are equally likely to interact at any position over 2π radians, the above equation can be integrated with the following results:

$$\bar{x} \cong d_p \cdot \left(1 + \frac{1}{R^2}\right)^{1/2} \qquad \frac{\sigma}{\bar{x}} \cong \frac{1}{2R\left(1 + \frac{1}{R^2}\right)}$$

where x-bar is the mean and σ/(x-bar) is the relative standard deviation. As long as the flow velocity is set to 1/5th or less of the tangential velocity (R≥5), the mean is within 2% of the true diameter, and the relative std. dev. is 10% or less. Given the 10-fold increase in tangential velocity of the CIS-100 over the CIS-1, a greater range of samples can now be measured accurately.

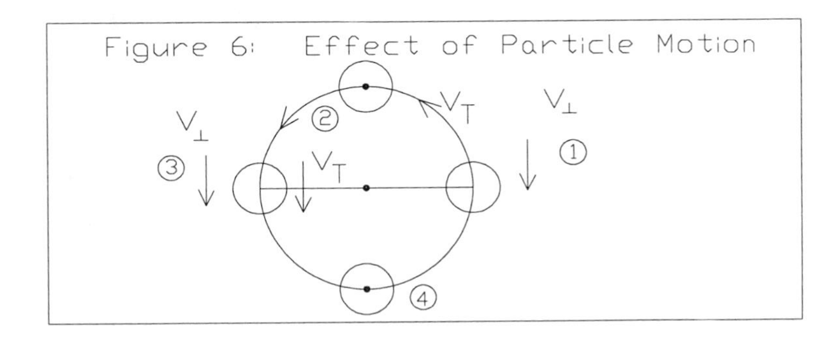

Figure 6: Effect of Particle Motion

Concentration Limitations. The lower limit of detection is one particle. For this reason, the TOT method is capable of measuring much lower concentrations than Fraunhofer, sieve, sedimentation, and field-flow fractionation devices; however, it is not as sensitive as a good single particle counter (SPC). A good SPC device counts all the particles in a precisely known volume, yielding the absolute concentration in various particle size classes. A SPC is sometimes employed more for its ability to measure the concentration than for its particle sizing accuracy or resolution.

The TOT method is not a suitable substitute for a good SPC for many reasons. Not all particles are counted in the TOT method. Some may be out of focus and rejected. The beam may transit across a chord and reject the particle. Finally, as will be seen below, a volume dependent on the particle size can be optically defined, and this is not as accurate as counting all the particles in a precisely known volume.

Nevertheless, if one assumes that the particles are not segregated by size as they are presented to the circling beam, then, on average, one can assume the calculated concentration is a reasonable, first-order approximation to the true concentration.

The upper limit of detection is set by the need to avoid overlapping particles which results in shifts to apparently larger sizes. Only one particle is allowed in a volume the shape of a rectangular parallelepiped. Two sides of the volume are set equal to twice the particle diameter since particles cannot get any closer than d_p apart. The long dimension is set to 4z, where z is determined by the criterion that the particle diameter $d_p = 2\ w(z)$. This optically defined volume leads to the following relationship when $d_p/2w_o) >> 1$:

$$\Phi_M = \frac{\lambda}{48 \cdot w_0}$$

where Φ_M is the maximum, theoretical volume fraction. Thus, for particle diameters larger than about 5 μ, this theoretical limit in water ($\lambda = \lambda_0 / n$) is 1.7% v/v. Measurements on 2 μ fat particles in milk have been made successfully in a specially designed flow cell up to 0.42% v/v which corresponds to 10^{+9} particles/cc. The path length in the cell was 1 mm, allowing the measurement of a milky looking sample. The standard path length is 10 mm. Such high concentration measurements are not possible with Fraunhofer devices, because, unlike the TOT method, the beam is not highly focused.

The above calculation for the maximum volume fraction is based on theory. It is a limiting value. Turbidity and multiple scattering, both complicated functions of concentration, particle size, and refractive indices, set the practical limit. As a first approximation for an unknown sample, a concentration of 0.2% v/v is a good starting point.

Nonspherical Particles. When the circling beam encounters a nonspherical particle, many of the interactions of the beam and particle will lead to asymmetric pulses which are rejected. Consider, as an extreme case, a long, thin rod *randomly* oriented as it passes through the circling beam. Interactions perpendicular to the length of the rod will lead to acceptable pulses; the size recorded will correspond to the width of the particle. There will be fewer acceptable interactions along the length of the particle, provided the length is small compared to the circling diameter.

The result, in this extreme case, is an apparent size distribution favoring sizes close to the rod's width, with the possibility of a second mode corresponding to the length of the rod. It may be argued that the TOT technique is not suitable for such highly nonspherical particles. The argument is correct, but no more so than for any other nonimaging device, where the size obtained is always a function of the particle shape. For globular particles the TOT technique is acceptable. For highly irregularly shaped particles, it is important to characterize the shape and size through image analysis. In such cases, the results obtained using the TOT method alone may not be sufficient to characterize the particle size distribution.

Interestingly, the simplicity of the TOT optical design allows images to be obtained by inserting optics at right angles to the laser beam. A synchronized flash and a CCD camera allow measurements from approximately 2 μ to 3,600 μ. Images provide not only the opportunity for particle sizing but also the opportunity to inspect the state of aggregation. An example of image analysis is given in this paper.

Measurements and Discussion

The TOT technique has been used for about a decade to solve both routine and complex problems in particle sizing. A few have been chosen to illustrate some of the capabilities of this technique.

Accuracy: BCR 67, Standard Quartz Powder. The results shown in Figure 7 demonstrate the basic accuracy of the TOT method. The sample, BCR 67, is a standard quartz powder (4) with a broad size distribution from 2.4 μ to 32 μ. The sample was prepared as a suspension in water using 0.1% sodium hexametaphosphate as a wetting/dispersing agent. Approximately 2 cc were placed in a standard, plastic cuvette. During the few minutes it takes to make the measurement, the particles were kept in suspension using a magnetic stir bar. Note the good agreement between the standard values obtained from sedimentation and the TOT results.

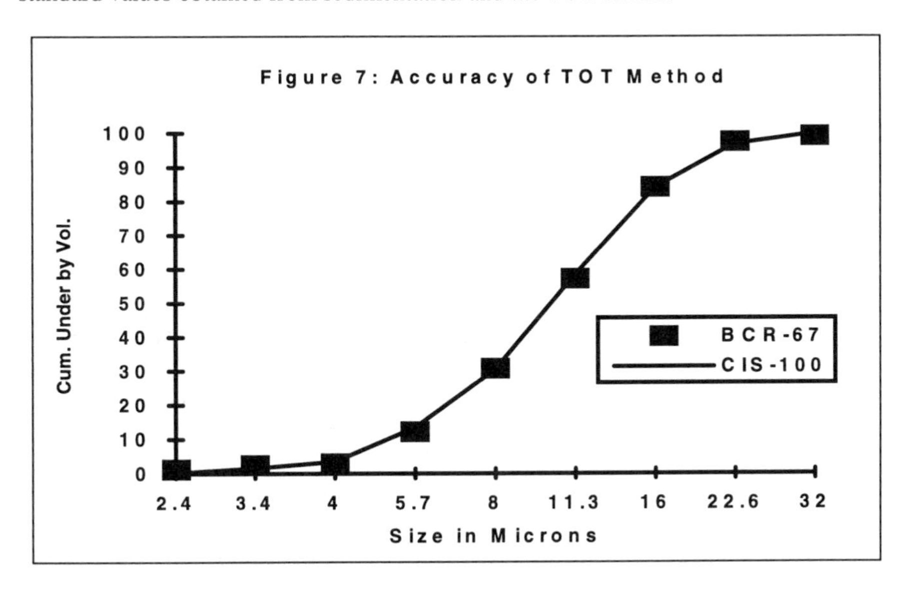

Large, Airborne Particles: Farina (Wheat). Farina is ground wheat, sieved in order to produce various fractions. The coarse fractions are used in making specialty foods like Cuscus, a Mediterranean delicacy. A fine fraction is used to make flour for bread. A super fine fraction is used in the production of baby foods. The production costs and selling price are functions of the particle size distribution. (Galam Ltd., Kibbutz Mahanit, Israel. Galam is a manufacturer of Farina.) Due to absorption, measurements cannot be made in liquid.

Figure 8 shows the overlap of three runs made on different dates: May 19, June 12, and June 19, 1995. The cumulative undersize distributions by volume are shown. The volume median diameter is around 500 μ for all three, with the curve for the run on May 19 shifted to distinctly smaller sizes, indicating this sample was suitable for use as a higher priced baby food fraction.

All three samples cover a broad range of sizes from a few 10's of microns to just under 1 mm. Measurements took approximately 2 minutes using a vibrating platform, similar to a riffler, to feed randomly these free-flowing particles into a 1 cm x 1 cm flow cell. Gentle suction is applied to assist the gravity feed while making the measurements.

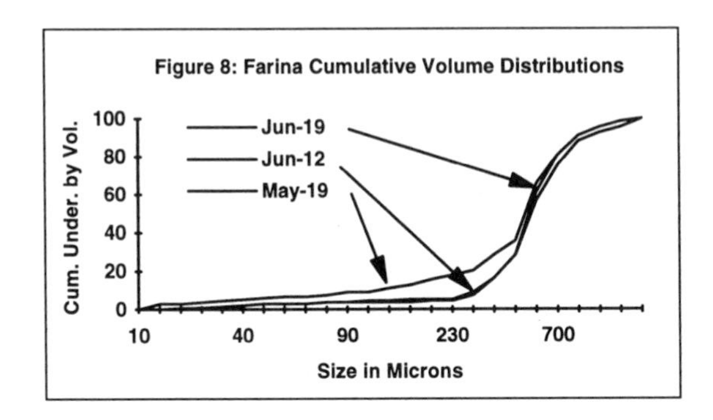

Figure 8: Farina Cumulative Volume Distributions

Oriented Rods: Mixture of Hardwood and Softwood Fibers. In the manufacture of paper, a mixture of hardwood (HW) and softwood (SW) fibers is used. Softwood is harder to shear, folds more easily, and is less expensive. Hardwood is easier to shear, folds less easily, and is more expensive. The proper mixture provides the desired properties at the right cost. During the milling process both HW and SW species are fed continuously into headboxes. It is desirable to adjust, in real-time, the ratio of HW to SW to get the final, desired properties.

Typical hardwoods used in paper making include the following: redgum, birch, aspen, oak, eucalyptus, and gmelina. Unprocessed, they vary in length from roughly 300 μ to 1,200 μ, and the widths vary from 15 to 30 μ. Typical softwoods used in paper making include southern yellow pine and western hemlock. Unprocessed, they vary in length from 2 to 8 mm, and the widths vary from 25 to 45 μ (*5*).

During processing the long fibers are chopped into a variety of lengths. Thus, the length distributions, and therefore the volume distributions, are broad and overlapping. Even though the width distributions overlap slightly, they characterize more sharply the differences between the HW and SW, because the widths remain nearly unchanged by the processing. By orienting the fibers in a flowing cell and measuring the fiber widths using the TOT method, one can reconstruct the volume distribution without ever measuring the fiber lengths.

Consider a long rod of diameter d_p and length L made to flow with speed $V_\perp$ perpendicular to the tangential speed V_T. As long as both $V_\perp$ and V_T are constant, and $V_T >> V_\perp$ the long rod will be cut into N_c pieces, each of length $\Delta \ell$. The volume V and length L of a rod are given by

$$V = \frac{\pi}{4} d_p^2 L \qquad where \quad L = N_c \Delta \mathtt{l}$$

The *surface area* of N_c *spheres*, each with a diameter of d_p is given by

$$A = \pi \, d_p^2 \, N_c$$

From these relationships one can show that the differential volume distribution is given by

$$\frac{dV}{dd_p} = \frac{\pi}{4} \cdot \Delta \mathtt{l} \cdot \mathrm{d}_p^2 \cdot \frac{dN}{dd_p}$$

where N is equal to the sum of N_c over all the rods. In addition, one can show that the differential surface area distribution of N total spheres with diameter d_p is given by

$$\frac{dA}{dd_p} = \pi \cdot d_p^2 \cdot \frac{dN}{dd_p}$$

where it should be emphasized that this differential surface area distribution formed by cutting equal slices from oriented rods, is not the surface area distribution of the rods themselves. Combining the above result with that for the differential volume distribution yields

$$\frac{dV}{dd_p} = \frac{\Delta \ell}{4} \cdot \frac{dA}{dd_p}$$

This rather interesting result says that the differential volume distribution with respect to the width of the rod is proportional to the differential surface area distribution of spheres whose diameters equal the widths of the rods. There is no need to know the rod lengths.

It turns out that the distributions obtained are not easily separable into two modes: one corresponding to the HW width and one to the SW width. However, by calibrating using a run of pure HW and pure SW, one can establish a working formula for calculating the ratio of HW:SW from the cumulative distribution obtained from a mixture. Figure 9 shows an example. Notice it is the cumulative undersize surface area distribution that is plotted as a function of particle size.

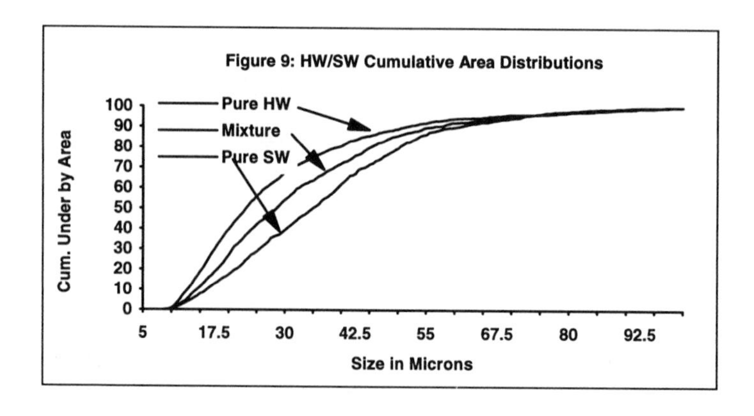

Using Pulse Shape to Discriminate: On-line Measurements of Oil/Sand Mixtures. Many single particle counters determine particle size by measuring the height of an electrical pulse and compare it to pulses generated using standards. In the TOT method it is the pulse width that determines particle size; calibration is not required. However, the pulse shape does play a role in deciding if the pulse is processed or not. Asymmetric pulses, pulses with long rise/fall times, and pulses with insufficient amplitude are normally rejected as corresponding to highly irregular shapes, chords, or out-of-focus particles. In special cases, with *a-priori* knowledge, asymmetric pulses are used to discriminate against two different particle types. This capability has been used to discriminate the particle size distribution of oil droplets remaining in sea water after processing in the presence of a background of sand particles.

Figure 10 shows two pulses. The more symmetric pulse is typical of an opaque oil droplet in sea water, the droplets remaining after most of the oil has been separated using a combination of holding tanks and hydrocyclones (*6*). The less symmetric pulse is typical of sand particles. The shape of the pulse from sand is believed to arise from internal reflections that are not present in a dark, strongly absorbing oil droplet. Since the pulse widths are roughly equal, it would be impossible to discriminate oil from sand based on size alone.

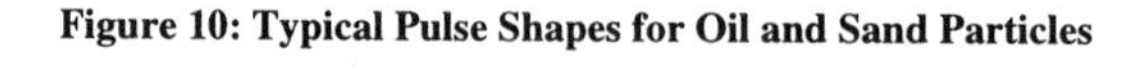
Figure 10: Typical Pulse Shapes for Oil and Sand Particles

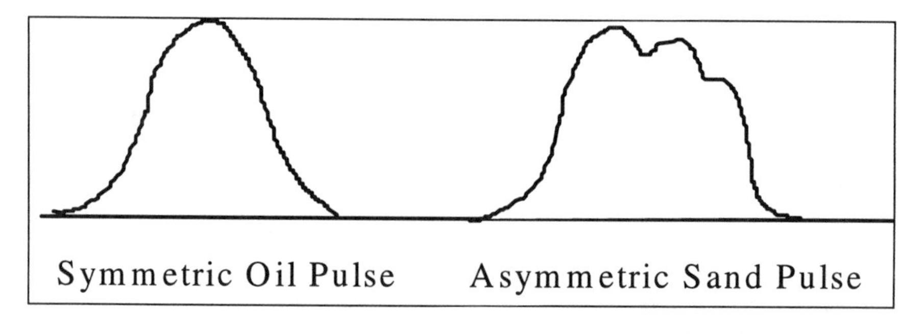

An obscuration ratio is defined as the ratio of the (rise time)/(fall time). Opaque oil droplets have characteristically higher obscuration ratios at the same particle size than do sand particles, and a threshold can be established as a function of particle size. Pulses with obscurations exceeding the threshold correspond to oil droplets and are processed further to determine the particle size; pulses with obscurations less than the threshold correspond to sand particles are not processed.

Figure 11 shows the overlay of two differential volume size distributions obtained on the same sample of oil droplets in salt water spiked with 20 μ sand particles. The distribution marked Standard, with the strong peak at 20 μ, was obtained using all the pulses. The distribution marked Special, without the peak, was obtained using the threshold criterion set by the obscuration. No trace of the sand peak is evident.

These measurements have been made with both a laboratory instrument and a dedicated, on-line, flow instrument with results equal to those shown here (*7*).

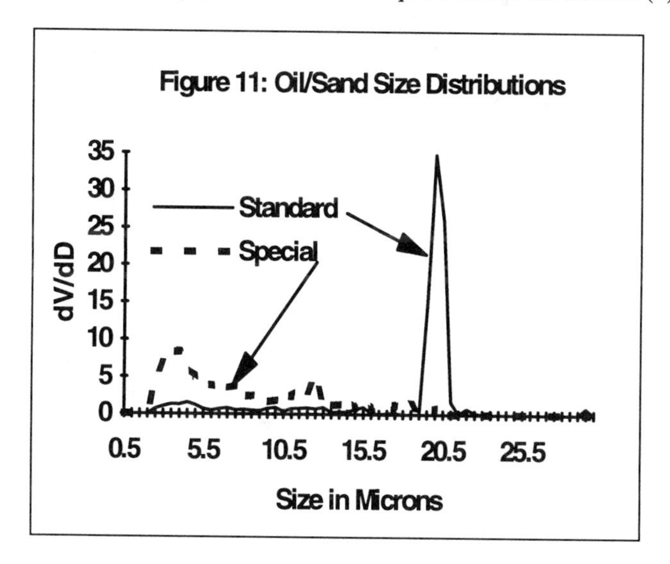

Dynamic Shape Analysis: Fly Ash/Cement Mixtures. Not all problems in particle sizing can be solved by assuming a spherical shape. Sometimes the information obtained from imaging is necessary. A good example of this consists of determining the amount of fly ash in a mixture with cement. Such a mixture is used in the making of concrete for surfacing roads. Fly ash is much less expensive than concrete, yet if it constitutes more than 20% of the mixture, the final properties of the concrete suffer (*8*).

Figure 12 shows a few typical renderings of fly ash and cement particles. Three things are evident: there is not a significant difference in size between the two types of particles; there is not a significant difference between the aspect ratios (length/width) between the two types of particles (both are globular); but the perimeter of fly ash particles is typically smoother than the perimeter of cement particles. This difference in perimeter can be exploited using the shape factor.

The shape factor, SF, is here defined as $4\pi \bullet Area/(Perimeter)^2$, where the Area and Perimeter are determined from the number and size of the pixels covering the particle and lying on the external contour. Using the definition, SF is exactly 1 for a circle, 0.785 for a square, and approaches zero for a long, thin rectangle. For smooth globular particles like Fly Ash, the SF is closer to 1 than for jagged-perimeter particles like cement with their longer perimeters.

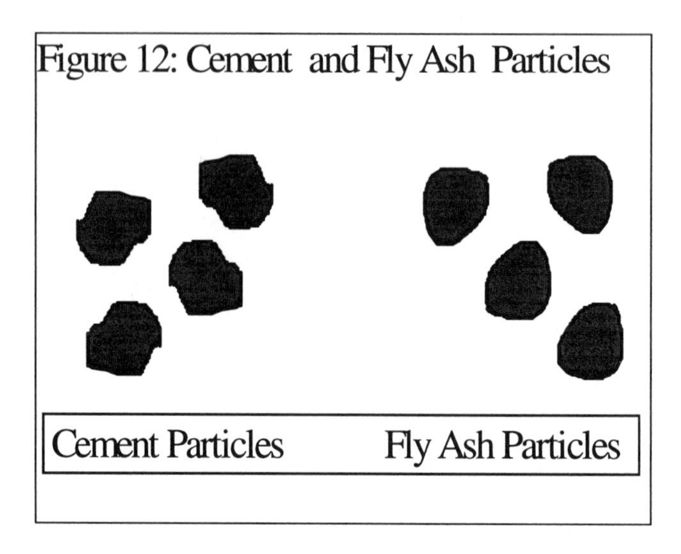

Figure 13 shows the cumulative SF distributions for cement and fly ash separately. As expected, the shape factors for cement are smaller, with approximately 100% less than 0.90. From the cumulative SF curve corresponding to the mixture, the amount of cement is set empirically as the percent corresponding to a SF of 0.90. Mixtures corresponding to less than 80% cement are rejected.

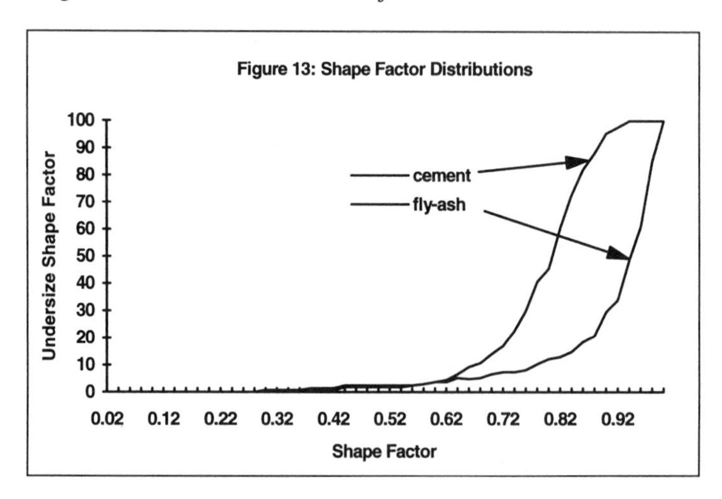

It should be noted that the images formed were obtained from particles falling through the measurement zone. As such, the images formed are dynamic, two dimensional images. This is different than forming the image from particles that have fallen onto a microscope slide (an alternate method possible for this instrument). For example, the stable position of platelet particles on a slide is face up, not standing on an edge. When analyzed, this static image will produce larger equivalent spherical sizes compared to an analysis of moving particles. Dynamic image analysis, also called dynamic shape characterization, offers the possibility of averaging size/shape information more realistically than traditional static image analysis. However, to date, there are no definitive studies on the differences. This remains a challenge for future investigation.

Summary

The TOT method presents an interesting blend of characteristics with the following disadvantages:

- It is relatively slow on broad distributions;
- Limited to particle sizes greater than ~ 0.5 μ ;
- Requires care in presenting large, dense particles in the measuring zone to avoid bias;

and the following advantages:

- Relatively high resolution from measurements on single particles;
- Straight forward data analysis;
- Can be made to operate over a wide size range;
- Very versatile, operating on particles in liquids, in air, or on transparent slides;
- Easily coupled to image analyzer, allowing size/shape analysis and visualization of shape and the state of aggregation.

Its unique capability of using pulse shape and height to ensure accuracy or to discriminate one type of particle from another, and pulse widths to determine particle size, make the TOT technique capable of solving some difficult, yet practical and interesting, particle sizing problems.

Literature Cited

1. Aharonson; Karasikov, N.; *J. Aerosol Science*, **1986**, *17*, 530-536.
2. Karasikov, N.; Krauss, M.; Barazani, G.; In *Particle Size Analysis*; Lloyd, P.J., Ed.; John Wiley & Sons: New York **1988**.
3. Melles-Griot Optics Catalogue; Melles-Griot: Irvine, California, **1995-1996**, 2-7.
4. BCR standard samples are available from the Commission of the European Communities, Community Bureau of Reference (BCR), Directorate General XII, 200 Rue de la Loi, B1049 Brussels, Belgium.

5. Kocurek, M.J.; Stevens, C.F.B.; In *Pulp and Paper Manufacture*; TAPPI: Atlanta, Georgia, 1989, Vol. 1.
6. Davies, R.H.S.; Palmer, A.J.; "Use of Hydrocylones for Solids Separation and Cleaning Applications", Presented at *4th International Conference on Water Management Offshore*, Aberdeen, Scotland, **1995**.
7. Tulloch, S.J.; "Evaluation of On-line Oil-in-Water Monitors and Particle Size Analysis In Relation to the Concentration and Drop Size Measurement of Dispersed Oil Droplets", *Project No. OWTC/90/059*; Orkney Water Test Center Ltd.: Orkney, U.K.
8. Gaskin, R.; *Laboratory Practice U.K.*, **1991**, *39*.

Chapter 9

Ultrafine Particle Size Measurement in the Range 0.003 to 6.5 Micrometers Using the Controlled Reference Method

Philip E. Plantz

Honeywell IAC, Microtrac Products, Mail Stop 191-1, 13350 U.S. Highway 19 North, Clearwater, FL 34624–7290

Dynamic light scattering (DLS) is a well established means of measuring particle size in the submicron range. DLS determines particle size from analysis of the Brownian motion of particles in suspension. Light that is incident on moving particles is scattered and undergoes a frequency shift. The frequency shift is the light equivalent of the well known Doppler shift of sound waves and can be analyzed to obtain information related to the particle size of the suspended particles. In many cases, the analysis is performed using the concepts embodied in photon correlation spectroscopy (PCS). This paper will describe a commercial DLS approach to analyzing the scattered light which we term the Controlled Reference Method (CRM) that allows measurement at high and low solids concentration without *a priori* assumptions of particle size distributions. Applications of this unique technology in industrial and pharmaceutical industries will be discussed using distribution graphs and tabular data. The CRM (heterodyne) approach will also be compared to conventional (homodyne) photon correlation spectroscopy.

Particle size measurement of submicron particles is difficult for several reasons, including limitations of light microscopy and electron microscopy. In the former case, particles in normal slurries cannot be confidently examined at sizes much smaller than 0.5 micrometers. In the latter case, considerable sample preparation, which often produces artifacts, is necessary and requires considerable expertise. Several approaches to measuring particle size from 0.003 micrometer (μm) to slightly above 1 μm have been attempted employing Dynamic Light Scattering (DLS) techniques. These methods embody monitoring the movement of particles as a result of solvent molecules colliding with the particles. These movements impart a frequency shift on the light that is incident on the particles. The light frequency shift is the counterpart of the well known Doppler frequency shift for sound waves.

Dynamic Light Scattering .

The earliest method for particle size determination using DLS was that of "homodyne" photon correlation spectroscopy. In PCS, there are several means of monitoring and analyzing the frequency shifts. Generally, these can be divided into two categories: multi- and single-angle scattering detection including selection of an acceptable algorithm and possibly *a priori* distribution assumptions. The multi-angle (goniometer) approach measures light at several angles ranging typically from 30 to 160 degrees to take advantage of the angular dependence of the light scattering signal. While the multi-angle approach is sometimes the preferred research method, it has difficulty with measuring high sample concentrations and requires specialized operator skills, measuring conditions and algorithms to yield useful results. A simplified version uses a single detection angle of 90°. The 90° single-angle photon correlation instruments provide a simplification of the multi-angle measurement, but usually require *a priori* assumptions concerning the particle size distribution. Both approaches embody analysis of light scattering intensity fluctuations, usually by concentrating on the time domain by the method of autocorrelation. Selection of one or more algorithms is needed to provide results in an intensity-weighted format and subsequent conversion to a volume-weighted format. Mathematical approaches have been developed to enhance the capability of PCS in measuring multimodal systems, but there is considerable controversy in the literature as to the best data treatment approach.(1). These issues have historically limited PCS application to well-behaved, dilute suspensions having relatively narrow monomodal distributions. This paper will describe an alternative approach to the use and analysis of Doppler shifted light that overcomes many of the inherent difficulties of measurement and data interpretation associated with Photon Correlation Spectroscopy. We refer to this approach as the Controlled Reference Method (CRM) known commercially as the Microtrac Ultrafine Particle Analyzer (UPA). In the scientific literature it is known as heterodyne dynamic light scattering.

Regardless of which measurement method is applied, DLS determines approximate particle size distributions from the analysis of the Brownian motion of suspended particles. Light which is incident on these particles is frequency shifted as a result of its being scattered by the moving particles. The shift in light frequency is analogous to that associated with the Doppler shift which occurs from a moving sound source in relation to a stationary observer or monitoring device (2). The moving particles give rise to a shift in the scattered light frequency relative to the incident light frequency that is related to the random diffusive movement and related velocity of the particles. The frequency shifted light carries information related to the size of the particles in suspension. Analysis of the frequency shift allows a computation of particle size distribution.

Photon Correlation Spectroscopy

The earlier approach to this analysis is conventional photon correlation spectroscopy (PCS), in which an autocorrelation function is developed from the scattered light photopulse signal produced by a photomultiplier tube detector. In this case the frequency shifted scattered light is self-referencing in that light scattered from one particle mixes or interferes with light scattered from all the other particles. This is termed homodyne scattering and is shown in Figure 1A. In both portions of Figure 1, a Lorentzian function of angular light frequencies, ω, is presented as the power spectrum. The homodyne mode is a "self-referencing" measurement as depicted by the power function being proportional to the square of only the scattered light intensity, I_s^2. When measuring in the homodyne mode, only the homodyne expression applies. The power spectrum is also a function of 2γ, where γ is a constant dependent upon the wavelength, scattering angle and the diffusion coefficient. Generally, the scattered light is measured at 90 degrees. The high optical frequency component is not detectable leaving only the combined frequency shifted light signal for analysis. In some devices, the analysis is performed in the time domain using an autocorrelator. From the autocorrelation function, one calculates a diffusion coefficient. The diffusion coefficient is inversely related to particle size through the Stokes-Einstein equation. In the simplest case of a monomodal, monodisperse system, analysis of the scattered intensity fluctuation signals yields an average decay constant, which is related to the mean diffusion coefficient and average intensity weighted particle size. In turn, the data are presented in the format of the mean diameter of an intensity-weighted distribution and a calculated breadth. Further analysis of the data is possible using a number of algorithms, including the method of cumulants and inverse Laplace transforms, such as, CONTIN. (1, 4-7) These approaches often invoke *a priori* assumptions that are very constraining, sometimes resulting in over-smoothing or are limited to multi-modal distributions of large size differences. (1,8) Discrimination of distribution tails or of multimodal distributions may not be possible except in special cases or by use of multi-angle PCS measurements. Current particle size analysts require greater definition and broader applicability than previously available capabilities. Such capabilities often can be obtained using the Controlled Reference Method, as described below.

Controlled Reference Method

Physics and Design. While conventional PCS measurement techniques usually employ homodyne scattering method, the Controlled Reference Method (CRM) utilizes heterodyne light scattering. In the heterodyne measurement (9), shown in Figure 1B, a portion of the light directed at the particulate suspension is diverted from the direction of the particles to act as a reference, or "local oscillator". The remaining light follows a path to the particulate suspension and becomes frequency shifted after being scattered by the particles. Light scattered at 180 degrees is combined coherently with the previously diverted light and follows a path to a silicon detector. The heterodyne measurement depends only upon γ, with the power spectrum proportional to the product of the scattered and reference intensities, I_s and I_o. The blending of I_s and I_o mixes the power spectra of the two Lorentzian widths γ and 2γ. However, by supplying a large enough reference intensity, the heterodyne component is made large compared to the homodyne component and dominates the mixed signal. The homodyne component, therefore, is negligible and non-interfering. The foregoing heterodyne approach is termed the Controlled Reference Method because of the use of a portion of the incident laser beam as a high intensity reference.

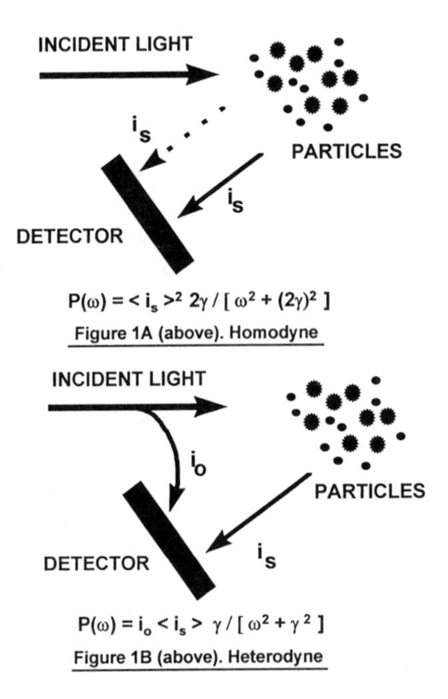

Figure 1. (A). Diagram showing self-referencing in homodyne scattering. (B). Diagram showing heterodyne scattering in which a portion of incident light is mixed with the scattered light. For (A) and (B), the related Lorentzian function is given below the corresponding drawing.

The combined scattered and reference (reflected) beams are detected using a silicon photodiode, the output of which resembles random noise. To allow analysis of the signals, the output is digitized to permit development of a frequency power spectrum using a digital signal processor and fast Fourier transform (FFT) algorithm. Power spectra for several different sizes of particles are shown in Figure 2. As the equations in Figure 1 show, the power spectrum takes the form of a Lorentzian function, in which the frequency shift is inversely proportional to size. Thus, power spectrum plots (Fig. 2) for different particle sizes show a shifts to higher frequencies as the size decreases. The power spectrum for a mixture of particles will have a power spectrum that is the sum of Lorentzian functions which is weighted by the volume concentration of each particle size. Additional weighting is performed by including the scattering efficiency for all the particles present in the mixture. The analysis routine deconvolves the combined power spectrum to produce a report of the particle size *distribution*.

The design is embodied in the Honeywell Microtrac Ultrafine Particle Analyzer (UPA) probe as shown in Figure 3. A waveguide is immersed in the particulate suspension. A portion of the input beam is reflected back to the detector as the result of the Fresnel reflection at the waveguide/suspension interface. The remaining light enters the sample, is frequency-shifted due to scattering by the particles, and is collected at 180 degrees. The reflected portion and the back-scattered portion are combined at the silicon detector to produce a stable, high intensity signal, having a frequency power spectrum which is related to the particle size distribution of the suspension. Proprietary computation is used to provide particle size distribution results.

Operational Advantages of the CRM. The CRM technique allows for full distribution measurement without special distributions assumptions. Thus multimodes, broad distributions, and mixtures can be measured directly without operator input of expected distribution characteristics or selection of algorithm type. In addition, the waveguide collects frequency shifted light over a very short pathlength, which allows the technique to demonstrate a very wide tolerance for variable sample concentration. As demonstrated in Figure 4 for measurement of several polystyrene samples, the sensitivity to concentration changes is far less, as compared to photon correlation spectroscopy (PCS) measurements. While the particle size reported by PCS is highly dependent upon polymer concentration, the Ultrafine Particle Analyzer(UPA) shows little response change over a wide concentration range. This comparison demonstrates that the precision of UPA data are little affected by concentrations up to 25 % latex solids. Depending upon material characteristics, this high concentration may be extendible to 40% solids. While evaluation of concentration effects should be performed for each material for a variety of concentrations, the Controlled Reference Method is designed to provide acceptably precise and accurate measurements at concentrations hundreds or thousands of times greater than those used in conventional PCS analysis. In addition, the graph shows the UPA to possess sensitivity to the lowest possible detection levels of PCS, while maintaining excellent precision. The combined use of probe technology and heterodyne mixing allows for unique opportunities in the particulate size measurement of non-diluted colloidal systems.

a = radius
η = viscosity
λ = wavelength
T = Absolute temperature
γ = frequency @ 1/2 power

$$\gamma = 8\pi kT / 3\lambda^2\eta a$$

$P(\omega)$ = Power as function of frequency

$$P(\omega) = I_o < I_s > \gamma / (\omega^2 + \gamma^2)$$

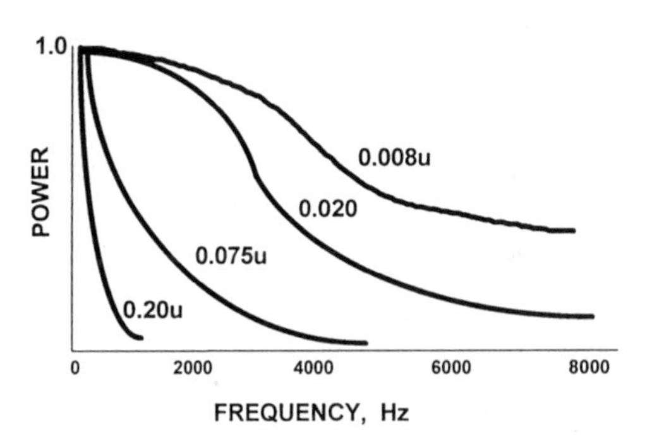

Figure 2. Equation describing power as a function of frequency with plots for several particle sizes.

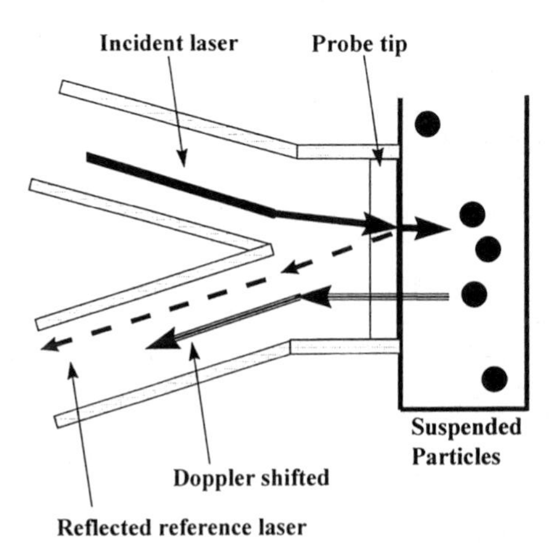

Figure 3. Design of the Ultrafine Particle Analyzer waveguide probe for laboratory and in-line process control applications.

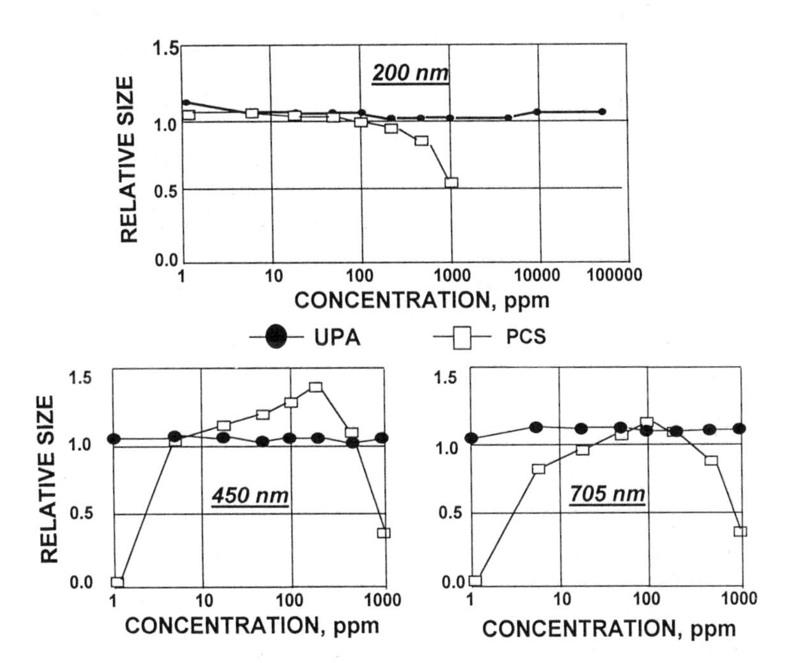

Figure 4. Response of Ultrafine Particle Analyzer and PCS to increasing concentration of 200 nm, 450 nm, and 705 nm polystyrene particles.

A special operational advantage of the UPA approach to ultrafine particle measurement is that interfering background scattering as a result of contaminating dust or other contaminating component is measured and is automatically eliminated from the power spectra. In contrast, PCS requires the use of highly purified suspension fluids to avoid the interfering effects of particulate contaminants. Without adherence to the use of such fluids, contaminating particles become part of the PCS particle size measurement or may act as an interference. In the case of the heterodyne (UPA) method, the reference beam acts as a light source to illuminate the fluid before being used for dilution of the sample (if required). Illuminated contaminating particles, therefore, are measured as background which is subtracted from the subsequent measurement of sample particle size. The combined attributes for concentration tolerance, contaminant background subtraction and measured distribution characteristics provide reliable, precise measurements of non-diluted particles in the range of 0.003 to 6.541 micrometers. General technical advantages of the UPA as compared to PCS are shown in Table I.

Table I. Comparison of the Technology Features of CRM and PCS.

Controlled Reference Method	PCS
1. Accepts wide range of concentrations	Concentration sensitive
2. Uses controlled (laser) reference	Moving particle reference
3. Background measure for contamination correction	Requires specially filtered fluid
4. Uses power spectrum for direct distribution determination	Uses autocorrelation function and *a priori* assumptions for distribution determination
5.Full volume and intensity-weighted distributions and summary values presented	May only provide mean and distribution breadth. Algorithms available for more in-depth analysis.
6.Automatic viscosity value compensation as a function as function of temperature.	Temperature control usually required
7. Commercially proven laboratory, on-line, at-line and in-line process control	Laboratory and on-line measurement only

Industrial Applications of the CRM. The forgoing discussion summarizes the concepts necessary to understand the functioning of the CRM immersible probe technology. Of particular interest is the application of the technology to the wide variety of materials that are produced within the operational size range of the commercial instrument named the Ultrafine Particle Analyzer. Emulsions of polymers and resins, biotechnology substances and microcrystalline ceramics all lend themselves to particle size measurement by this technology. Examples of these applications are discussed below.

Biotechnology Applications. Lipoproteins are complex macromolecules that, in combination with other molecules, segregate tissue into definable structures called cells. The contents of the cells are contain structures called organelles whose boundaries are composed of other structural lipoproteins. The organelles allow for the efficient interplay between the biochemicals that occur in living systems. In addition, specialized forms termed serum lipoproteins provide the transport mechanism for lipids through blood systems for catabolic and anabolic reactions. In animals, several basic types exist and are classed according to the lipid composition and the relative density. Generally they are termed chylomicrons, low-density lipoprotein (LDL), very-low density lipoprotein (VLDL) and high-density lipoprotein(HDL). Other types of lipid-protein complexes exist to provide special structure-function relationships, but of special importance to humans are relatively elevated blood serum low-density lipoprotein concentrations which may be an indicator of potential or ongoing cardiovascular disease. Other disease states may be diagnosed according to the presence, absence or elevated concentration of other serum or cellular proteins. The CRM offers the opportunity for evaluation and measurement of these substances without the concerns noted in Table 1.

At present, special medical significance is given to HDL and LDL. Potential rapid screening of blood serum for these substances may result in less expensive preliminary testing, with less operator dependence and greater speed. To evaluate the ability of the UPA to measure the size of lipoproteins and other proteins of varying molecular weight and size, several experiments were conducted. Samples of human serum VLDL, LDL and HDL were measured using the UPA in the volume measurement mode. The materials were suspended in a 0.01% pH 7.4 EDTA buffer prepared in isotonic saline. The particle size distribution histograms of the HDL and LDL samples are shown in Figures 5 and 6, respectively. Similarly, samples of the proteins thyroglobulin, cytochrome c, lysozyme, bovine serum albumin (BSA), and human Immunoglobulin G (IgG) were measured. Distribution data are shown for the latter two species in Figures 7 and 8.

While the distribution plots are useful, it is also important to compare the experimental size information to accepted values. To this end, Ultrafine Particle Analyzer data were compared to literature values (9,10) obtained from conventional photon correlation spectroscopy and are shown in Figures 9A and 9B. Figure 9A (PCS data from Reference 10) depicts the relationship of the measured values for six proteinaceous substances. A strong linear correlation (correlation constant 0.986) was found comparing UPA size data to literature values for PCS data. Agreement was not expected since the approaches to the measurement use different detection systems, physics (heterodyne Vs homodyne) and mathematics. Since the suspensions for the UPA measurement were prepared according to literature citation, molecular association due to concentration or some other effect resulting in increased particles size values would not seem to be a causative factor. Macromolecular or particulate size is generally accepted to increase as a function of molecular weight for non-associated species. As such, one should assume that a linear relationship between protein size and molecular weight will occur. Molecular weight for four of the proteins was plotted as a function of the size reported by the two methods as shown in Figure 9B. (9) It is interesting to note that the correlation coefficients for the two methods are similar, but the y-intercept values are considerably different which suggests that the UPA provides more realistic DLS size information. Particle distribution breadth affects the reported values for mean and median size and may also be influential. The UPA collects such breadth

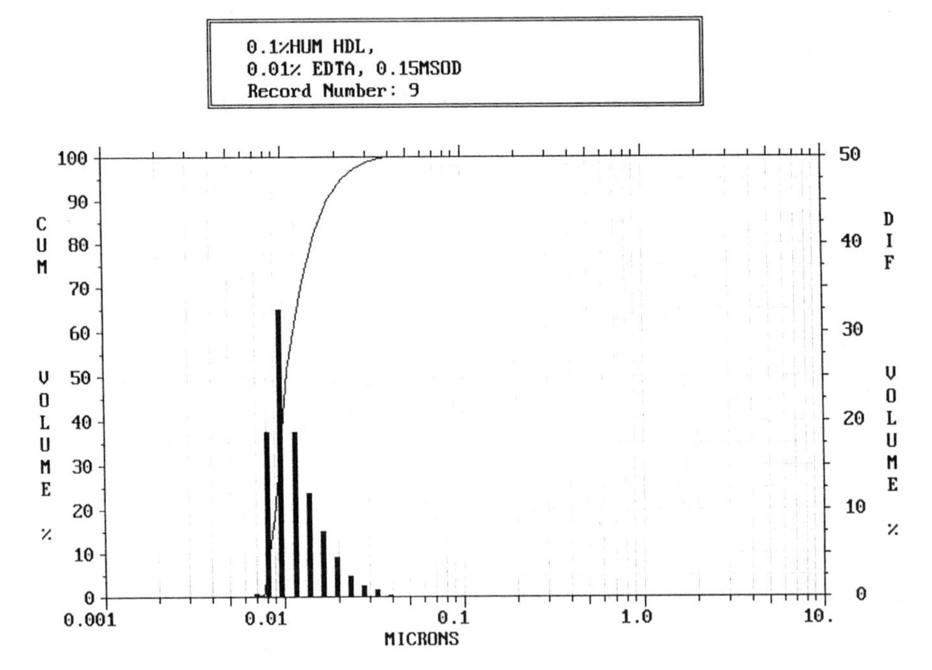

Figure 5. Particle size distribution histogram of human high-density lipoprotein measured under the conditions given in the text.

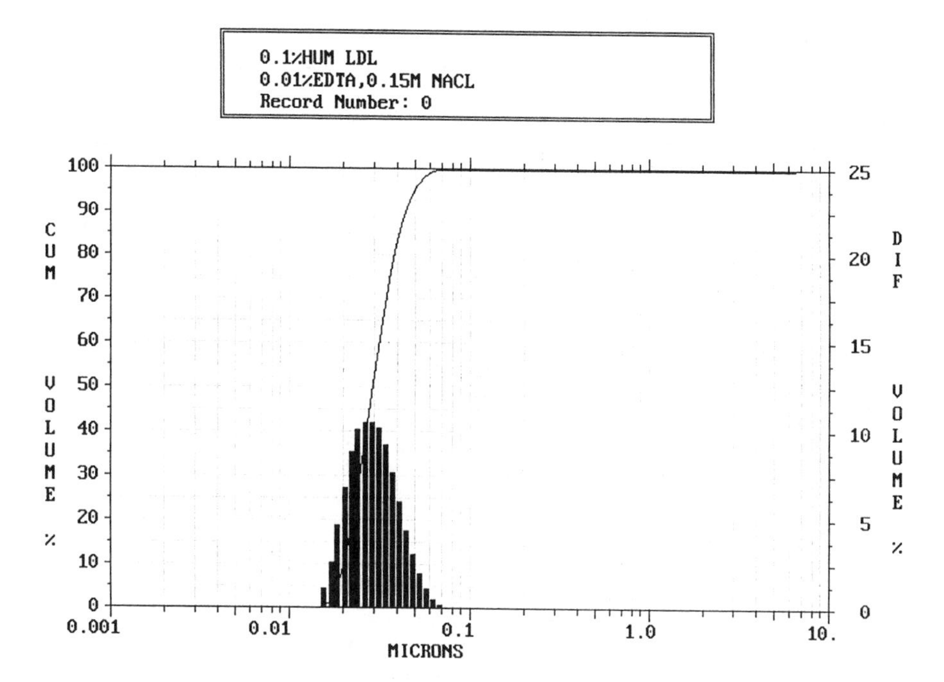

Figure 6. Particle size distribution histogram of human low-density lipoprotein measured under the conditions given in the text.

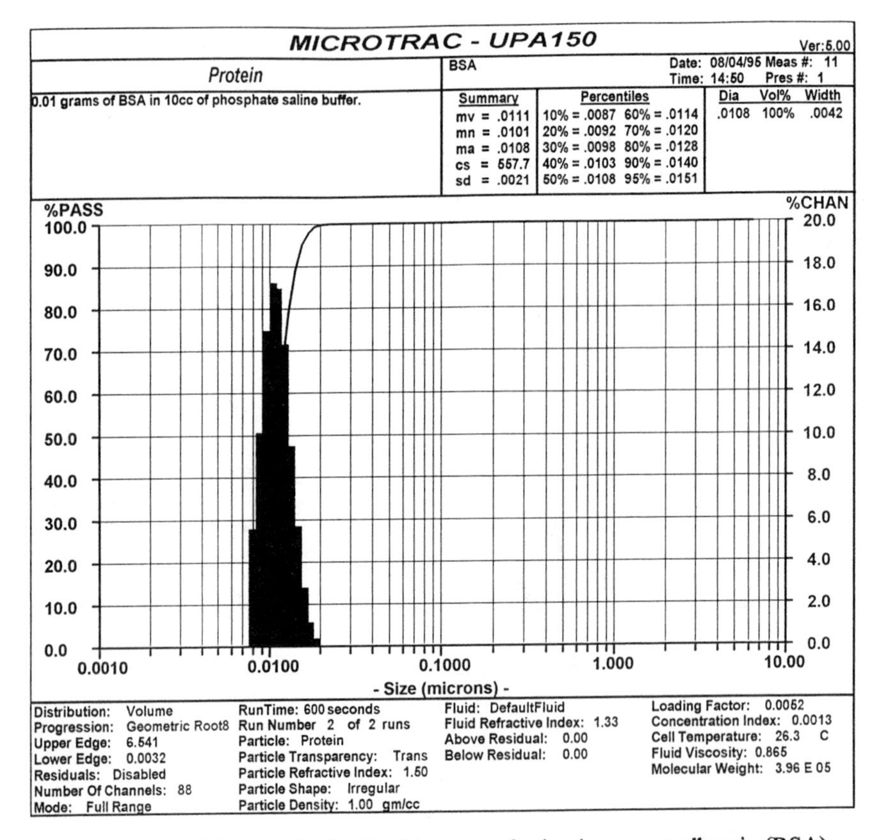

Figure 7. Particle size distribution histogram for bovine serum albumin (BSA) measured as described in the text.

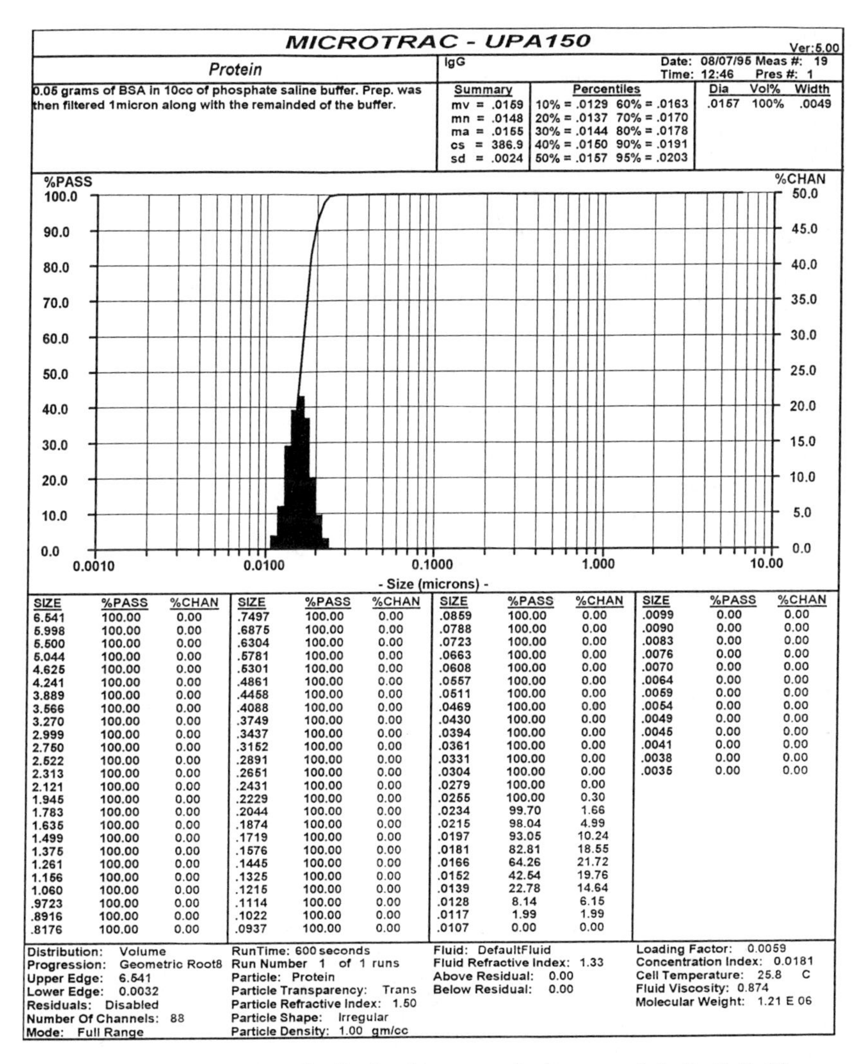

SIZE	%PASS	%CHAN	SIZE	%PASS	%CHAN	SIZE	%PASS	%CHAN	SIZE	%PASS	%CHAN
6.541	100.00	0.00	.7497	100.00	0.00	.0859	100.00	0.00	.0099	0.00	0.00
5.998	100.00	0.00	.6875	100.00	0.00	.0788	100.00	0.00	.0090	0.00	0.00
5.500	100.00	0.00	.6304	100.00	0.00	.0723	100.00	0.00	.0083	0.00	0.00
5.044	100.00	0.00	.5781	100.00	0.00	.0663	100.00	0.00	.0076	0.00	0.00
4.625	100.00	0.00	.5301	100.00	0.00	.0608	100.00	0.00	.0070	0.00	0.00
4.241	100.00	0.00	.4861	100.00	0.00	.0557	100.00	0.00	.0064	0.00	0.00
3.889	100.00	0.00	.4458	100.00	0.00	.0511	100.00	0.00	.0059	0.00	0.00
3.566	100.00	0.00	.4088	100.00	0.00	.0469	100.00	0.00	.0054	0.00	0.00
3.270	100.00	0.00	.3749	100.00	0.00	.0430	100.00	0.00	.0049	0.00	0.00
2.999	100.00	0.00	.3437	100.00	0.00	.0394	100.00	0.00	.0045	0.00	0.00
2.750	100.00	0.00	.3152	100.00	0.00	.0361	100.00	0.00	.0041	0.00	0.00
2.522	100.00	0.00	.2891	100.00	0.00	.0331	100.00	0.00	.0038	0.00	0.00
2.313	100.00	0.00	.2651	100.00	0.00	.0304	100.00	0.00	.0035	0.00	0.00
2.121	100.00	0.00	.2431	100.00	0.00	.0279	100.00	0.00			
1.945	100.00	0.00	.2229	100.00	0.00	.0256	100.00	0.30			
1.783	100.00	0.00	.2044	100.00	0.00	.0234	99.70	1.66			
1.635	100.00	0.00	.1874	100.00	0.00	.0215	98.04	4.99			
1.499	100.00	0.00	.1719	100.00	0.00	.0197	93.05	10.24			
1.375	100.00	0.00	.1576	100.00	0.00	.0181	82.81	18.55			
1.261	100.00	0.00	.1445	100.00	0.00	.0166	64.26	21.72			
1.156	100.00	0.00	.1325	100.00	0.00	.0152	42.54	19.76			
1.060	100.00	0.00	.1215	100.00	0.00	.0139	22.78	14.64			
.9723	100.00	0.00	.1114	100.00	0.00	.0128	8.14	6.15			
.8916	100.00	0.00	.1022	100.00	0.00	.0117	1.99	1.99			
.8176	100.00	0.00	.0937	100.00	0.00	.0107	0.00	0.00			

Distribution: Volume
Progression: Geometric Root8
Upper Edge: 6.541
Lower Edge: 0.0032
Residuals: Disabled
Number Of Channels: 88
Mode: Full Range

RunTime: 600 seconds
Run Number 1 of 1 runs
Particle: Protein
Particle Transparency: Trans
Particle Refractive Index: 1.50
Particle Shape: Irregular
Particle Density: 1.00 gm/cc

Fluid: DefaultFluid
Fluid Refractive Index: 1.33
Above Residual: 0.00
Below Residual: 0.00

Loading Factor: 0.0059
Concentration Index: 0.0181
Cell Temperature: 25.8 C
Fluid Viscosity: 0.874
Molecular Weight: 1.21 E 06

Figure 8. Particle size distribution histogram for Immunoglobulin G (IgG) measured as described in the text.

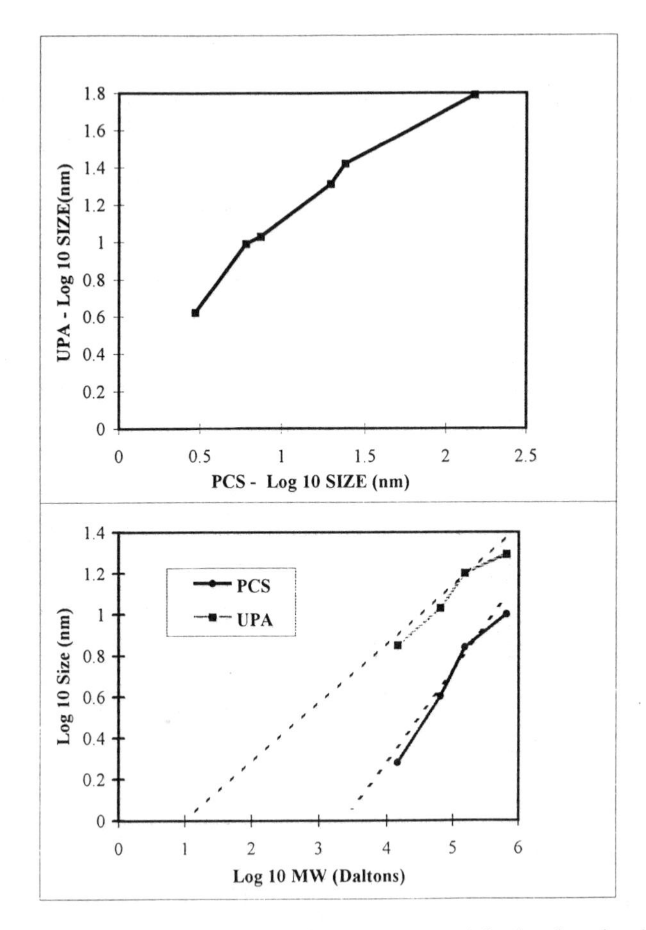

Figure 9. (A). Comparison of UPA and PCS mean particle size data for six biochemical molecules: very-low-density lipoprotein, low-density lipoprotein, thyroglobulin, bovine serum albumin, high-density lipoprotein, and cytochrome c. (B). Comparison of literature values for molecular weight with measured particle size measurements from the Ultrafine Particle Analyzer (heterodyne) and PCS (homodyne) methods for lysozyme, bovine serum albumin, Immunoglobulin G and thyroglobulin.

information during the measuring stage and calculates the mean and median from the volume distribution. For informational and comparative purposes, accepted size and molecular weight information for many biological substances is presented in Table II.

Table II. Accepted Particle Size Data for Biochemicals

Substance	Size(nm)	Molecular Weight (Daltons)
Liposomes	50 - 10,000	-------
Micelles	5 - 20	20,000
BSA	8 - 10	65,000
IgM	40	150,000
Carbonic Anhydrase	5	29,000
Thyroglobulin	19	670,000
Cytochrome c	3	11,000
HDL Lipoprotein	6	------
LDL Lipoprotein	25	-------
VLDL	140	-------
IgG	6.9	-------
Egg white lysozyme	7	27,000

An experiment was designed to test the UPA response to a mixture of proteins and to preliminarily evaluate the ability to distinguish between the two substances. Cytochrome c (3 nm) and thyroglobulin (19 nm) were prepared in buffered saline and mixed in a ratio of 90/10 ratio respectively. The final total protein concentration was not monitored. A measurement period of 10 minutes was selected. The results shown in Figure 10 provide information on the relative amounts of the proteins as well as their respective sizes. Notable is the separation of the two modes representing the two proteins. The slight shift in particle size for each (cytochrome c larger and thyroglobulin smaller than expected) is suggestive of molecular interaction or, possibly, small errors in the FFT analysis. Further study is required to elucidate the mechanism causing the shift, but it is notable that the CRM is capable of distinguishing the two proteins without the necessity of selecting a particular mathematical solution. The Controlled Reference Method (CRM) technique provided information within acceptable, limited error on the relative amounts of the two proteins in suspension as well as their respective sizes (Figure 10).

Other biomedical applications. Other pharmaceutical and medical applications include the areas of bacterial motility, antigen-antibody interactions and vesicle formation. Two other very promising applications include evaluation of protein and liposome stability. Both materials are included in advanced forms of drug delivery and drug targeting. Drug delivery systems may use proteins as carriers of specific pharmaceuticals, while liposomes are being investigated as a means of providing therapeutic agents to specific anatomical or physiological sites. In both cases, the shelf-life is of extreme importance. Shelf-life can be evaluated using wet and spectrochemical biochemical techniques, but considerable laboratory skill is normally required. Therefore, of considerable interest are techniques that indicate stability by means other than biological activity or chemical studies. One such method is size stability

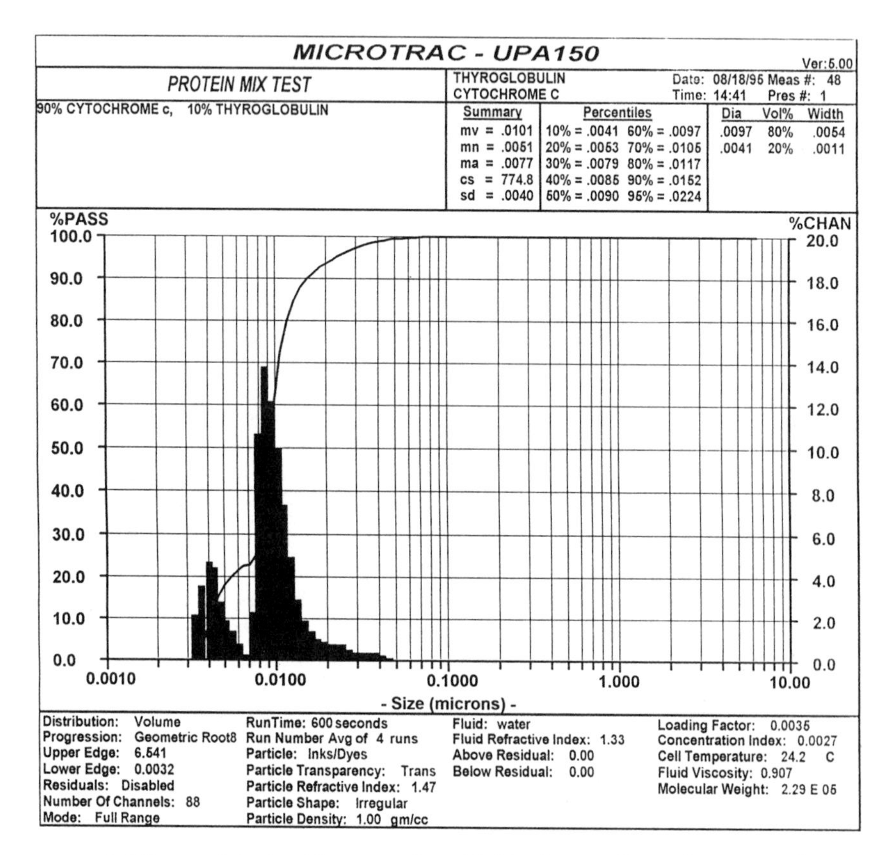

10. UPA particle size histogram for a 90:10 mixture of cytochrome c and thyroglobulin.

determination. Proteins when destabilized normally will unfold or form bundles which are measurable in the UPA size range. The increase in size relative to the known starting size of the nascent or modified protein is an indication of loss of structural stability, biological activity or perhaps loss of suitability to pass through the blood to retard disease progression. A similar effect on size would be expected for destabilized liposomes.

The particle size of egg white lysozyme in phosphate buffer water was measured in 0% and 8% sodium chloride. It is well established that egg white lysozyme precipitates at 8% sodium chloride concentrations. The data in Figure 11 clearly show the effect of sodium chloride concentration on a lysozyme solution. The initial value for lysozyme size is reported by the UPA to be 8.2 nanometers while literature values suggest 7 to 8 nanometers. Increasing the salt concentration of the lysozyme sample produced the expected effect of increased size. Two modes appeared at the conclusion of the experiment. Interestingly the size of the first mode lies at approximately 28 times the original value. A second mode lies at 44 times the original size. This information may be indicative of the mechanism of protein agglomeration from the molecular to the precipitated state. Further studies are required to determine the value of this approach in understanding the mechanism as it is applied to this technology, but suffice to state that destabilized protein solutions can be monitored by the Controlled Reference Method.

Liposomes were prepared in the laboratory by mixing 5 ml of a commercially available vegetable oil transferred to 95 ml deionized water. The mixture was treated with ultrasonic energy for three periods of 30 seconds, 2 minutes and 5 minutes. At the conclusion of each treatment period, the resulting emulsion of liposomes was measured without further dilution or other treatment. As shown in Figure 12 each succeeding treatment period caused a shift in the reported particle size. The middle peak shows a slight tail or bias at the coarse end of the distribution suggestive of residual larger oil droplets which disappear following further treatment at 5 minutes. Whether the small tail in the distribution of the third treatment period (smallest size) would disappear with longer treatment was not determined. This experiment suggests that the UPA instrument can be applied to monitoring and measurement of the stability of liposomes in which the size would progressively increase. Figure 13 shows such a size change after mixing with sucrose.

Aside from the above applications potential in the ever-widening fields of pharmaceuticals and biotechnology, there are many other industrial uses of ultrafine particle size measurement. The industries of ink manufacture, polymer production, diamond dust formation, emulsion technologies and ceramics offer many opportunities for applications. Industrial manufacture of commercially applied polymers requires measurement at high solids concentration. As discussed previously, size measurement of polymer emulsions at concentrations above 20% solids is possible in several instances and is presently finding industrial use and acceptance. Light scattering data may be presented as a volume-weighted format, which may differ from an intensity-weighted format because particles of different size can scatter light at different

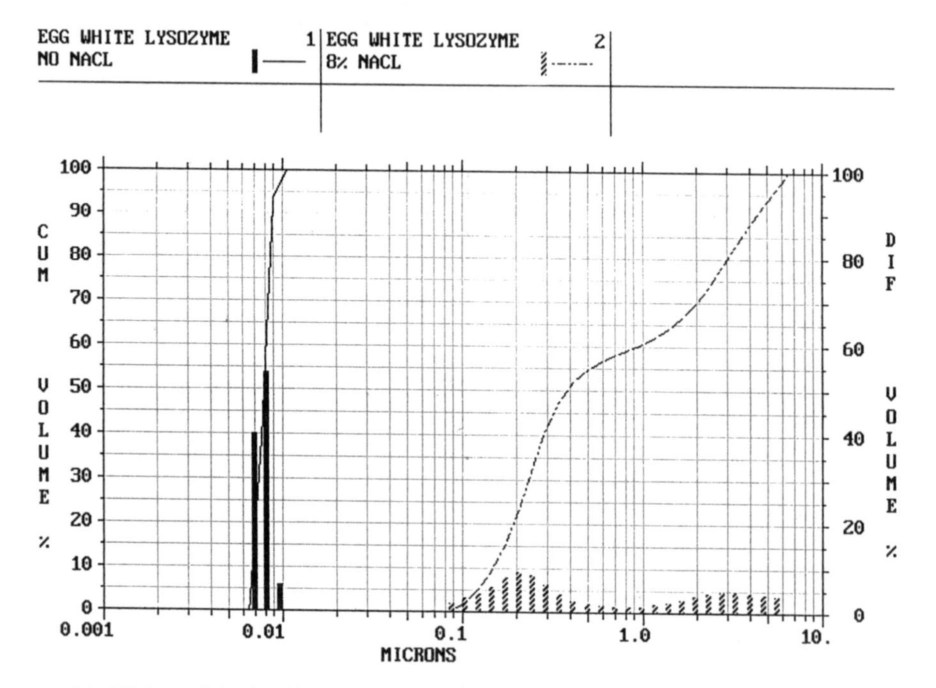

11. UPA particle size histograms of lysozyme before and after treatment with 8% sodium chloride used to induce protein precipitation for simulating protein destabilization.

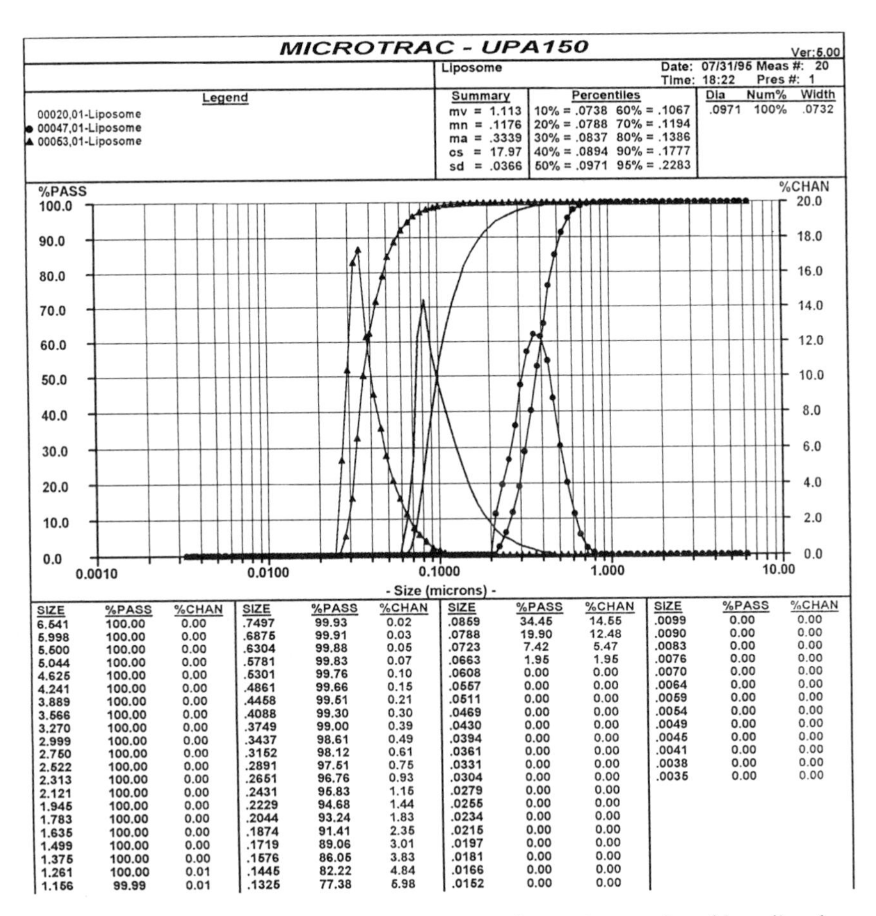

MICROTRAC - UPA150 Ver:5.00

Liposome Date: 07/31/95 Meas #: 20
Time: 18:22 Pres #: 1

Legend
00020,01-Liposome
● 00047,01-Liposome
▲ 00053,01-Liposome

Summary	Percentiles		Dia	Num%	Width
mv = 1.113	10% = .0738	60% = .1067	.0971	100%	.0732
mn = .1176	20% = .0788	70% = .1194			
ma = .3339	30% = .0837	80% = .1386			
cs = 17.97	40% = .0894	90% = .1777			
sd = .0366	50% = .0971	95% = .2283			

SIZE	%PASS	%CHAN	SIZE	%PASS	%CHAN	SIZE	%PASS	%CHAN	SIZE	%PASS	%CHAN
6.541	100.00	0.00	.7497	99.93	0.02	.0859	34.45	14.55	.0099	0.00	0.00
5.998	100.00	0.00	.6875	99.91	0.03	.0788	19.90	12.48	.0090	0.00	0.00
5.500	100.00	0.00	.6304	99.88	0.05	.0723	7.42	5.47	.0083	0.00	0.00
5.044	100.00	0.00	.5781	99.83	0.07	.0663	1.95	1.95	.0076	0.00	0.00
4.625	100.00	0.00	.5301	99.76	0.10	.0608	0.00	0.00	.0070	0.00	0.00
4.241	100.00	0.00	.4861	99.66	0.15	.0557	0.00	0.00	.0064	0.00	0.00
3.889	100.00	0.00	.4458	99.51	0.21	.0511	0.00	0.00	.0059	0.00	0.00
3.566	100.00	0.00	.4088	99.30	0.30	.0469	0.00	0.00	.0054	0.00	0.00
3.270	100.00	0.00	.3749	99.00	0.39	.0430	0.00	0.00	.0049	0.00	0.00
2.999	100.00	0.00	.3437	98.61	0.49	.0394	0.00	0.00	.0045	0.00	0.00
2.750	100.00	0.00	.3152	98.12	0.61	.0361	0.00	0.00	.0041	0.00	0.00
2.522	100.00	0.00	.2891	97.51	0.75	.0331	0.00	0.00	.0038	0.00	0.00
2.313	100.00	0.00	.2651	96.76	0.93	.0304	0.00	0.00	.0035	0.00	0.00
2.121	100.00	0.00	.2431	95.83	1.15	.0279	0.00	0.00			
1.945	100.00	0.00	.2229	94.68	1.44	.0255	0.00	0.00			
1.783	100.00	0.00	.2044	93.24	1.83	.0234	0.00	0.00			
1.635	100.00	0.00	.1874	91.41	2.35	.0215	0.00	0.00			
1.499	100.00	0.00	.1719	89.06	3.01	.0197	0.00	0.00			
1.375	100.00	0.00	.1576	86.05	3.83	.0181	0.00	0.00			
1.261	100.00	0.01	.1445	82.22	4.84	.0166	0.00	0.00			
1.156	99.99	0.01	.1325	77.38	5.98	.0152	0.00	0.00			

12. Comparison UPA particle size histograms for a mixture of cooking oil and deionized water successively treated with ultrasonic energy for 30 seconds, 2 minutes and 5 minutes.

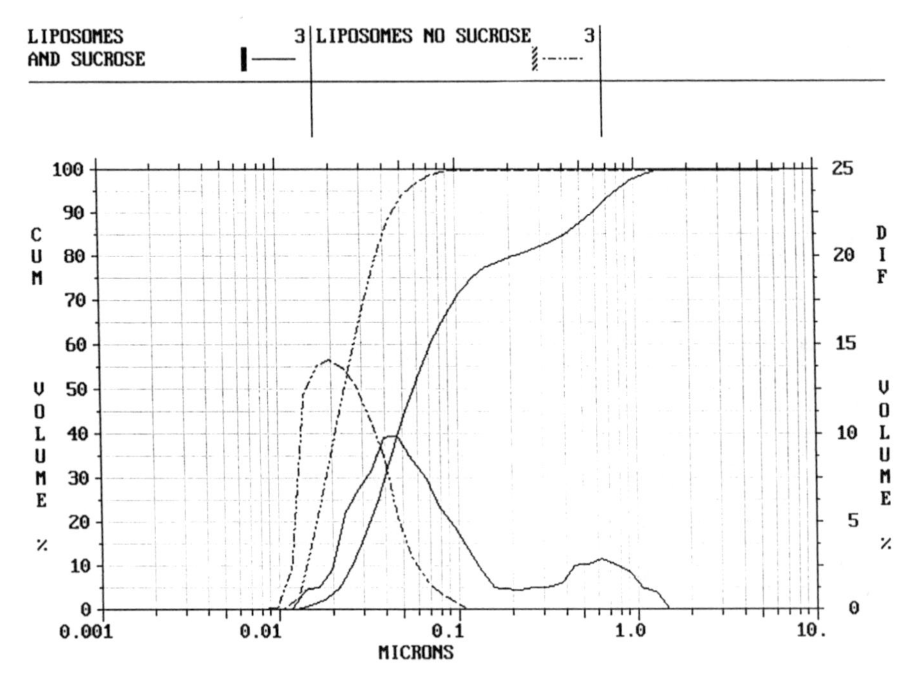

13. Comparative UPA particle size histograms for liposomes measured at room temperature before and after mixing with sucrose used as a destabilizing chemical.

intensities. The intensity format provides a distribution that is not corrected for the non-linear relationship between size and scattering efficiency and represents merely relative intensity as a function of size and not the volume(weight) amount of particulate present. Correction for the size/scattering efficiency relationship using Mie concepts (11) allows data presentation in a "real-life", volume format.

Special Data Considerations. To produce a volume distribution, scattering efficiency corrections must be considered. Thus while a particular size may be present in significant volume or mass, low scattering ability can preclude its being reported in a intensity-weighted distribution. Often sensitivity to the presence of specific particle size ranges can be enhanced by selecting the volume-weighted format. The value of reporting data in the volume-weighted format is shown in Figures 14 and 15. A volume-weighted histogram for a 50/50 mixture of 0.305 and 0.050 micrometer polystyrene is shown in Figure 14.. This example simulates the presence in an industrial sample of undersized or incompletely polymerized particles. According to the intensity-weighted distribution shown in Figure 15, the unpolymerized polymer (50 nm) represents approximately 16% of the entire product distribution. In order to enhance its presence and gain a truer perspective of the "real-life" distribution, recalculation to a volume format demonstrates that the volume(weight) percent of the unpolymerized material is 48.5%. A value in this range may exceed specification and demand process intervention and alterations. In contrast, the prominent large particle size mode may be viewed with greater sensitivity by reporting the data in the intensity-weighted format. Thus selection of distribution presentation can enhance selective sensitivity over the entire measuring range. By measuring the sample at manufacturing concentrations using these presentations, monitoring for coagulation, over-processing and under-processing can be performed without the common deleterious effects of dilution on the polymer suspension.

Polishing Computer Hard Disk Drives. The fine grinding and milling necessary for the manufacture of computer hard disk drives dictate the use of diamond abrasives and industrial oxides of very small size. Typically the size range of the various grades is from 0.1 to 4 microns. Figure 16 shows the measurement of six sizes of diamond dust abrasive between 100 and 2000 nanometers. To ensure stable large particle suspensions, each sample was prepared in an aqueous 60% glycerol solution to retard settling. Correction for the effect of higher viscosity (lower particle velocity) on Brownian motion was accomplished automatically through software selection and automatic temperature compensation. Each sample was measured at 1% solids. In each case the median particle diameter accuracy was reported within 4 percent of that expected by the manufacturer.

Ink Applications. In addition to the above mineral applications, ink manufacture for jet-type printers, ball-point pens and other industrial printing operations are highly dependent upon size and distribution. Figure 17 shows the effect of one extremely high concentration ink sample measured at both 15 and 1% wt solids. An apparent slight shift in data occurs and is believed to be due to particle-particle interactions. Charge or other surface chemical effects are believed to cause the concentrated particles to diffuse

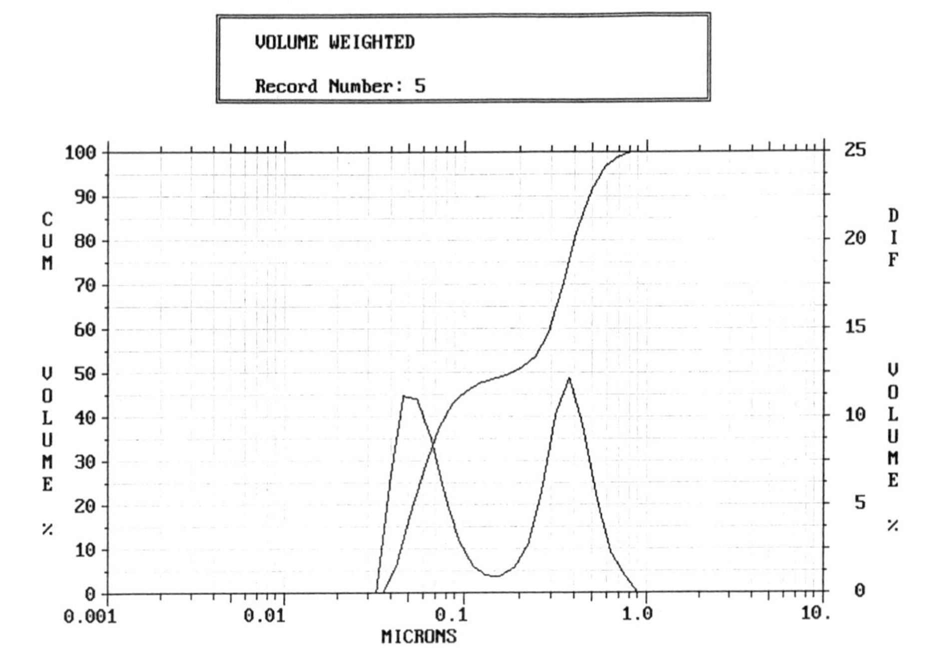

14. Bimodal distribution of polystyrene measured on the UPA, heterodyne method. Histogram is presented as the volume-weighted distribution.

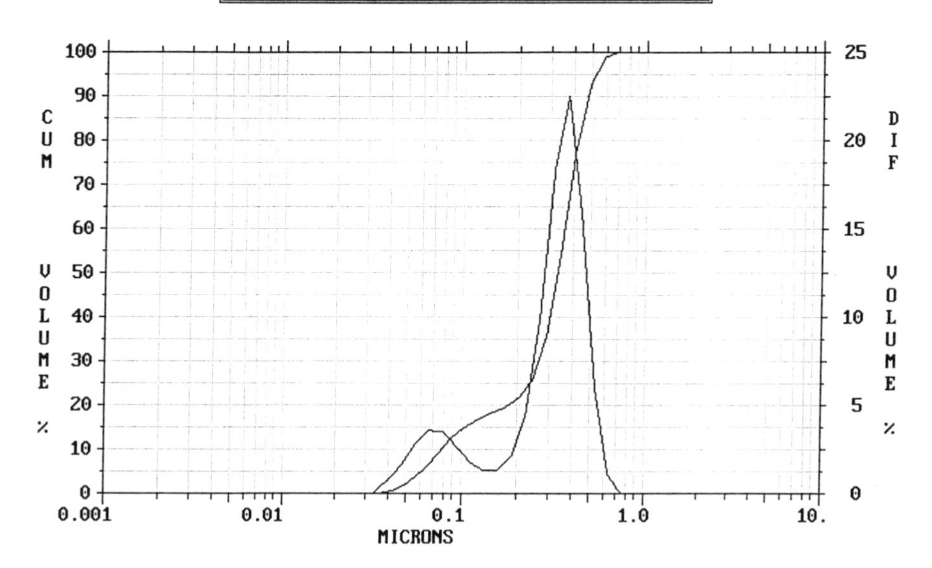

15. Data from Figure 14 presented as the intensity-weighted distribution.

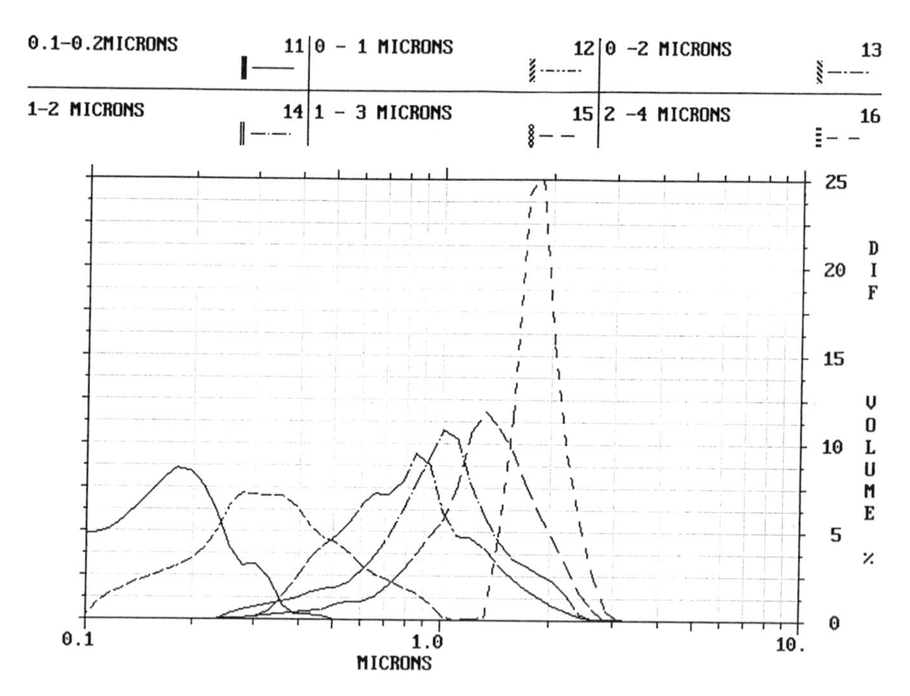

16. UPA comparative distributions of six sizes of diamond dust used for computer hard disk polishing.

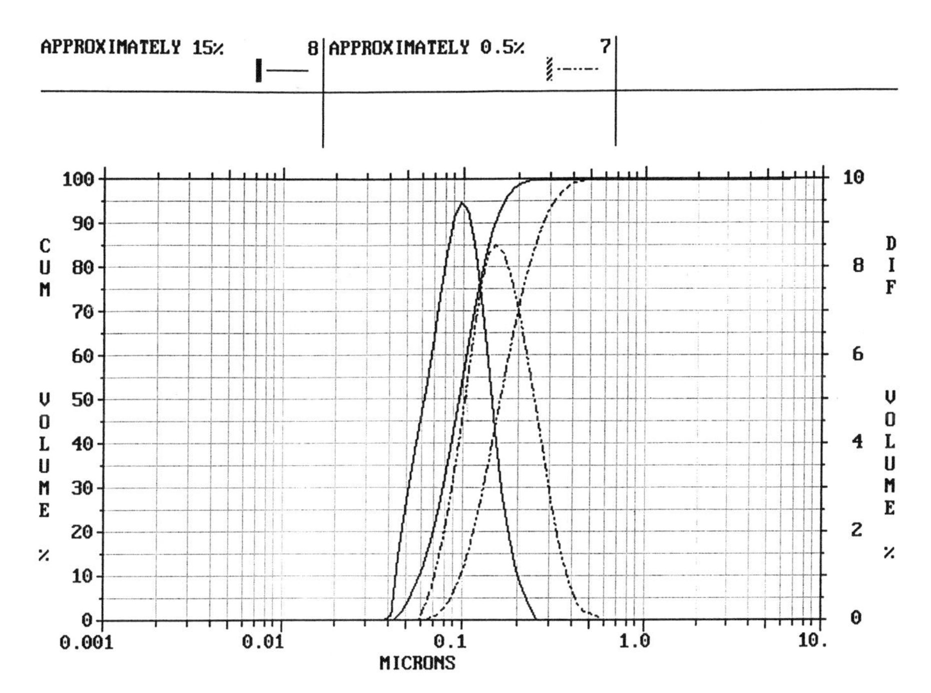

17. Histograms showing the effect of ink concentration on the data reported by the UPA Controlled Reference Method of dynamic light scattering.

at higher rates causing the apparent shift to smaller size. Modification or disappearance of this shift in size can be used as a new predictor of inappropriate surface chemistries, coagulation issues or under-milled particulate. The effect of the slight shift is far outweighed by the potential effects of extreme dilution necessary to complete PCS measurements. Such dilutions can result in changes in particle dispersion stability that may lead to reports of over-sized or agglomerated particulate which may falsely indicate out-of-spec product. Thus, the ability to measure particles at very high concentrations can provide more useful data while avoiding the undesirable effects of dilution.

Conclusions

The concept and implementation of the heterodyne Controlled Reference Method of dynamic light scattering were explained and compared to Photon Correlation Spectroscopy. Subtle to mild differences in the capture of light scattering information from particles exhibiting Brownian motion between the CRM (UPA) and PCS methods result in apparent greater performance capability than heretofore reported. Application of the CRM method were reviewed and data presented to establish the validity of the technique when measuring sub-micron species in dilute or concentrated suspension. Measurements at previously unknown high concentration demonstrated a slight shift explainable by particulate repulsion. Some substances were measurable to 40% solids without dilution. The UPA was shown to be designed as a small but sturdy tool that is applicable to a variety of substances including biologics, pharmaceuticals, inorganics and inks.

References

1. Vaidya, R.A., Mettile, M.J., and Hester, R.D. In *Particle Size Distribution: Assessment and Characterization,* Provder, T., Ed.; ACS Symposium Series No 332; Amer. Chem. Soc: Wash, D.C. **1987,** Chapt 4.

2. Pecora, R. *Dynamic Light Scattering. Applications of Photon Correlation Spectroscopy*; Plenum Press: 1985.

3. Trainer, M.N.; Freud, P.J.; Leonardo, E.M. *Amer. Lab.*, **1993,** *Vol 69* , pp 42-45.

4. Provencher, S.W. *Makromol. Chem.* **1979,** 180, pp 201.

5. Provencher, S.W. *Computer Physics Communications.* **1982**, 27, pp 229.

6. Koppel, D.E. *J.Chem.Phys.* **1972**, 57, pp 4814-20.

7. Gulari, E., Gulari, E., Tsunashima, T., Chu, B., J.Chem.Phys., **1979**, 70, pp 3965-72.

8. Bott, S.E., In *Particle Size Distribution: Assessment and Characterization*; Provder, T., Ed.; ACS Symposium Series No 332,. Amer. Chem. Soc: Wash, D.C., 1987, Chapt. 5

9. Claes, P., Dunford, M.,Kenney, A.,and Vardy, P. In *Laser Light Scattering in Biochemistry;* Harding S.E., *Sattelle. D.B., and Bloomfield, V.A.;* Ed.; Royal Society of Chemistry; 1990.

10. Elings, V.B. and Nicoli, D., *Amer. Lab.*, **1984**, *Vol 60*, pp34-39.

11. Mie, Gustave, *Annalen der Physik.* **1908**. *vol 25,* pp. 377-445.

Chapter 10

Optimization of Particle Size Measurement from 40 Nanometers to Millimeters Using High-Angle Light Scattering

Michael N. Trainer

Honeywell Industrial Automation and Control, 1100 Virginia Drive, Fort Washington, PA 19034

Conventional static angular scattering has demonstrated excellent size accuracy and reproducibility down to 300 nanometers. High angle scattering measurement extends this size range to 40 nanometers, while preserving the reproducibility and speed of angular scattering. To cover a large range of scattering angles, two optical detector arrays are multiplexed among three lasers to measure light scattered over three different angular ranges. These three angular scattering distributions are combined to form a single continuous distribution, from 0.02 to 165 degrees, which is inverted by an iterative algorithm to produce the particle size distribution. The measurement of high angle scattering demonstrates superior small particle size resolution and accuracy when compared to other techniques that measure at lower angles or use polarization effects. Results for both spherical particles and non-spherical samples such as diamond powders are shown.

Optical scattering techniques are the method of choice for particle size measurement in laboratories and process control. These methods are accurate and repeatable over a wide range of operator skill level. Optical scattering also covers a size range of 6 orders of magnitude, from nanometers to millimeters in liquid or gas dispersants. This wide size range is divided between two scattering methodologies, static and dynamic scattering. In dynamic light scattering (*1*), the Doppler spectral broadening of the scattered light due to Brownian motion of the particles is measured through either the power spectrum or autocorrelation of the scattered light intensity. This method works well for particle sizes below 1 micron where the 1/f noise and statistical accuracy of the power spectrum estimation are favorable. Above 1 micron, static angular scattering is preferred due to higher reproducibility and shorter measurement time. Since dynamic scattering is a stochastic process, at least one minute data collection time is required to accurately estimate the spectral characteristics of the scattered light. On the other hand, static angular scattering is a deterministic measurement, where excellent accuracy and repeatability are obtained in a 5 second measurement. Also other factors, such as dispersant viscosity, sample flow, and temperature instability can affect dynamic scattering results, while having a negligible effect on angular scattering.

Progress in the development of static angular scattering methods has continued to improve performance for measuring submicron particles. Polarization methods (*2*) near the 90 degree scattering angle have extended static measurements down to 100 nanometers, but with poor accuracy and high sensitivity to particle composition and shape. Differential polarization shows poor resolution and accuracy for non-spherical particles, which are not consistent with the polarization model. We have abandoned this method in preference for high angle scattering which provides better accuracy and less sensitivity to particle composition and shape. By measuring light scattering at angles up to 165 degrees, the useful size range is extended down to 40 nanometers, without the need for polarization measurements.

Measurement of Scattered Light with a Focal Plane Array

A typical light scattering configuration is shown in Figure 1. A beam of light illuminates a group of particles which may be dispersed in a liquid or gas stream. In both cases the particles may be confined by a sample cell with optical windows. The incident beam of light and light scattered by the particles are focused by a lens. An array of optical detectors, in the focal plane of the lens, measures the angular distribution of scattered light. Notice that the bundle of light scattered at angle θ is focused to a point on the detector array. Therefore, each point on the detector array corresponds to a single scattering angle from all the particles in the beam and angular resolution is independent of the sample volume size. The detector array light distribution, which is the sum of all the distributions from individual particles in the ensemble, must be mathematically inverted to obtain the particle size distribution. This inversion process relies upon an accurate model for the scattering process. The most widely used model is based on Mie theory, which solves Maxwell's equations exactly for the boundary conditions of a spherical particle. We use the Mie model for inversion of scattered light from spherical particle samples; but the unclosed form of the Mie solution does not offer insight into the scattering process and system optimization.

Fraunhofer Approximation and System Optimization

The most widely used closed-form solution, the Fraunhofer approximation, is not appropriate for non-absorbing particles because this model does not account for light transmitted by the particle. Optical interference, between light transmitted and light scattered by the particle, creates secondary angular resonances which are also not described by this approximation. However, since the general form of the Fraunhofer scattering function shows sufficient agreement with the exact Mie model for particles larger than 0.4 microns, this approximation can provide a basis for optimization of static scattering systems.

In the Fraunhofer approximation, for n_l particles of diameter D in a light beam of irradiance E_0 and wavelength λ, the scattered optical intensity $I(\theta)$ as a function of scattering angle θ is:

$$I(\theta) = \frac{n_l E_0 \lambda^2 \alpha^4}{16\pi^2}\left[\frac{2 J_1(\alpha\theta)}{\alpha\theta}\right]^2 \quad \text{where } \alpha = \frac{\pi D}{\lambda}$$

The scattering angle θ is measured relative to the incident light beam and J_1 is a first order Bessel function of first kind. For particles of total volume v, the scattered intensity becomes:

$$I(\theta) = \frac{3 E_0 v}{8\lambda}\frac{J_1^2(\alpha\theta)}{\alpha\theta^2}$$

Figure 2 shows the normalized scattered intensity $I(\theta)$ for 20, 40, and 80 micron particles. As in most scattering phenomena, smaller objects create scattering over a larger angular range. The size information is contained in the region where the scattering functions have the highest slope. Measurement of particles larger than 100 microns, requires scattering angles below 0.1 degree with spatial filtering of the incident light beam to reduce background scattering from laser optics.

Optimum Angular Scaling and Resolution

The optimum angular scaling for this inversion problem depends upon the distribution type (*3*). In the case of particle a volume-vs.-diameter distribution, the optimum scaling for the detector array is logarithmic. The width of each detector element is constant, but the element length progresses logarithmically with increasing scattering angle. The scattering angle corresponding to the upper edge of the nth detector element is given by θ_n where β is a constant.

$$\theta_n = \beta^n \theta_0$$

For this detector scale we can define logarithmic angle and size parameters, x and x_1.

$$x = \ln\theta \text{ and } x_1 = -\ln\alpha$$

Then the optical flux integrated over each logarithmic detector element is a sampled value of the function f(x).

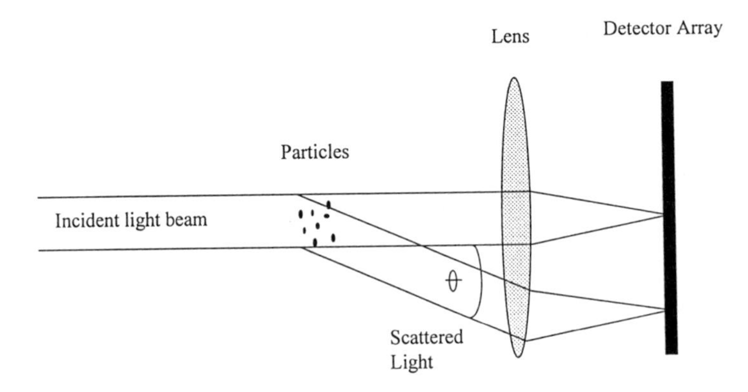

Figure 1. Scattering configuration with focal plane detector array.

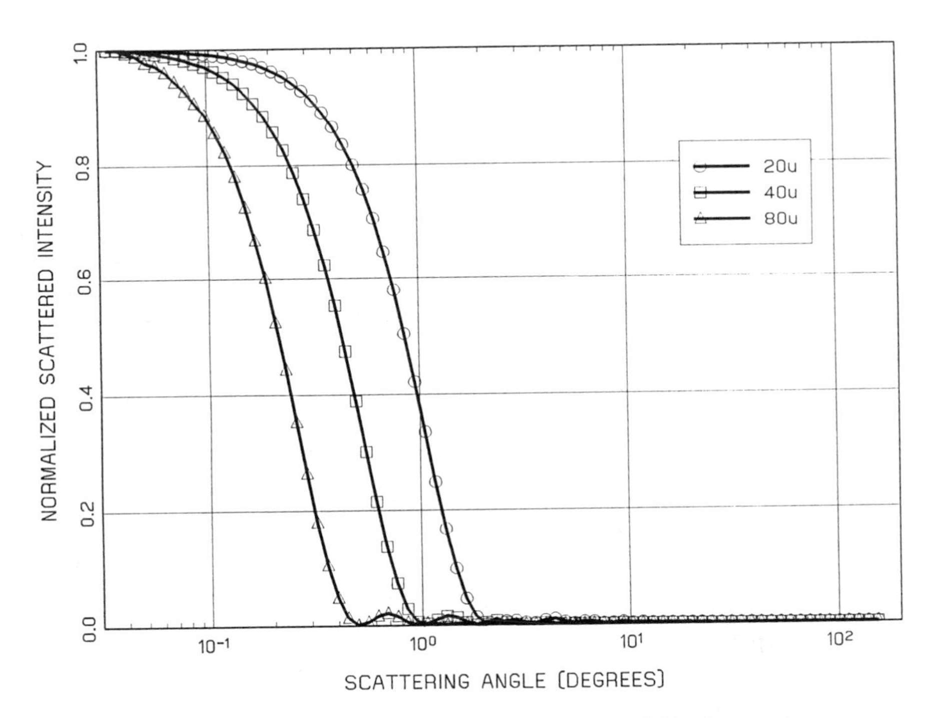

Figure 2. Normalized scattered intensity for 20, 40, and 80 micron spheres.

$$f(x) = Kv\frac{J_1^2(e^{(x-x_1)})}{e^{(x-x_1)}}$$

$$\text{where } K = \frac{3E_0}{8\lambda}$$

Notice that f(x) is shift-invariant to the size parameter x_1. This shift invariance of the Fraunhofer approximation is also shown by the Mie model for polystyrene spheres in Figure 3. The scattering functions maintain the same shape and shift along the logarithmic axis as size changes. Integrating f(x), weighted by the logarithmic volume distribution $v(x_1)$ over all x_1, we obtain f(x) for an arbitrary size distribution.

$$f(x) = K\int_{-\infty}^{\infty} v(x_1)\frac{J_1^2(e^{x-x_1})}{e^{x-x_1}}dx_1$$

f(x) is a convolution of the impulse response h(x) with the volume distribution v(x). h(x) is the impulse response of the linear system which describes the Fraunhofer diffraction process.

$$f(x) = K * v(x) \otimes h(x)$$

$$h(x) = \frac{J_1^2(e^x)}{e^x}$$

This formulation of the scattering problem offers two insights into optimization of the inversion problem. First, the logarithmic detector scale provides a scattering function with a convolution form which can be inverted by robust iterative deconvolution techniques (*4*). Second, the convolution form for f(x) has a useful correspondence in Fourier space.

$$v(x) \otimes h(x) \leftrightarrow V(s) * H(s)$$

V(s) and H(s) are Fourier transforms of the logarithmic functions v(x) and h(x) as functions of the spatial frequency, s, in the detector plane. The frequency response of this linear system, H(s), determines the maximum particle size resolution. This is evident after inspection of the Fourier solution for V(s).

$$V(s) = F(s)/H(s)$$

F(s) is the Fourier transform of the flux distribution f(x). Band-limitation of H(s) determines the reduction of resolution in v(x) due to noise in F(s). In particular, at frequencies where H(s) is small, the noise in F(s) will produce large errors in V(s) and v(x). Filtering of f(x) will reduce these size errors but with loss of size resolution. This analysis demonstrates that the improvement in size resolution

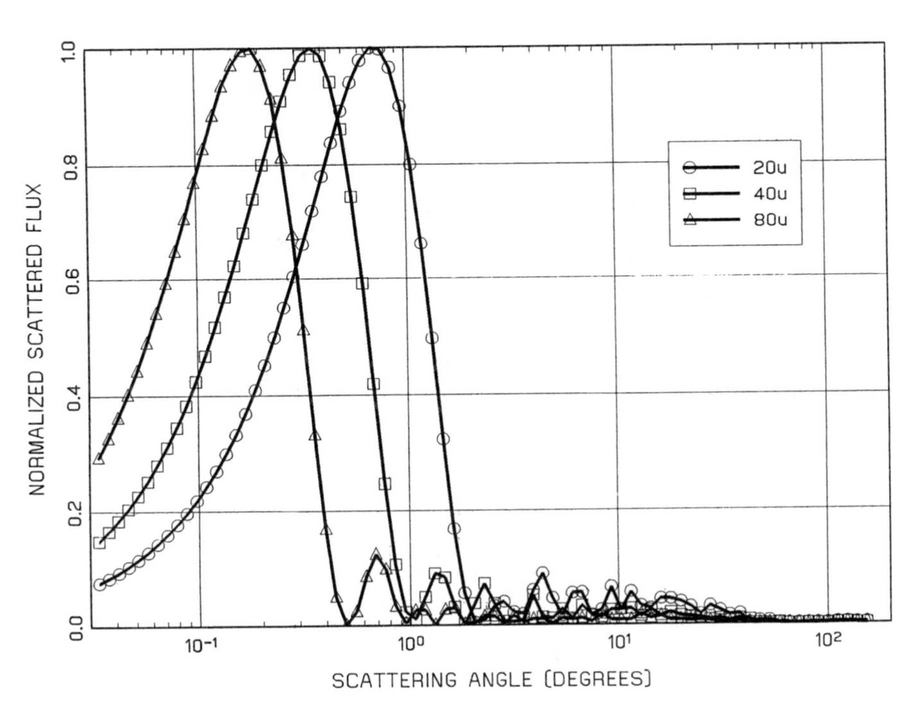

Figure 3. Normalized scattered flux for a logarithmic detector array. (Reproduced with permission from ref. 8. Copyright 1997 Instrument Society of America)

diminishes as the number of detectors in the array is increased. For typical radiometric measurement accuracy, the optimal logarithmic scale (*5*) is approximately $\beta = 1.2$. More detectors add noise in the frequency region where H(s) is attenuated, providing poor signal to noise in V(s) and marginal improvement in resolution.

Deviation From the Convolution Form

The principal convolution form is maintained over a large size range. Figure 4 shows the normalized f(x) contours of Mie calculations for polystyrene spheres in water. The straight contour lines indicate the shift invariant nature of the convolution. Secondary resonances start at diameters below 10 microns and substantial deviation from the convolution form is evident below 0.5 microns. Figure 5 shows the progression of this deviation down to 40 nanometers, for light polarized perpendicular to the scattering plane. The convolution form of the scattering function disappears below 0.2 microns and the inversion problem becomes more ill-conditioned. Radiometric accuracy is sufficient to provide accurate size determination down to 40 nanometers. However, since the deconvolution algorithm no longer applies, a generalized linear system model must be employed. This ill-conditioned problem is solved by introducing an *a priori* positivity constraint, which is implemented through an iterative procedure (*4*). Positivity of particle volume is the only assumption. The functional form of the particle size distribution is not assumed. Each point of the volume-vs.-diameter distribution is determined independently as part of a constrained linear system.

Multi-laser Design

As shown in Figure 5, measurements at scattering angles up to 165 degrees are required to obtain a favorable condition number for this linear system down to 40 nanometers. To extend the angular range to 45 degrees, with low optical aberration, a second lens and detector array are added as shown in Figure 6 (compare to Figure 1). However, additional lenses and detector arrays are an impracticable approach for further extension to 165 degrees. A better solution is to multiplex the dual array system among many optical beams (*6*) as shown in Figure 7. Since the scattering angle is defined from the direction of the incident light beam, the angular scattering range of the detector arrays is different for each beam. For example, beam 3 provides scattering angles up to 165 degrees.

This three laser system is the basis for the Honeywell Microtrac X100 particle size analyzer. Three laser diodes are collimated and projected through the particle sample region at three different angles of incidence. Each laser is turned on and off in sequence to multiplex the two detector arrays over three different ranges of angular scattering. These three scattering distributions are combined into one continuous angular distribution, from 0.02 to 165 degrees, which is inverted as a constrained linear system.

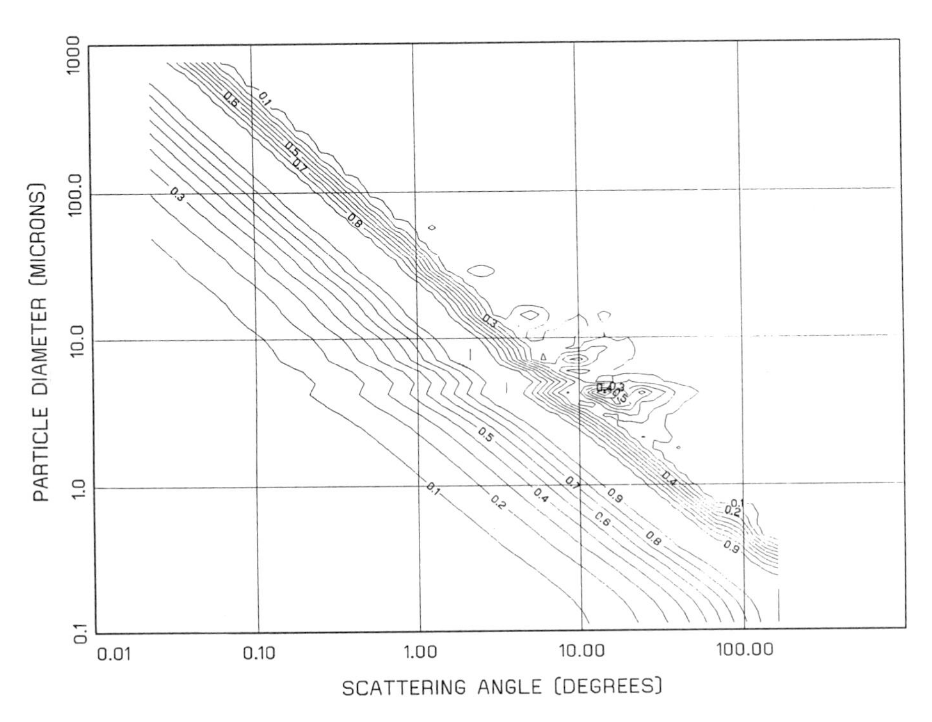

Figure 4. Normalized scattering response function contours for polystyrene spheres.

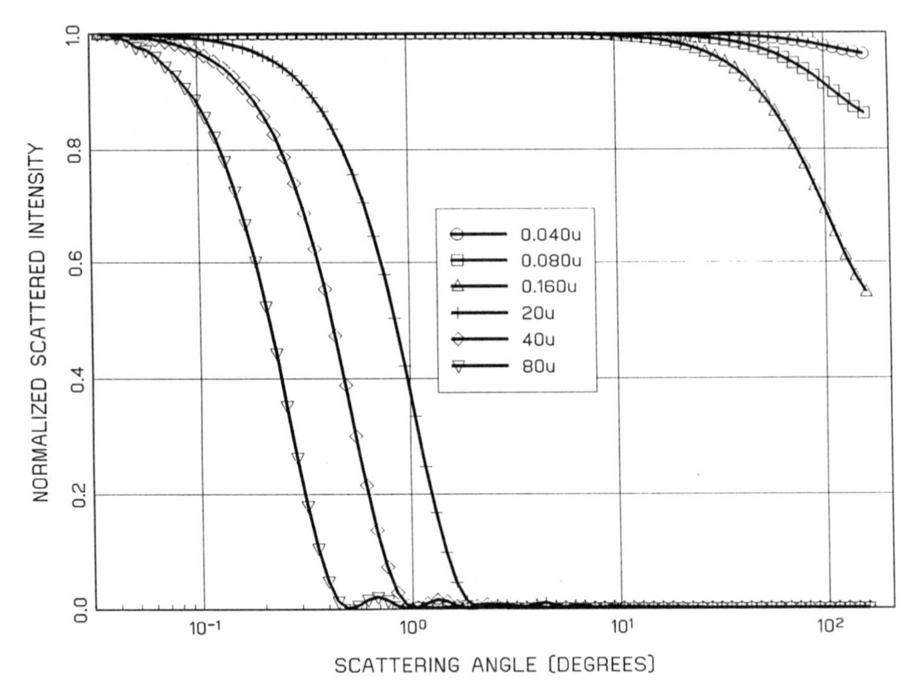

Figure 5. Normalized scattered intensity for polystyrene spheres.

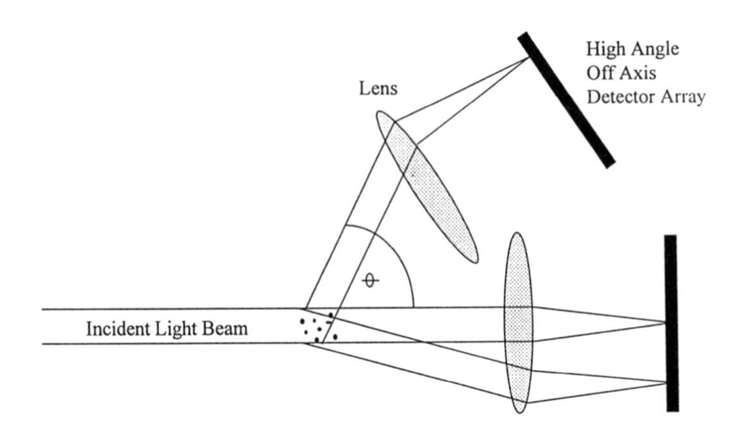

Figure 6. Scattering configuration with an off axis detector array.

Comparison of Dynamic and Static Scattering

Size measurement accuracy is determined by comparison with appropriate referee methods. Below 1 micron, the best referee methods are electron microscopy and dynamic light scattering (photon correlation spectroscopy). The following results use polystyrene size standards (electron microscopy) and samples of diamond powder which are characterized with dynamic scattering using the Honeywell Microtrac Ultrafine Particle Analyzer (UPA).

The UPA waveguide interferometer measures scattered light which is Doppler broadened by Brownian motion (*4*). The scattered light is heterodyned with the incident light to produce a down-shifted optical spectrum with high signal to noise. This waveguide interrogates a small scattering volume. The short optical path reduces multiple scattering to provide accurate size measurements at concentrations from ppm to 10% volume concentration (*7*).

Figures 8 and 9 show comparisons between static and dynamic scattering for 105 and 69 nanometer polystyrene, respectively. These results show excellent agreement between dynamic scattering (UPA) and wide angle static scattering (X100) in a size region where static scattering has typically failed. Low noise measurement at angles up to 165 degrees provides the linear system conditioning and high radiometric accuracy required for accurate size determination. Further evidence is shown in Figure 10 which shows static scattering results for 10 polystyrene samples between 69 and 1260 nanometers. Good resolution and accurate mean diameter are maintained throughout the size range. However, as the mean diameter decreases below 100 nanometers, the mode width and resolution approach limiting values due to poorer conditioning of the linear system. This condition number limitation is a consequence of diminishing size discrimination for particle size below 0.1 microns, as shown by the scattering functions in Figure 5.

Non-spherical Particles

The Mie scattering model accurately describes the optical scattering resonances of spherical particles. These resonances are strongly attenuated for non-spherical particles because the random orientation of particles, without rotational symmetry, destroys the interference structure of their coherently combined scattered fields. In addition, Mie theory does not accurately describe the light transmitted by non-spherical particles, which do not focus the transmitted light. The Mie theory model requires an input of particle and dispersant refractive indices, which are sometimes adjusted to produce acceptable results for non-spherical particles. This use of spherical particle models to invert scattered light distributions from non-spherical particles has no sound theoretical basis and essentially requires that the instrument user provide the size distribution to the instrument, negating the measurement need.

A more accurate scattering model for non-spherical particles is developed by analysis of the differences between scattering distributions from spherical and non-spherical particles. For example, most non-spherical particles, such as diamond and garnet, have angular forms with flat surfaces. In general, these particles generate an additional scattering component which can be determined empirically from

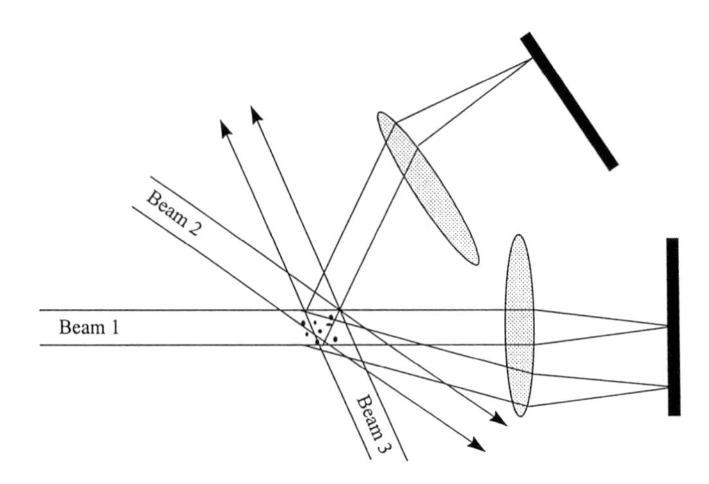

Figure 7. Three laser system with wide scattering angle range.

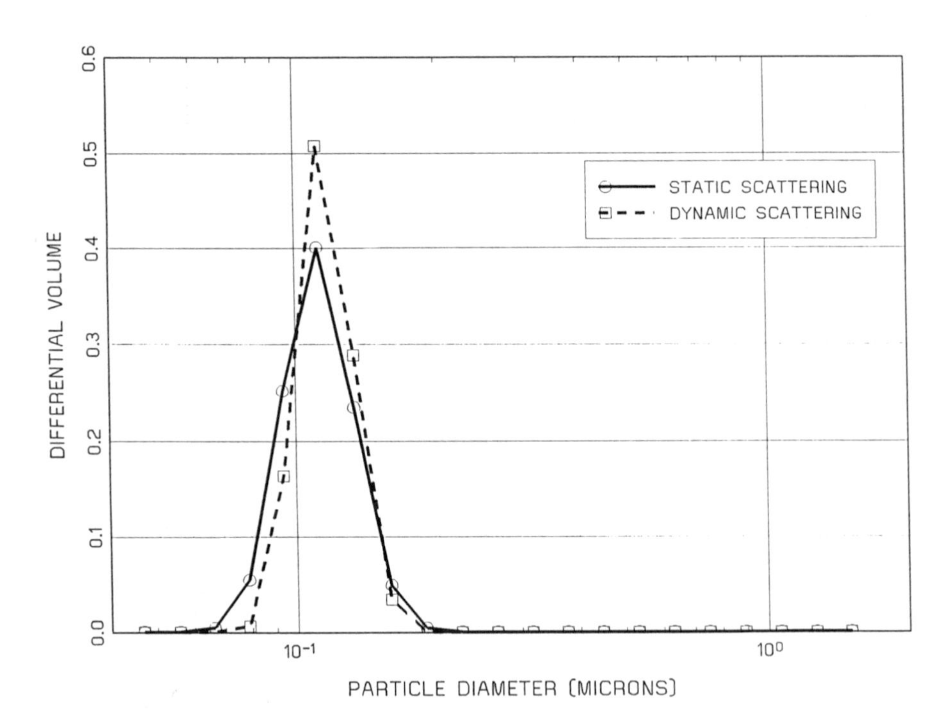

Figure 8. Differential volume distribution for 105 nanometer polystyrene spheres: static vs. dynamic scattering.

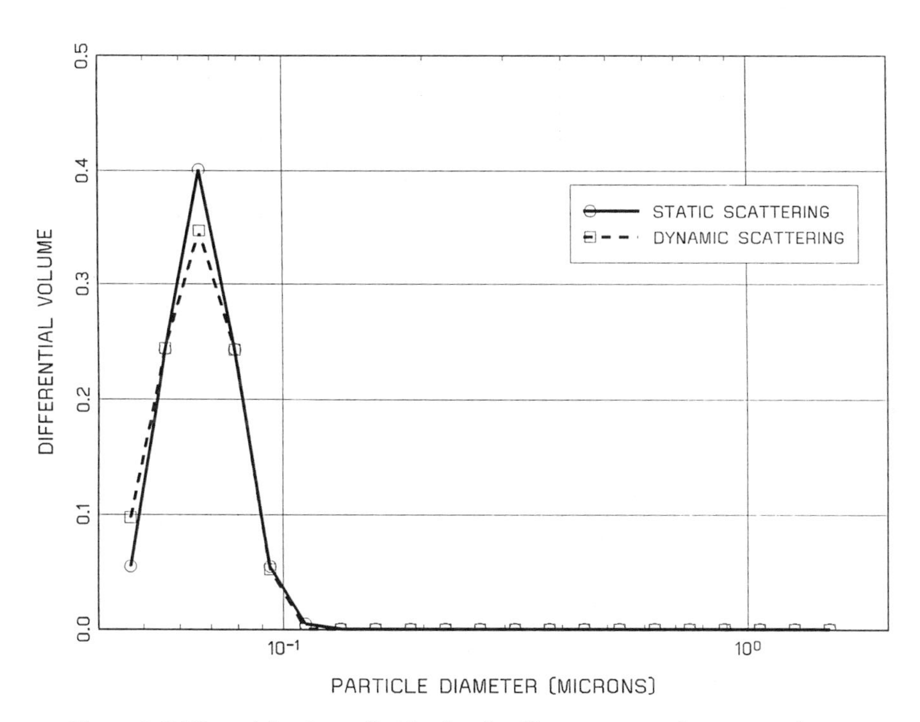

Figure 9. Differential volume distribution for 69 nanometer polystyrene spheres: static vs. dynamic scattering. (Reproduced with permission from ref. 8. Copyright 1997 Instrument Society of America)

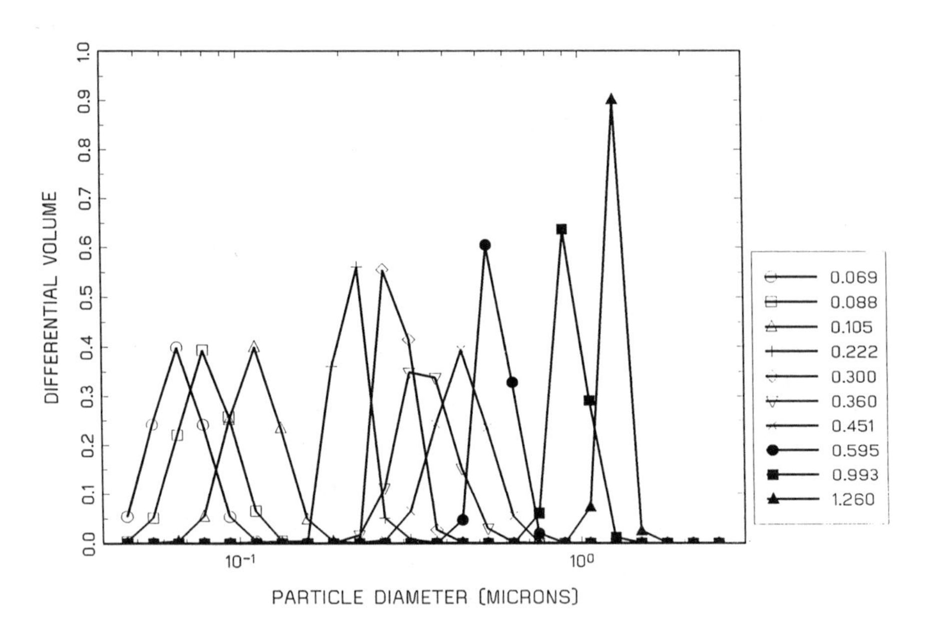

Figure 10. Static scattering differential volume distribution for narrow distributions of polystyrene spheres with mean diameters between 69 to 1260 nanometers.

comparison of size measurements based upon static and dynamic scattering. The equivalent size distribution is determined by the referee, which in our case is a "Brownian" equivalent sphere. This procedure is successful for the class of particle shapes which have a reasonable correlation between "Brownian" and "optical" equivalent diameters.

Dynamic and Static Scattering Results for Diamond Particles

This additional scattering component was calculated for narrow distributions of diamond powders with mean sizes between 0.1 and 10 microns. Then the static scattering model was corrected to account for this component. The results for three diamond powders are shown in Figures 11, 12, and 13. Reasonable agreement is shown between dynamic and static scattering for 0.12 micron diamond in Figure 11. In this size range, the spectral estimation for dynamic scattering is optimal but the static scattering model is only moderately conditioned. The condition number for static scattering improves for larger particles. Figures 12 and 13 show excellent agreement with dynamic scattering for 0.27 and 0.40 diamond samples, respectively. Usually dynamic scattering methods are preferred below 0.1 microns due to superior condition number and higher signal to noise. Above 1 micron, the spectral estimation of lower frequency components degrades dynamic scattering performance, but the conditioning of the static scattering problem is excellent.

Larger Particles

Angular scattering is capable of measuring particles from 40 nanometers to millimeters in diameter. Figure 14 shows a size distribution measured by the X100 with a mixture of five particle samples: 0.4, 2.4, 27, and 200 micron silicon dioxide and 8 micron aluminum oxide. All five modes are resolved with accurate mode position, demonstrating the wide size range provided by wide angle measurements and the ability of iterative algorithms to accurately invert complicated scattering distributions.

Conclusions

In conclusion, the method of wide angle static scattering provides accurate and reproducible size distribution results over a large size range, from 40 nanometers to millimeters. The optimal angular scale and resolution are determined from analysis of the Fraunhofer approximation, which predicts a convolution relationship between the particle size distribution and logarithmic scattering distribution. Deviation from this convolution relationship below 0.4 microns creates the need for a constrained linear system solution. Measurements at scattering angles up to 165 degrees are required to measure particles down to 40 nanometers. This large angular range is best handled by using multiple laser sources. Excellent agreement with dynamic light scattering is demonstrated down to 69 nanometers for polystyrene spheres. A non-spherical model provides good agreement with dynamic scattering for diamond powders. Wide size range, size accuracy, measurement speed, and excellent

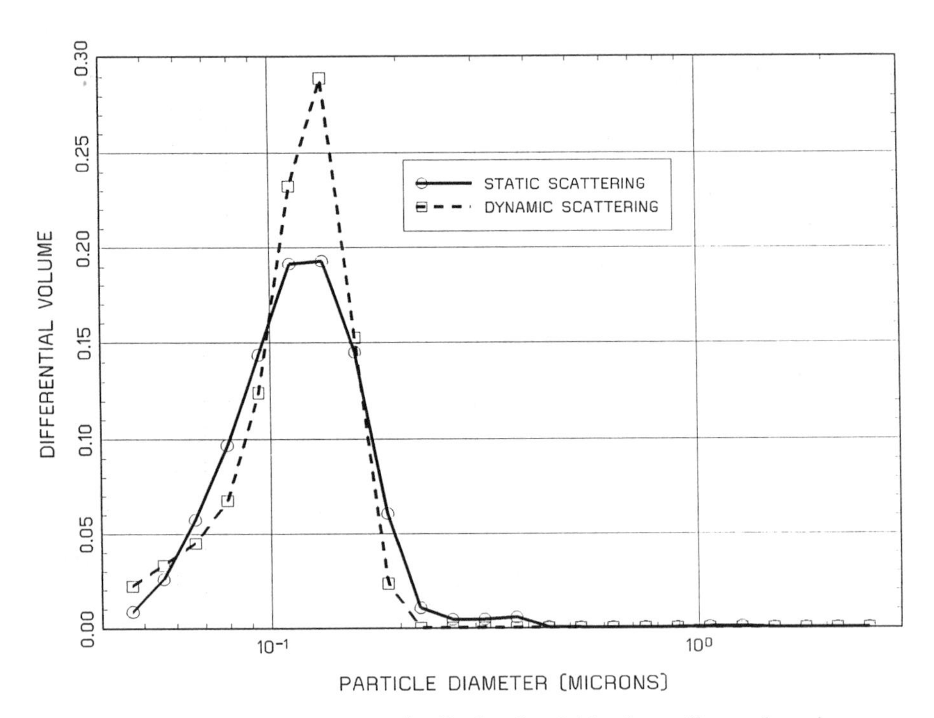

Figure 11. Differential volume distribution for 0.12 micron diamond: static vs. dynamic scattering.

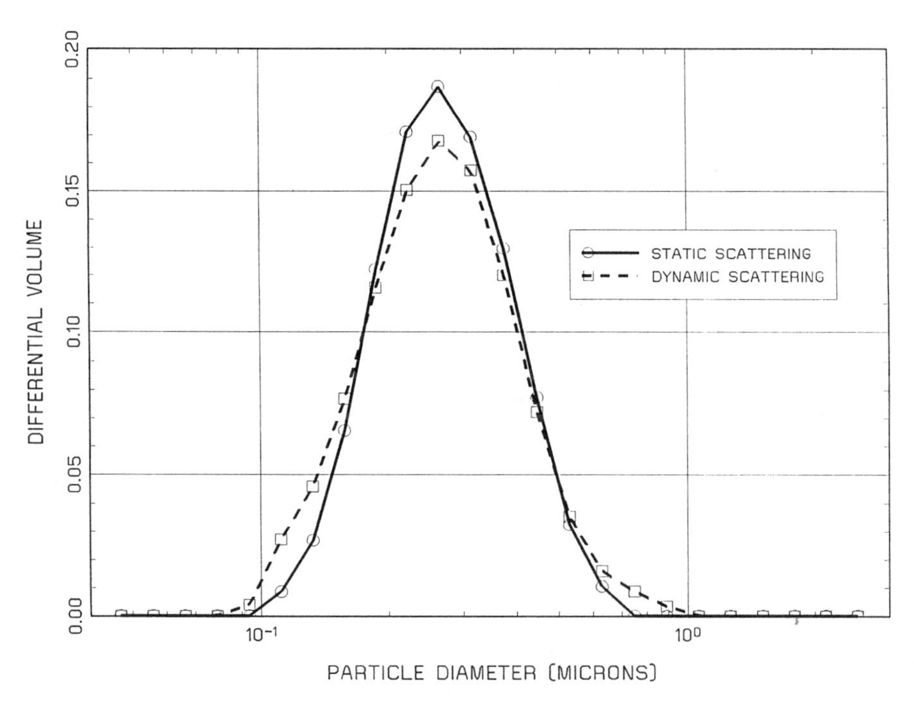

Figure 12. Differential volume distribution for 0.27 micron diamond: static vs. dynamic scattering.

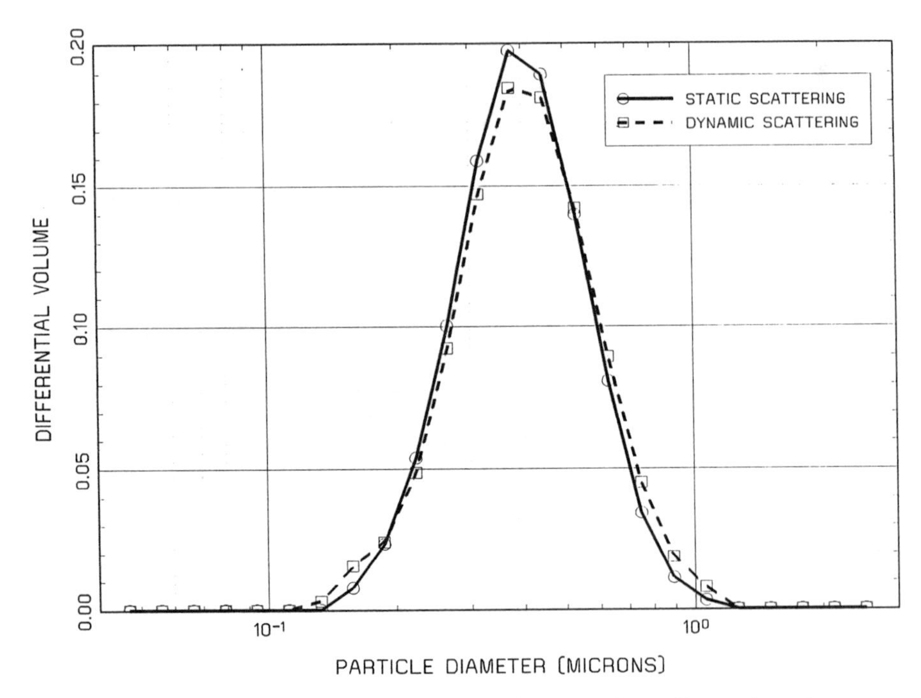

Figure 13. Differential volume distribution for 0.40 micron diamond: static vs. dynamic scattering.

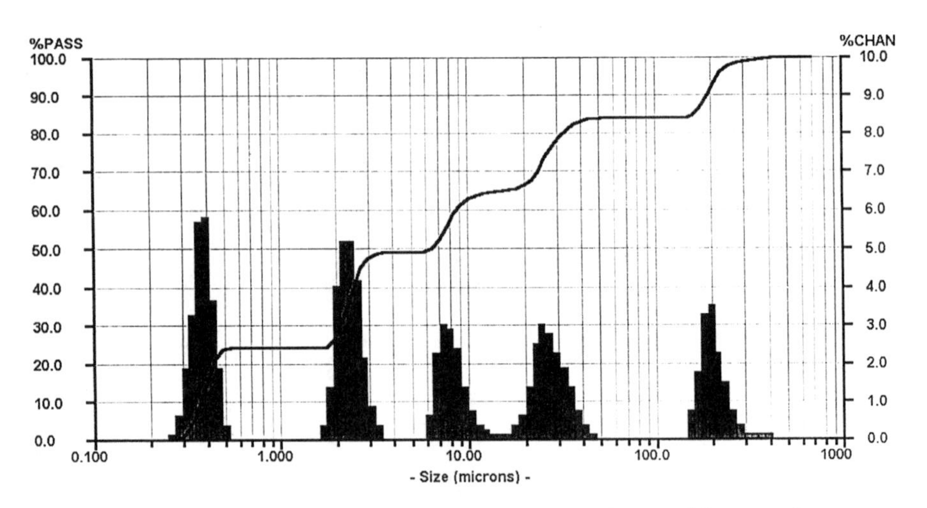

Figure 14. Differential volume distribution for a mixture of five samples between 0.4 to 200 microns. (Reproduced with permission from ref. 8. Copyright 1997 Instrument Society of America)

reproducibility will continue to promote the application of static angular light scattering in both the laboratory and process control environments.

Literature Cited

1. Pecora, R. *Dynamic Light Scattering: Applications of Photon Correlation Spectroscopy*, **1985**, Plenum Press.
2. Wertheimer, A.L.; Trainer, M.N., *Proceedings of the Annual Meeting of the Optical Society of America*, Rochester, New York, October 1979
3. Hirleman, E.D. , *Part. Chract.* **1987**, vol 4, pp 128-133.
4. Trainer, M.N., et al., U.S. Patent 5094532, March 10, 1992
5. Trainer, M.N.; Freud, P.J.; Weiss, E.L.; and Wilcock, W.L., *Proceedings of the 2nd International Congress on Optical Particle Sizing*, Tempe, Arizona, March 6,1990, pp 169-177
6. Trainer M.N., U.S. Patent 5416580, May 16, 1995
7. Trainer, M.N.; Freud, P.J.; and Leonardo E.M., *Amer. Lab.*, **1992**, vol 24 #11
8. Tenney, A.S.; Trainer, M.N.; *Proceedings of ISA TECH/97,* Anaheim, California, October 7-9, 1997

Chapter 11

Absorbed Layer Thicknesses of Associative Polymers on TiO_2 Particles

S. Sa-nguandekul, M. S. El-Aasser, and Cesar A. Silebi

Chemical Engineering Department and Polymer Interface Center, Lehigh University, Bethlehem, PA 18015

Particle sizes of TiO_2 and polystyrene dispersions in distilled-deionized water and in solutions of water soluble associative polymers (AP) were measured by dynamic light scattering. The associative polymers used consist of a water soluble backbone, with molecular weights ranging from 34,200-100,400, that have been capped with two hexadecyl-end groups. The adsorption isotherms of these polymers on TiO_2 and polystyrene particles were determined by serum replacement. The conformation of the adsorbed associative polymers on TiO_2 particles is slightly different from their conformation on polystyrene latexes. Adsorption isotherms suggest that polymer molecules with hydrophobic end groups form multilayers as the associative polymer concentration increases. The adsorbed layer of associative polymers with and without hexadecyl end-groups were measured by dynamic light scattering and the results were compared with the adsorbed layer on polystyrene latexes. The adsorbed layer of associative polymers on TiO_2 is approximately three times the radius of gyration of the polymer molecule. The adsorbed layer of the same polymer on polystyrene particles was twice as thick as the layer adsorbed on TiO_2. This indicates that the extend of stretching of the hydrophilic segment of the AP depends significantly on its affinity to the particle surface, therefore, the polymer may be assuming a random coil conformation with an extension of the chain into the aqueous medium.

Soluble polymers added to colloidal dispersions can either stabilize or flocculate a dispersion. The earliest known role, as stabilizer, aids or preserves the dispersion through adsorption of the polymer onto the surfaces of the particles due to a steric repulsion interaction between particles in close proximity. However, under some conditions, adsorption of polymer flocculates otherwise stable dispersion, due to the bridging of particles by adsorbed polymer molecules (*1*). In addition, even in the absence of adsorption, a soluble polymer can cause flocculation by the osmotic attraction induced by the exclusion of polymer from the gap between two particles (*2*). The range and strength of the steric repulsive interaction between particles with adsorbed polymers is determined to a great extend by the dimension of the adsorbed layer. The adsorption of polymer molecules on colloidal particles can influence the stability of a dispersion due to the particle-polymer interaction and the polymer-solvent interaction. In order to understand the mechanism of stabilization, characterizing the conformation of polymers on particle surfaces is very important. In general, the configuration of an adsorbed polymer layer on particle surfaces can be categorized as trains, loops and tails (*3*). The conformation is affected by factors such as polymer chemical structures, surface properties of colloidal particles, and solvent type. The conformation of an adsorbed layer of a homopolymer depends on the affinity of the polymer molecules toward the particle surface and the solvent. On the other hand, a block copolymer composed of hydrophobic and hydrophilic segments, can adsorb onto particle surfaces in different ways; for example, one segment can be attractive to the particle surface while the other can be repulsive to the particle surface. Several theories have been proposed to predict the adsorbed polymer layer conformation distributions or the transition between conformations for different types of copolymers, such as the scaling theory (*4-5*) and the mean field theory (*6-7*).

Model associative polymers, which are block copolymers consist of a hydrophilic backbone with two hydrophobic end groups, have unique chemical structures that are capable of forming dynamic network junctions in an aqueous phase and adsorbing on colloidal particles. Associative polymers with several hydrophobic groups grafted to the hydrophilic backbone are typically used as rheological modifiers for latex paints, and the interactions between associative polymers and other components in the formulation will help the stabilization of latex paints. It has been shown that the adsorption of associative polymers on polystyrene latexes affects the stability and rheological properties of latex dispersions *(8)*. Ou-Yang and Gao *(9)* observed a pancake to brush conformation (trains to loops and tails conformation) for model associative polymers adsorbed on polystyrene latexes as a first order phase transition. At very low polymer concentration in solution, every polymer molecule can attach to the many available adsorption sites on the polystyrene surface in a pancake-like conformation, with the polymer molecules lying down on the polystyrene surfaces. As the polymer concentration increases, the number density of polymer chains at particle surfaces increases. Hydrophobic end groups of the chain, that favorably attach to the surface will replace the backbone to lower the overall free energy of the system, and drive the polymer backbone to stretch out into the solution, forming a brush conformation. Both pancake and brush conformations exist during the transition, and only the brush conformation is observed at the saturated adsorption.

In this work, we study the conformation and adsorption of these associative polymer molecules on another component in the latex paint formulation; i.e., TiO_2 particles which are hydrophilic in nature. The adsorbed layer thickness can be investigated by many methods; e.g., dynamic light scattering, sedimentation, electrophoresis and viscometry (*10-15*). In the present work, the dynamic light scattering and viscosimetric methods were used.

Materials

Nonionic Hydrophobically Ethoxylated Urethane water-soluble associative polymers (HEUR), kindly provided by Union Carbide Co., were used as model associative polymers. The molecular structure of associative polymers is

$$R\text{-}O\text{-}(DI\text{-}PEG)_y\text{-}DI\text{-}O\text{-}R$$

Where DI is isophorone diisocyanate, PEG (Polyethylene glycol) is CARBOWAX™ 8,000 having a nominal molecular weight of 8,200, y is an index with even numbered value from 2 to 12, and R is either hydrophobic end group ($C_{16}H_{33}$) or hydroxyl terminated end group(OH). Four different molecular weights of associative polymers, ranging from 34,200 to 100,400, were examined to study the molecular weight effect on the adsorbed layer thickness. Based on the chemical structure of the associative polymer, one would expect two hydrophobes per chain. Yekta et al (*16*) investigated on these associative polymer, and observed approximately 1.4-1.7 hydrophobes per chain upon the purification of the associative polymer. According to this result, most of the associative polymer molecules contain two hydrophobes. Moreover, this supported by the significant enhancement in the viscosity of the model associative polymer solutions.

P25-TiO_2(Degussa) particles were utilized as the inorganic pigment particles for the adsorption isotherm experiments. The particle size as received is about 25 nm in diameter which is too small for all experiments, so the particles were agglomerated by calcination at 875° C for 2 hrs. This calcination causes the particles to transform from the 80% rutile/20% anatase form to the 100% rutile crystal form, which is commonly used in commercial paints. Since TiO_2 particles agglomerate after calcination, TiO_2 particles were milled with small size glass beads to deagglomerate and obtain the appropriate particle size. The final particle size, determined by a Nicomp Submicron Particle size Analyzer, was approximately 345 nm. The specific surface area of TiO_2 determined from the Brunauer Emmet and Teller (BET) nitrogen adsorption/desorption isotherm is 6.8 m^2/g.

Methods

Adsorption Isotherms. To understand the interaction at polymer/particle surfaces, the adsorption of associative polymers on TiO_2 particles was carried out by using a serum replacement technique (*17*). Associative polymer solution of known concentration was fed into a serum replacement cell containing 0.3 wt% TiO_2 dispersion, allowed to mix, then the concentration of associative polymers in the exit stream was detected by a differential refractometer. The effluent was collected and weighed to determine the cumulative associative polymer mass flux through the serum cell. The adsorbed amount of associative polymers onto TiO_2 particles was determined through a mass balance of associative polymers over the serum replacement cell. The desorption experiment was performed by using DDI water as a feed stream.

Adsorbed Layer Thickness. Two techniques were employed to determine the adsorbed layer thickness of associative polymers on TiO_2 particles: dynamic light scattering and viscosimetry.

Dynamic light scattering experiments were performed using a Nicomp Submicron particle size analyzer with a polarized light source of wavelength 632.8 nm and the scattering angle of 90 degrees at 25°C. The particle size of bare particles in distilled deionized water was measured as the reference particle size. The thickness of adsorbed layer is one half the difference between a measured particle size in the presence of associative polymers and the bare particle. Layer thicknesses of associative polymers on 91 nm and 357 nm polystyrene latexes were also determined to elucidate the effect of polydispersity and the effect of particle curvature. Two mixtures of the two polystyrene latexes (91 nm and 357 nm) were employed to study the effect of the particle size distribution on the accuracy of the measurements.

A viscosimetry experiment to determine the adsorbed layer thickness was carried out by a Ubbelohde viscometer in a controlled water bath at 25°C. The time required for a TiO_2 dispersion containing associative polymers to flow through a capillary tube was compared to the supernatant associative polymer solution, to determine the relative viscosity of the suspension. Based on the modified Einstein equation, the relative viscosity of a suspension can be related to the volume fraction of solid particles in the dispersion, and the adsorbed layer thickness of adsorbed polymers on particles (*18,19*).

$$\frac{\eta}{\eta_o} = 1 + 2.5\phi(1 + \frac{\delta}{r})^3 \quad (1)$$

Where η and η_o are viscosities of a suspension and a medium, respectively, ϕ is the volume fraction of particles in a suspension, δ is the adsorbed layer thickness of polymers on particle surfaces, and r is the bare particle radius.

Result and Discussion

Adsorption Isotherms. The adsorption/desorption isotherm of associative polymer ($C_{16}H_{33}$ hydrophobic end group, Mw ~ 34,200) on TiO_2 particles is shown in Figure 1. Associative polymer molecules start to adsorb onto the particle surfaces, and the adsorbed amount reaches a small plateau which probably corresponds to a monolayer of adsorbed polymer. As the associative polymer concentration increases, the adsorbed amount of polymers continues to increase, which may indicate multilayer adsorption. Some of associative polymer molecules were washout from particles after DDI-water was fed into the serum replacement cell, suggesting the desorption of additional layers. Lundberg and Glass (*20*) observed no adsorption of HEUR associative polymers on TiO_2 particles in an alkali media due to the negative charge of the particle surface. However, in our case, TiO_2 particles were simply dispersed in DDI-water, and the pH of the solution is around 6.5-7 which is close to the isoelectric point of the pigment (pH ~ 6.5). Hence, the adsorption interaction is likely to be due to the hydrophilic interaction between the EO group of the associative polymer and the OH group on the hydrate layer of TiO_2 particles.

On the other hand, the adsorption of associative polymer on TiO_2 particles is different from that on polystyrene latex particles investigated by Jenkins (*8*). Due to the hydrophobic surface of polystyrene latex particles, the adsorption isotherm was found to follow a sigmoidal shape (Langmuir type 4). The associative polymer molecules slowly

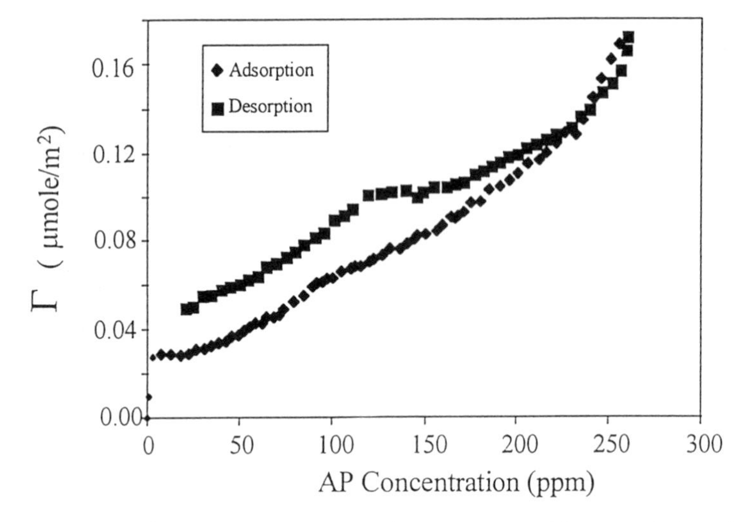

Figure 1. Adsorption isotherm of associative polymers with a nominal molecular weight of 34,200 $gmol^{-1}$, on TiO_2 particles.

adsorb onto particle surfaces at very dilute polymer concentration, and the adsorption rapidly rises to a plateau region as more polymers are added.

Since our major concern in this study is the interaction at polymer/particle interfaces and its layer thickness, we focused our investigation on the small plateau occurring at low polymer concentration. Figure 2 shows the adsorption/desorption isotherms of associative polymers (Mw~ 34,200) and the hydroxyl terminated polymers (Mw ~ 33,400), the same hydrophilic chain length, at low polymer concentration. Hydroxyl terminated polymers adsorb and reach the small plateau at 0.007 μmole/m^2 (0.234 mg/m^2). The adsorption continues to increase which may result from the further adsorption of polymer molecules on adsorbed layer due to the affinity of the polymer-polymer interaction. The desorption of polymers from particles is not observed that at this concentration, which indicates that the adsorbed amount may be in the equilibrium. In the case of the polymers with hydrophobic end groups, a small plateau of the adsorbed polymer is roughly 0.011 μmole/m^2 (0.37 mg/m^2) which is higher than that of the hydroxyl terminated polymer. Since the difference between these two polymers are their end groups, it is possible that end groups of associative polymers do not absorb onto particle surfaces but, instead, extend from the adsorbed layer so that more polymer molecules can attach to these end groups.

Adsorbed Layer Thickness Measurements. The adsorbed layer thickness of polymer molecules can be measured and compared to the radius of gyration of the polymer in

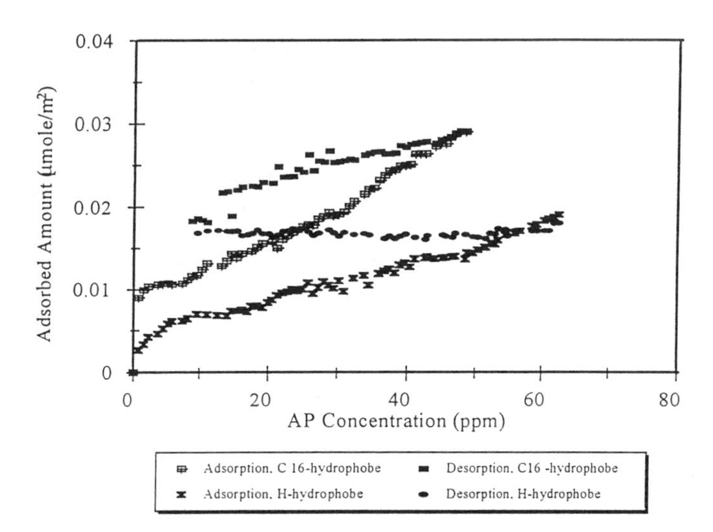

Figure 2. Adsorption isotherms of studied polymers, hydrophobic end group polymers (Mw ~34,200 gmol^{-1}) and hydroxyl terminated polymers (Mw ~ 33,200 gmol^{-1}), on TiO_2 particles.

solution. The radius of gyration of the associative polymer in solution can be determined from dilute solution viscometry using Flory's equation (*21*).

$$[\eta] = \frac{\phi < r^2 >^{3/2}}{Mw} \tag{2}$$

$$R_g = \frac{<r^2>^{1/2}}{\sqrt{6}} \tag{3}$$

Where $[\eta]$ is the intrinsic viscosity of the associative polymer solution, ϕ is $2.1 \pm 0.2 \times 10^{21}$ (dl/g· mol· cm^3), Mw is the molecular weight of the associative polymer, $<r^2>^{1/2}$ is the root mean square end-to-end distance of associative polymer, and R_g is the radius of gyration. The radius of gyration of these associative polymers obtained by Jenkins (*8*) are summarized in Table I.

Table I. The Radius of Gyration of Associative Polymer in Water

Mw ($gmol^{-1}$)	R_g (*nm*)
34,200	6.7
51,000	8.1
100,400	11.5

The thickness of an adsorbed layer is affected by the size of bare particles, since the curvature of particles is expected to play an important role in the conformation of polymer molecules on surfaces. Figures 3 and 4 illustrate the effect of polystyrene particle

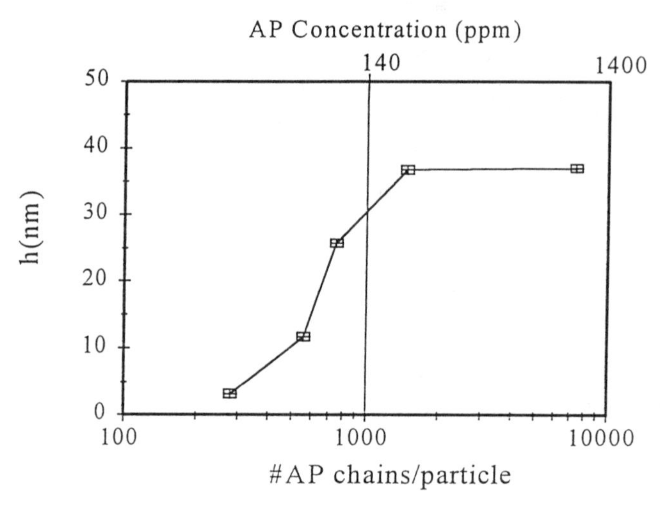

Figure 3. The layer thickness, h, of the associative polymer with a nominal molecular weight of 34,200 $gmol^{-1}$, on 0.1 wt% PS latex (D~ 91 nm)

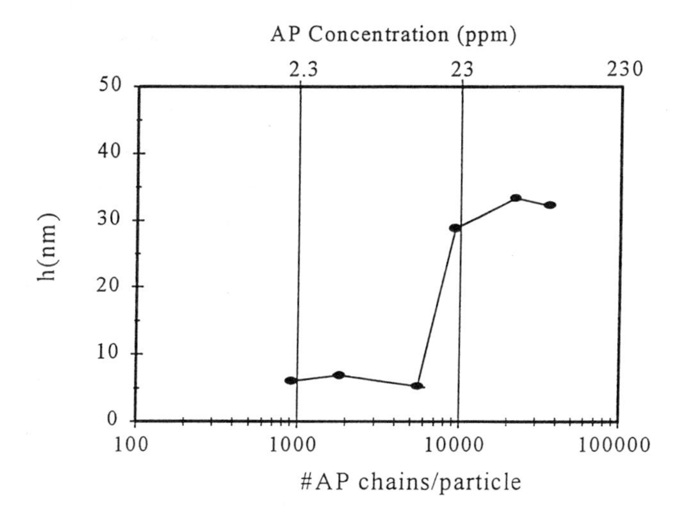

Figure 4. The layer thickness, h, of the associative polymer with a nominal molecular weight of 34,200 $gmol^{-1}$, on 0.1 wt% PS latex (D~ 357 nm)

sizes on the adsorbed layer thickness of associative polymers with nominal molecular weight of 34,200 and a radius of gyration of 6.7 nm. These results indicate that at low polymer concentration, the layer thickness is smaller than the radius of gyration, suggesting a pancake polymer conformation on the surface. As the polymer concentration increases the layer thickness becomes much higher and is about six times greater than the radius of gyration. Previous results by Gao (*22*) showed a similar profile of the layer thickness of associative polymers on polystyrene latex particles and they concluded that at higher polymer concentrations, the adsorbed polymer assumed a brush type conformation. In comparison, as the particle size of polystyrene latexes increases, the layer thickness of adsorbed polymers on 357 nm polystyrene latexes at full coverage is about 32 nm which is approximately 3 nm less than the layer thickness on 91 nm polystyrene latexes. The difference between the adsorbed layer thickness is small, so the difference in the particle size is not significant enough to show the effect of curvature. The adsorbed layer thicknesses at full coverage for these two latexes from two mixtures of them are summarized in Table II. These results show that although the mixtures are not a uniform size we were still able to determine the adsorbed layer thickness by dynamic light scattering.

Table II. The Adsorbed Layer Thickness of Associative Polymers on Polystyrene Latex Solutions at Full Coverage

PS Mixture Sample	*Particle Ratio 91 nm/357 nm (weight ratio)*	*Weight Average Particle Diameter of Mixture (nm)*	*Layer Thickness (nm)*
1	1:0	91	35
2	1:1	250	33
3	1:10	341	36
4	0:1	357	32

The adsorbed layer thickness of associative polymers, Mw ~ 34,200, on TiO_2 particles as a function of associative polymer concentration is shown in Figure 5. The adsorbed layer thickness is consistently smaller than that found for the adsorption on polystyrene latexes, and furthermore it does not show a sharp transition as the associative polymer concentration is increased. It appears that on TiO_2 particles both the hydrophobic end groups and the hydrophilic backbone of the polymer do adsorb on the TiO_2 surfaces, effectively reducing the extent to which the polymer molecule can stretch into the aqueous medium. Although the hydrophilic backbone of the associative polymer has a high water solubility, its affinity to the particle surface also has a significant effect on adsorption, as determined through adsorption isotherm studies with the homopolymer. Thus, the layer thickness will rely on the competition between these two factors.

Since the thickness of the adsorbed layer of associative polymer on TiO_2 particles is dependent on its hydrophilic backbone, the effect of polymer molecular weight was also investigated. Table III summarizes the adsorbed layer thickness of associative polymer on TiO_2 and polystyrene particles as a function of associative polymer molecular weight. The solid content was kept constant at 0.15 wt%, requiring a polymer concentration of 0.1% to fully cover the particle surface. To assure that the polymer concentration is high

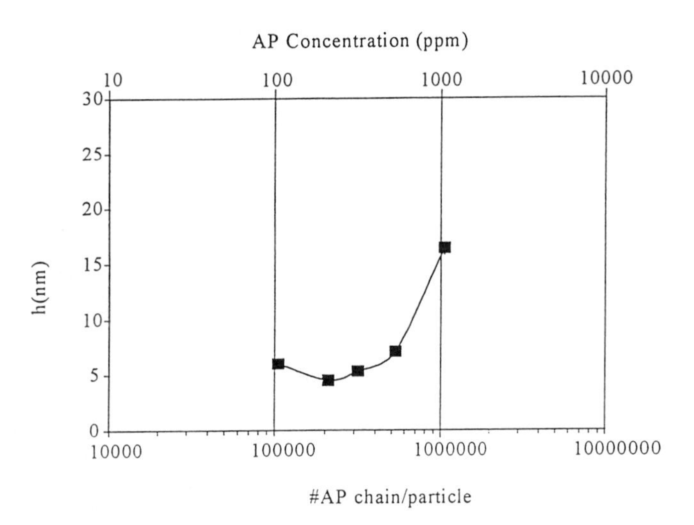

Figure 5. The layer thickness, h, of the associative polymer with a nominal molecular weight of 34,200 $gmol^{-1}$, on 0.15 wt% TiO_2 dispersion (D~ 345 nm)

enough to cover particle surfaces, the concentration of TiO_2 particles was also reduced to 0.03 wt%. As expected, for both concentrations of TiO_2 , the layer thickness increases as the molecular weight increases (which indicates a longer hydrophilic backbone), and is roughly three times the radius of gyration of free polymers. For a free random coil conformation, the layer thickness is less than twice the radius of gyration; so, the conformation of associative polymers on TiO_2 appears to be a random coil with the chain also extending out into the solution. The layer thickness of the adsorbed polymers on polystyrene particles is approximately 1.5 to 2.5 times greater than the layer thickness on TiO_2 particles, suggesting that the affinity of the hydrophilic backbone for the TiO_2 surface is sufficiently strong to limit its extending into the aqueous phase.

Table III. Layer Thicknesses of Hexadecyl End Group Associative Polymers on TiO_2 and PS Latex Particles.

Mw of AP Polymer	*Layer Thickness (nm)*		
(0.1 wt%)	*0.15 wt% TiO_2*	*0.03 wt% TiO_2*	*PS Latex*
34,200	16.4	13.1	32.4
51,000	20.8	17.5	48.6
100,400	39.6	42.1	59.2

In addition, the adsorption of the hydroxyl terminated polymer on TiO_2 and polystyrene particles was studied as reported in Table IV. In spite of the hydrophobicity of the surfaces, hydroxyl terminated polymers adsorb on polystyrene latexes and the layer thickness increases as a function of polymer molecular weight. Other workers have observed the same effect of polyethylene oxide (PEO) molecules adsorbed on polystyrene particles (*23-25*). In the case of TiO_2 particles (0.03 wt%), the layer thickness of low molecular weight hydroxyl terminated polymers is significantly greater than that of the associative polymers with the same molecular weight. These results suggest that more polymer molecules adsorb on particles and form multilayers on particle surfaces. For a 0.15 wt% TiO_2 dispersion, since the layer thickness of hexadecyl end group associative polymers is somewhat less than that of hydroxyl terminate polymers, it is possible that the hydrophobic end groups also attract the extended polymer toward the particle surfaces.

Table IV. Layer Thicknesses of Hydroxyl Terminated Associative Polymers on TiO_2 and PS Latex Particles

Mw of AP Polymer	*Layer Thickness (nm)*		
(0.1 wt%)	*0.15 wt% TiO_2*	*0.03 wt% TiO_2*	*PS Latex*
33,400	26.0	41.5	29.8
50,200	17.5	32.7	32.9
100,400	46.2	37.6	67.0

To confirm the hydrodynamic layer thicknesses determined by the light scattering method, viscosimetric analysis was carried out (Table V). The TiO_2 concentration is around 4.0 wt% ($\phi \sim 0.0099$) which is sufficiently dilute so that all the interactions between particles can be neglected. Therefore, it is appropriate to use the modified Einstein equation to calculate the layer thickness of polymer molecules on particle surfaces. The layer thickness determined from the viscosity measurements is somewhat smaller than that measured by dynamic light scattering. At low molecular weight, the difference between the measurements was significant, but a very small difference was observed at high molecular weight. Since the layer thickness of low molecular weight polymers is very small, an uncertainty of viscosity can result in a large deviation of the layer thickness. Also, since associative polymers adsorb on particles, the medium viscosity might not be the true viscosity for the calculation. However, the trend of the measurements was the same confirming the reliability of the light scattering method.

Table V. Layer Thicknesses of Associative Polymers on 4.0 wt% TiO_2 Dispersions Determined by Light Scattering Method and Viscosimetry Method.

Mw of AP	*Layer thickness (nm)*		*Difference*
(0.1 wt%)	*Light Scattering*	*Viscometry*	*(%)*
34,200	27.8	17.1	38.5
51,000	15.2	9.0	40.0
100,400	21.4	20.6	3.7

Summary and Conclusions

Adsorption isotherms of associative polymers on TiO_2 particles show an increase in adsorbed amount of polymers as the associative polymer concentration increases, which indicates multilayer conformation occurs at high associative polymer concentration. The layer thickness of associative polymers on TiO_2 particles were determined based on the first small plateau, supposedly monolayer, from the adsorption isotherm.

Due to the hydrophilic characteristic of the TiO_2 surfaces, the conformation of associative polymers on TiO_2 particles is slightly different from their conformation on polystyrene latexes. From the experimental results, a loosely coil conformation of the associative polymer chain simply lies on particle surfaces. The extent of polymer chain stretching depends on the affinity of the hydrophilic backbone to the particle surfaces and its solubility in water. Furthermore, the thickness of the adsorbed layer also depends on the hydrophobic end groups of the associative polymer which can also on the particle surface.

Literature Cited

1. Vincent, B., *Adv. Colloid & Interface. Sci.*, **1974**, 4, 193
2. Napper, D.H., *Polymeric Stabilization of Colloid Dispersion*, Academic Press, **1983**
3. Fleer, G.J., Cohen Stuart, M.A., Scheutjens, J.M.H.M., Cosgrove, T., and Vincent, B., *Polymers at Interfaces*, Chapman & Hall, Cambridge, UK, **1993**, 31
4. de Gennes, P.G., *J. de Physique*, **1976**, 37, 1461
5. de Gennes, P.G., *Macromolecules*, **1980**, 13, 1069

6. Scheutjens, J.M.H.M. and Fleer, G.J., *J. Phys. Chem.*, **1979**, 83, 1619
7. Joanny, J.F. and Johner, A., *J. Phys. II France*, **1996**, 6, 511
8. Jenkins, R.D., *Ph.D. Dissertation*, Lehigh University, Bethlehem, PA, **1990**
9. Ou-Yang, H.D., and Gao, Z.M., *J. Phys. II France*, **1991**, 1, 1375
10. van der Beek, G.P. and Cohen Stuart, M.A., *Langmuir,* **1991,** 7, 327
11. Bohmer, M.R., Koopal, L.L., Janssen, R., Lee, E.M., Thomas, R.K., and Rennie, A.R., *Langmuir*, **1992**, 8, 2228
12. Hedgus, C.R. and Kamel, I.L., *J. Coating Tech.*, **1993**, 65(821), 49
13. Levitz, P. Van Damme, H. and Karavis, D., *J. Phys. Chem.*, **1984**, 88, 2228
14. Siffert, B. and Li, S.F., *Colloids & Surfaces*, **1992**, 62, 307
15. Beckett, R., Ho, J., Jiang, Y. and Giddings, J.C., *Langmuir*, **1991**, 7, 2040
16. Yekta, A., Xu, B., Duhamel, J., Adiwidjaja, H., and Winnik, M.A., *Macromolecules*, **1995**, 28, 956
17. Ahmed, S.M., El-Aasser, M.S., Pauli, G.M., Poehlein, G.H. and Vanderhoff, J.W., *J. Colloids & Interfaces Sci.*, **1980**, 72(2), 388
18. Krieger, I.M., *Adv. Colloid Interface Sci.*, **1972**, 3, 111
19. Goodwin, J.W., *Colloidal Dispersions*, J.W., Goodwin, Ed., The Royal Society of Chemistry, UK, **1982**, 165
20. Lundberg, D.J., and Glass, J.E., *J. Coating Tech.*, **1992**, 64(807), 53
21. Flory, P.J., *Principle of Polymer Chemistry*, Cornell University, NY, **1978**, 611
22. Gao, Z.M., *PhD. Dissertation*, Lehigh University, Bethlehem, PA, **1994**
23. Kato. T., Nakamura, K., Kawaguchi, M. and Takahashi, A., *Polym. J.*, **1981**, 13(11), 1037
24. Polverari, M. and van de Ven, T.G.M., *Colloids & Surfaces A; Physicochemical & Eng. Aspect*, **1994**, 86, 209
25. Killman, E. and Sapuntzjis, P., *Colloids & Surfaces A; Physicochemical & Eng. Aspect*, **1994**, 86, 229

Chapter 12

Characterization of Polybutyl Acrylate Latex Particles Stabilized by a Reactive Surfactant

Chorng-Shyan Chern and Yu-Chang Chen

Department of Chemical Engineering, National Taiwan University of Science and Technology, Taipei 106, Taiwan, Republic of China

In semibatch emulsion polymerization of butyl acrylate, the concentration of the reactive surfactant sodium dodecyl allyl sulfosuccinate (JS-2) in the initial reactor charge is the most important parameter in determining the final latex particle size (d_p). The JS-2 stabilized latex particles were characterized by soap titration and experiments of coagulation kinetics. The saturated particle surface area occupied by one surfactant molecule used in the soap titration experiment increases with increasing particle surface polarity for the JS-2 stabilized latex products. The fraction of the chemically incorporated JS-2 buried inside the particles increases with increasing d_p. At constant surfactant concentration, the JS-2 stabilized latex shows poorer chemical stability than its counterpart stabilized by conventional sodium dodecyl sulfate. This result can be attributed to the effect associated with the buried JS-2 molecules.

Conventional surfactants such as sodium dodecyl sulfate (SDS) are small and mobile. These surface active species tend to migrate toward the surface layer of the coating formed from a latex product. This phenomenon can have a negative effect on the application properties of the polymeric film (e.g., adhesion and water resistance of a pressure sensitive adhesive).

Such a surfactant migration problem can be eliminated by means of surfactant-free emulsion polymerization (*1, 2*). However, the latex particles, stabilized only by the sulfate end-groups derived from the persulfate initiator, are relatively unstable because of their very low particle surface charge density. As a result, a significant amount of coagulum forms during the reaction. Limited flocculation often observed in semibatch surfactant-free emulsion polymerization of butyl acrylate (BA) makes the task of particle size control even more difficult (*1*). Both factors cause significant problems in the production of latex products. Incorporation of a small amount of functional monomer (e.g., acrylic acid or methacrylic acid) into the particles can greatly improve their colloidal stability and, thereby, retard limited flocculation (*2*). However, the particle size (d_p) of the latex product is still quite large (normally $d_p >$ 300 nm in diameter).

Another promising approach to alleviate such application problems associated with residual surfactant in the latex product is to use a reactive surfactant such as sodium dodecyl allyl sulfosuccinate (Eleminol JS-2) (38% active, Sanyo Chemical Industries) as a stabilizer in emulsion polymerization (*3-6*). The surfactant JS-2 can be chemically incorporated into the latex particles in the course of polymerization. Thus the immobilized JS-2 molecules are no longer capable of diffusing toward the surface layer of the coating and deteriorating the film properties. Furthermore, the sulfonate group covalently coupled onto the particle surface can greatly enhance the potential energy barrier against flocculation and, hence, retard the annoying limited flocculation.

The objective of this study was to characterize the JS-2 stabilized polybutyl acrylate latices produced in a semibatch reactor. The soap titration technique (*7-9*) was used to determine the saturated particle surface area covered by one molecule of soap used in the titration experiment (A_m). The parameter A_m provides information on the particle surface polarity and, therefore, it serves as an indicator that shows the relative level of JS-2 attached onto the particle surface during polymerization. This technique was also used to estimate the distribution profile of the covalently bonded JS-2 in the particles by a mass balance (*6*). Information on the location of chemically incorporated JS-2 molecules is crucial in evaluating the effectiveness of the reactive surfactant since only those JS-2 molecules attached onto the particle surface are capable of stabilizing the particles.

Experiments of coagulation kinetics (*10-12*) were used to study the colloidal stability of the JS-2 stabilized latex products toward added sodium salt. The coagulation kinetic data were then employed to determine the critical coagulation concentration (CCC), diffuse potential (ψ_δ) and Hamaker constant (A) according to DLVO theory (*13, 14*). The results obtained from this work can help one gain a better insight into the microscopic features of the polybutyl acrylate latex particles stabilized by JS-2. Such information should be useful to the manufacturers of tapes and labels.

Experimental

Materials. The chemicals used in this work include butyl acrylate (BA) (Formosa Plastics Co.), sodium dodecyl sulfate (SDS) (Henkel Co.), sodium alkyl allyl sulfosuccinate (Eleminol JS-2) (38% active, Sanyo Chemical Industries), sodium chloride (Riedel-de Haen), and deionized water (Barnsted, Nanopure Ultrapure Water System, specific conductance $< 0.057\ \mu$ Scm^{-1}). The monomer BA was distilled under reduced pressure before use. All other chemicals were used as received.

Polymerization Process. Semibatch emulsion polymerization was carried out in a 1-liter reactor equipped with a 4-bladed agitator, a thermometer, and a reflux condenser. A typical recipe designated as P1 is shown in Table I. The polymerization process involves addition of water along with initial surfactant and monomer to the reactor at room temperature. The initial reactor charge was purged with nitrogen for 10 min to remove dissolved oxygen while the reactor temperature was brought to 80 °C. The reaction was then initiated by adding the initiator solution to the reactor. After 15 min, the monomer emulsion was fed to the reactor over 3 h by an FMI pump. Polymerization temperature and agitation speed were kept constant at 80 °C and 400 rpm, respectively, throughout the reaction. After monomer emulsion feeding was complete, the reaction system was kept at 80 °C for 30 min to reduce the residual monomer to an acceptable level. The theoretical total solids content at the end of polymerization is 40%.

Table I. Typical Recipe Designated as P1 for Semibatch Polymerization of Butyl Acrylate Stabilized by Polymerizable Surfactant JS-2.

	Chemicals	Weight (g)
Monomer Emulsion Feed	H_2O	79.65
	JS-2 (38% active)	18.93
	BA	294.00
Initial Reactor Charge	H_2O	372.55
	JS-2 (38% active)	0.15
	BA	16.93
Initiator Solution	H_2O	15.00
	$Na_2S_2O_8$	0.97
Total Weight		798.18

Conditions: Theoretical Total Solids Content = 40 %; Monomer Feed Rate = 2.26 g/min; Agitation Speed = 400 rpm; T = 80 °C

Characterization. Particle size (d_p) data were obtained from the dynamic light-scattering method (Otsuka Photal LPA-3000/3100). Based on the cumulant method, the latex particle size was obtained from the average value of the correlation function standardized for every measurement. The accumulation times was set at 50 throughout this work. The particle size distribution (d_w/d_n) data were determined according to the histogram method.

The zeta potential (ζ) of the latex particles was determined by Malvern's Zetamaster. For appropriate measurements, the latex product with a volume of 0.05 ml was diluted with 100 ml water to adjust the number of photons counted per second (cps) to a proper value of 5000-15000. The dilution water has the same pH and conductivity as the original latex product. The pH and conductivity of the dilution water were adjusted by using HCl, NaOH, and NaCl. Thus, after sample preparation the latex particles should be exposed to a similar aqueous environment as compared to the original latex product. Five measurements were made for each latex sample and the average of these five measurements was reported as the ζ of the latex product.

The soap titration method was described in the literature (*7-9*). Before the start of the soap titration experiment, the latex sample was dialyzed (molecular weight cutoff = 12,000-14,000 g/mol) against deionized water for two days to remove the unreacted JS-2 and other impurities. The soap titration experiment was conducted at 25 °C using a 0.01 M solution of JS-2. The total solids content of the latex sample was in the range of 0.5%-2.5%. The surface tension of the sample was measured by using a surface tension meter (Face CBVP-A3).

Experiments of coagulation kinetics, monitored at 540 nm by a UV spectrophotometer (Shimadzu UV-160A), were used to determine the critical coagulation concentration (CCC), Hamaker constant (A), and diffuse potential (ψ_δ) of the latex sample (final total solid content = 0.01%) at various values of pH and 25 °C. The pH and ionic strength of the latex sample were adjusted by using NaOH, HCl, and NaCl.

Results and Discussion

Particle Size Control. Control of the final latex particle size is the key to guarantee the quality of the latex products. For example, the rheological property of the latex product is dependent on the particle size and size distribution. Changes in the viscosity due to batch-to-batch variations might cause production problems to the tape and label manufacturers. For trade paint applications, the rheological properties are generally adjusted by a rheological modifier (e.g., a high molecular weight cellulose derivative, a poly(carboxylic acid) latex, or an advanced associative thickener) to achieve the desired non-Newtonian viscosity profile required in the coating process. The resultant rheology of the coating system containing an associative thickener is strongly dependent on the particle size and size distribution of the latex particles (*15, 16*). In addition, the primary functions of trade paints and industrial coatings are protection and decoration. The particle size and size distribution play an important role in the application properties of these latex products (e.g., film formation, mechanical properties, chemical resistance, and optical properties). Thus it is very important to be able to control the particle size of the latex products.

Table II summarizes the formulae and the final latex particle size (d_p) data for the latex products investigated in this work. The final latex particle size decreases rapidly with increasing surfactant concentration in the initial reactor charge ($[S]_i$) for the JS-2 or SDS stabilized latices (see the d_p data for the JS series or SDS series in Table II). This is because the parameter $[S]_i$ controls the number of primary particles formed during the particle nucleation period. The value of d_p for latex PBA, prepared by

Table II. Recipes and Particle Size Data for Latices Chosen for This Study.

Latex	$[S]_i{}^a$ (%)	$[S]_i$ (M)	$[S]_f{}^b$ (%)	$Na_2S_2O_8$ (%)	d_p (nm)	d_w/d_n
P1	0.015		2.34	0.25	234.6	1.01
P3	0.48		3.23	0.25	86.1	1.05
PBA[c]	-		-	0.13	532.3	1.02
JS-2-1	1.495	0.0349	2.5	0.25	68.7	1.13
JS-2-2	0.149	0.00347	2.5	0.25	144.9	1.11
JS-2-3	0.0148	0.00034	2.5	0.25	237.7	1.04
SDS-1	1.00	0.0349	2.5	0.25	67.7	1.10
SDS-2	0.10	0.00347	2.5	0.25	136.9	1.01
SDS-3	0.01	0.00034	2.5	0.25	235.6	1.02

a Surfactant concentration in the initial reactor charge.
b Surfactant concentration in the monomer emulsion feed.
c Prepared by a semibatch surfactant-free process.

semibatch surfactant-free emulsion polymerization, is the largest (532.3 nm). In general, the particle size distribution (see the d_w/d_n data in Table II) is broader for the polymerization system containing a higher $[S]_i$ because it has the longest particle nucleation time. These experimental data show that semibatch emulsion polymerization stabilized by a polymerizable surfactant such as JS-2 can be used to prepare latex products with d_p ranging from ca. 50-250 nm, which is much smaller than those prepared by a surfactant-free technique.

Soap Titration. Latices P1 and P3 (both stabilized by JS-2) and PBA were then chosen for the soap titration studies. Figure 1 shows typical changes in the surface tension (γ) with the progress of soap titration for latex P1 with total solid content being in the range of 0.5%-2.5%. The soap titration profiles for both the latices P3 and PBA also exhibit similar trends. At constant γ , the parameter A_m can be calculated according to the following equations (*9*):

$$X_s/(1-C_s) = C_s/(1-C_s) + [E_s - C_s/(1-C_s)]\, X_p \quad (1)$$

$$E_s = 6MW_s/(d_p \rho_p A_m) \quad (2)$$

where C_s is the weight of surfactant per unit weight of water, X_s is the weight fraction of surfactant in the latex sample, X_p is the weight fraction of polymer particles in the latex sample, E_s is the weight of adsorbed surfactant per unit weight of polymer particles, and MW_s is the molecular weight of surfactant. The parameter A_m can be obtained from the slope and intercept of the X_s versus X_p plot at constant γ according to equations 1 and 2.

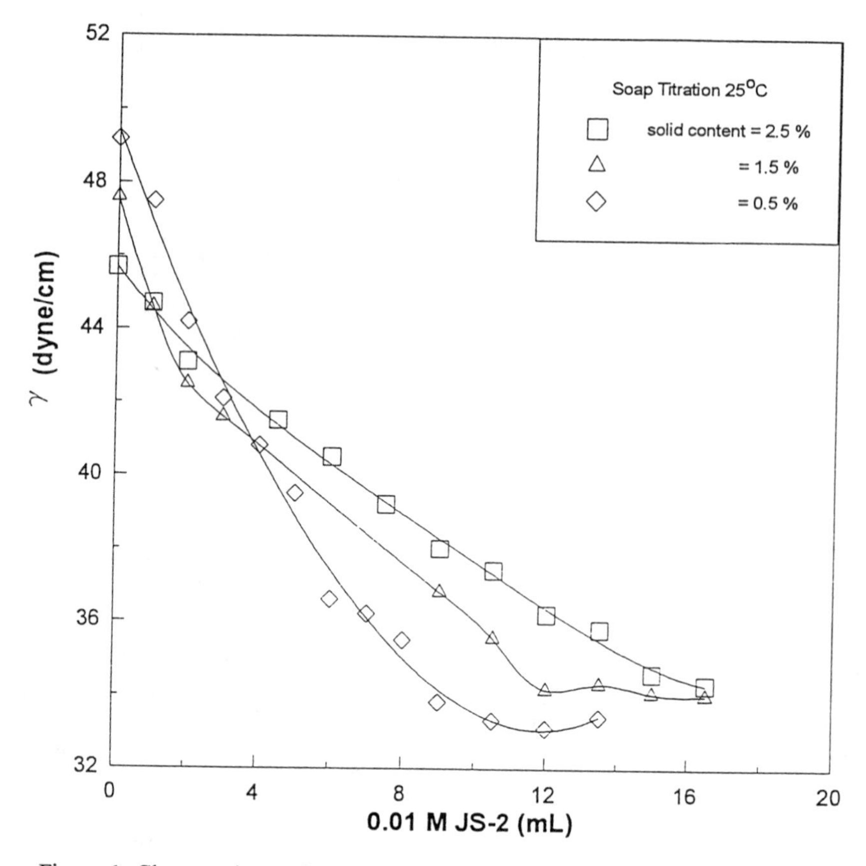

Figure 1. Changes in surface tension with progress of soap titration process for latex P1 (Reproduced with permission from reference 6).

Figures 2-4 show the X_s versus X_p plot at various values of γ for latex PBA, P1, and P3, respectively. As shown in Table III, the parameter A_m first decreases significantly and then levels off with the progress of soap titration. The saturated particle surface area covered by one molecule of JS-2 in the decreasing order is: P3 >

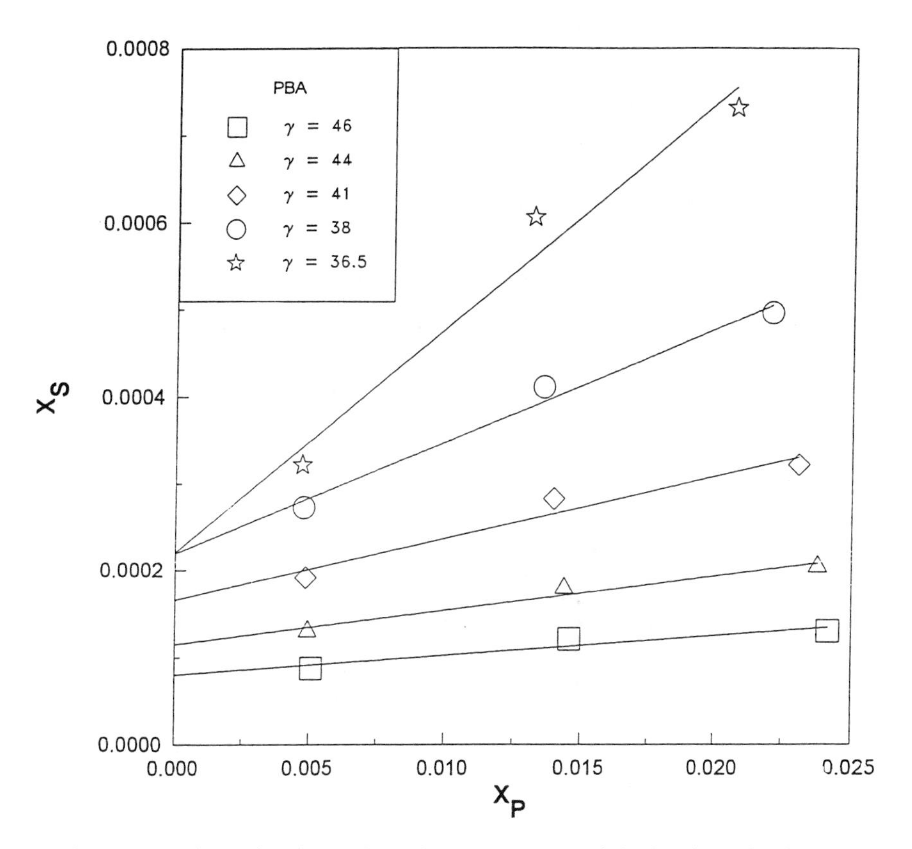

Figure 2. Weight fraction of surfactant versus weight fraction of polymer particles in latex sample for latex PBA.

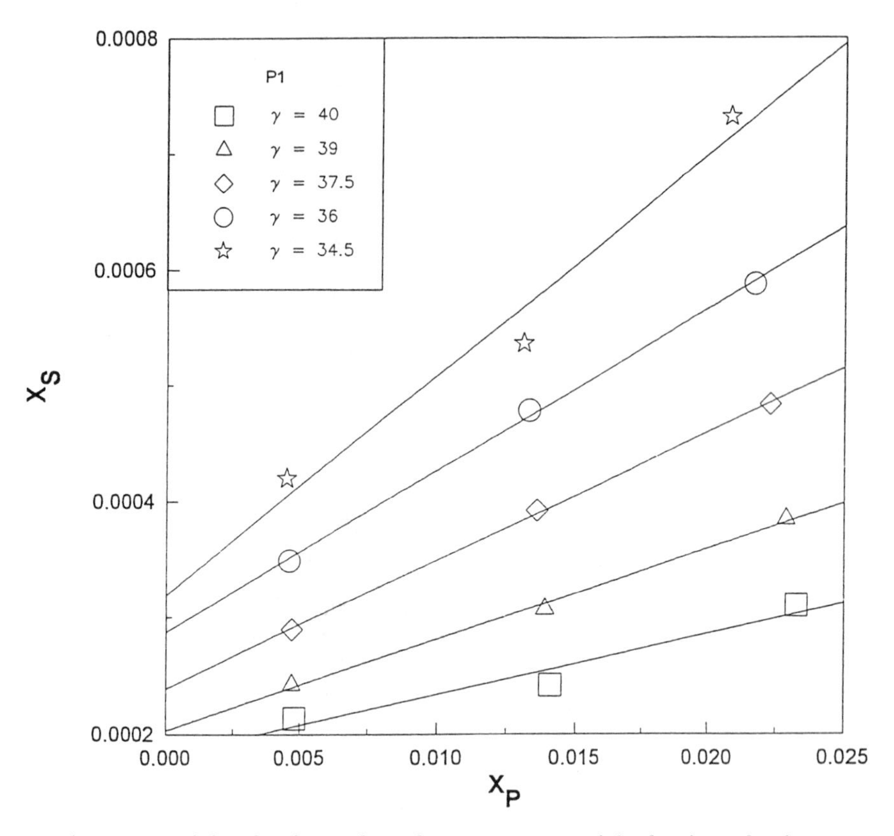

Figure 3. Weight fraction of surfactant versus weight fraction of polymer particles in latex sample for latex P1 (Reproduced with permission from reference 6).

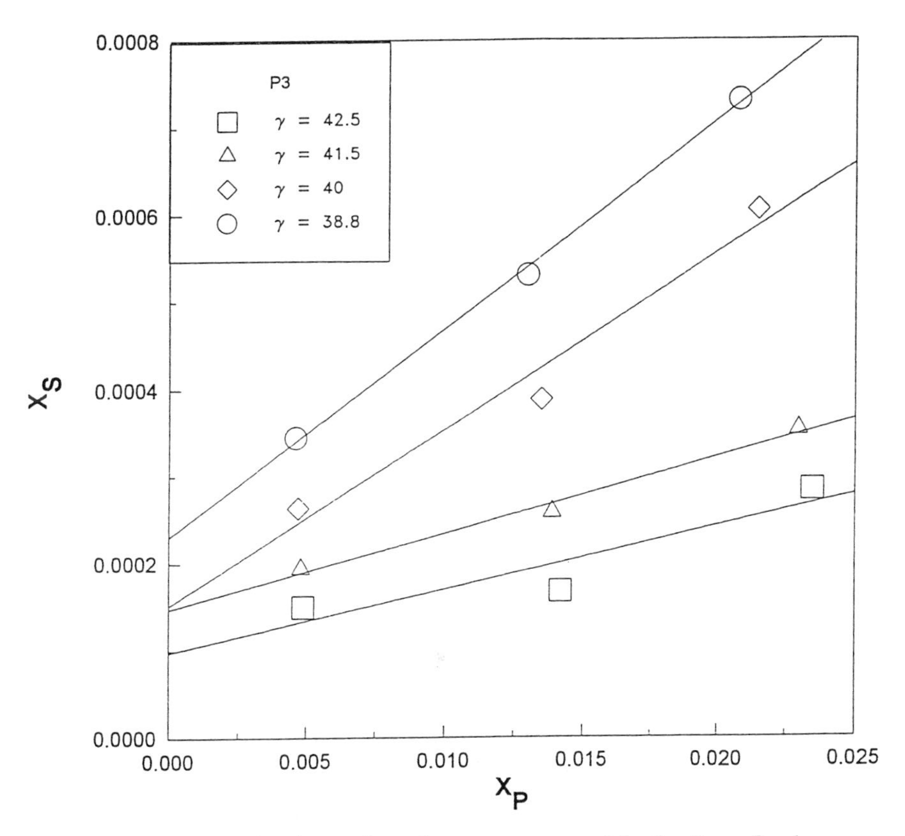

Figure 4. Weight fraction of surfactant versus weight fraction of polymer particles in latex sample for latex P3.

P1 > PBA. This observation is perhaps related to the fact that P3 particles containing the highest level of covalently bonded JS-2 should exhibit the most hydrophilic particle surface (see the recipe in Table II). Thus the higher the level of chemically incorporated JS-2 on the particle surface, the less is the amount of soap (i.e., JS-2 in this work) that can be physically adsorbed on the particle surface in the soap titration experiment. This factor can lead to the largest value of A_m for latex P3 (2.29 nm^2). On the other hand, PBA particles stabilized only by the sulfate end-group derived from the persulfate initiator are extremely hydrophobic. Therefore, these relatively hydrophobic particles can physically adsorb a significant number of JS-2 molecules in the soap titration experiment. This argument then leads to the smallest value of A_m for latex PBA (0.36 nm^2). These results are in agreement with the work of Paxton (*8*) and Ali et al. (*9*).

The fraction of chemically incorporated JS-2 buried inside the P1 or P3 particles can be estimated according to the procedure outlined in the appendix (*6*). The estimated fraction of the buried JS-2 is 56% and 18% for latex P1 (d_p = 234.6 nm) and latex P3 (d_p = 86.1 nm), respectively. This is reasonable because the probability for JS-2 molecules to be buried inside the particles should be greater for the much larger P1 particles. This result further supports the above finding that P1 particles are more hydrophobic than P3 particles. Therefore, the saturated value of A_m for latex P3 is larger than that for latex P1.

Table III. Caculated Particle Surface Area Coverd by One Molecule of JS-2 as a Function of Surface Tension for Latices P1, P3, and PBA (Reproduced with permission from reference 6).

PBA		P1		P3	
γ (dyne/cm)	A_m (nm^2)	γ (dyne/cm)	A_m (nm^2)	γ (dyne/cm)	A_m (nm^2)
46	4.17	40	3.81	42.5	7.54
44	2.38	39	2.54	41.5	6.25
41	1.30	37.5	1.80	40.0	2.69
38	0.72	36	1.42	38.8	2.29[a]
36.5	0.36[a]	34.5	1.04[a]		

a Saturated particle surface area covered by one molecule of JS-2

Coagulation Kinetics. The formulae and d_p data for the JS-2 series (JS-2-i; i = 1, 2, 3) selected for this study are listed in Table II. For comparison, the corresponding SDS stabilized series (SDS-i; i = 1, 2, 3) with the same molar concentration of surfactant in the initial reactor charge are also included in this study. In this manner, latex JS-2-i and the corresponding latex SDS-i show comparable values of d_p (e.g., the d_p data for latices JS-2-1 and SDS-1 are 68.7 and 67.7 nm, respectively, as shown in Table II). A typical set of absorbance versus time data at various NaCl concentrations ([NaCl]) for latex JS-2-1 at pH 5-6 is shown in Figure 5 (*17*). At constant [NaCl], the absorbance of the coagulated latex sample increases linearly with time during the early stage of coagulation experiment. Furthermore, the initial slope of the absorbance versus time plot (i.e., the rate of particle coagulation) first increases and then levels off when [NaCl] is increased.

According to Fuchs (*18*), the stability ratio (W) is defined as the ratio K_r/K_s, where K_r is the rapid flocculation rate constant and K_s is the slow flocculation rate constant. The parameter W (e.g., at [NaCl] = 0.7 M) can be estimated as the ratio of the asymptotic slope of the absorbance-vs.-time curve at zero time (e.g., the initial slope as [NaCl] → 1.3 M in Figure 5) to the slope of the absorbance-vs.-time curve at zero time (e.g., the initial slope at [NaCl] = 0.7 M in Fig. 5) because the rate constant K_r or K_s is proportional to the initial slope of the absorbance-vs.- time curve. The log W-vs.-log [NaCl] profiles for the JS-2 and SDS series at pH 5-6 are shown in Figures 6 and 7, respectively. The discontinuous point in the log W-vs.-log [NaCl] profile is then identified as the critical coagulation concentration (CCC) of the latex sample. In addition, the diffuse potential (ψ_δ) and Hamaker constant (A) of the latex product can be calculated according to the following equations (*19*):

$$-d \log W/d \log [NaCl] = 2.15x10^{7}\, r \gamma^{2} \quad (3)$$

$$A = [1.73x10^{-36} (d \log W/d \log [NaCl])/r^2 z^2 CCC]^{1} \quad (4)$$

where r is the particle radius, $\gamma = \tanh(ze\psi_\delta/2kT)$, z is the valence of counterions, e is an electron charge, k is Boltzmann constant, and T is the absolute temperature.

The parameters CCC, A and ψ_δ thus obtained are summarized in Table IV. The value of ψ_δ should depend on those parameters such as the particle surface charge density, particle size and ionic strength, etc. For a particular latex sample, the parameter ψ_δ increases with increasing pH. At relatively low pH, in addition to Na^+ ions, H^+ ions also can contribute to the ionic strength of the solution. When pH is decreased from 6 to 3, the increased ionic strength can compress the electric double layer of the particles and, hence, lead to a decrease in ψ_δ. However, it is quite difficult to explain the maximum ψ_δ data observed at pH 11. This is probably caused by ionization of the carboxyl groups derived from hydrolysis of the sulfate end-groups and subsequent oxidation of the generated hydroxyl groups (*20*). The

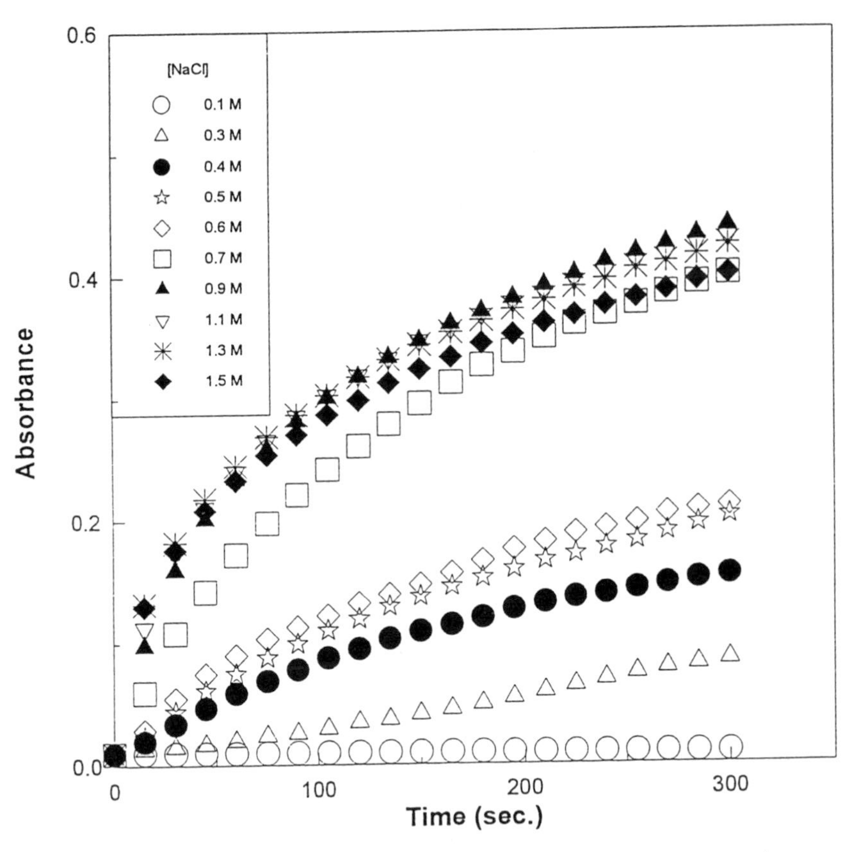

Figure 5. Absorbance of coagulated latex JS-2-1 at pH 5-6 versus time at various concentrations of NaCl (Reproduced with permission from reference 17).

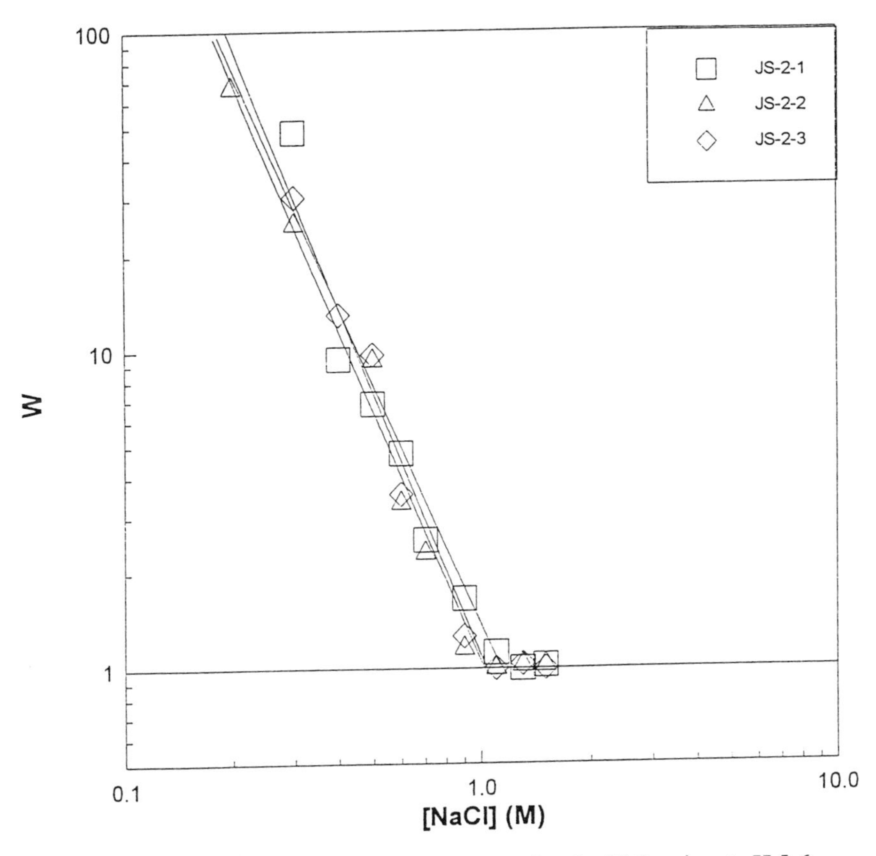

Figure 6. Stability ratio versus NaCl concentration for JS-2 series at pH 5-6.

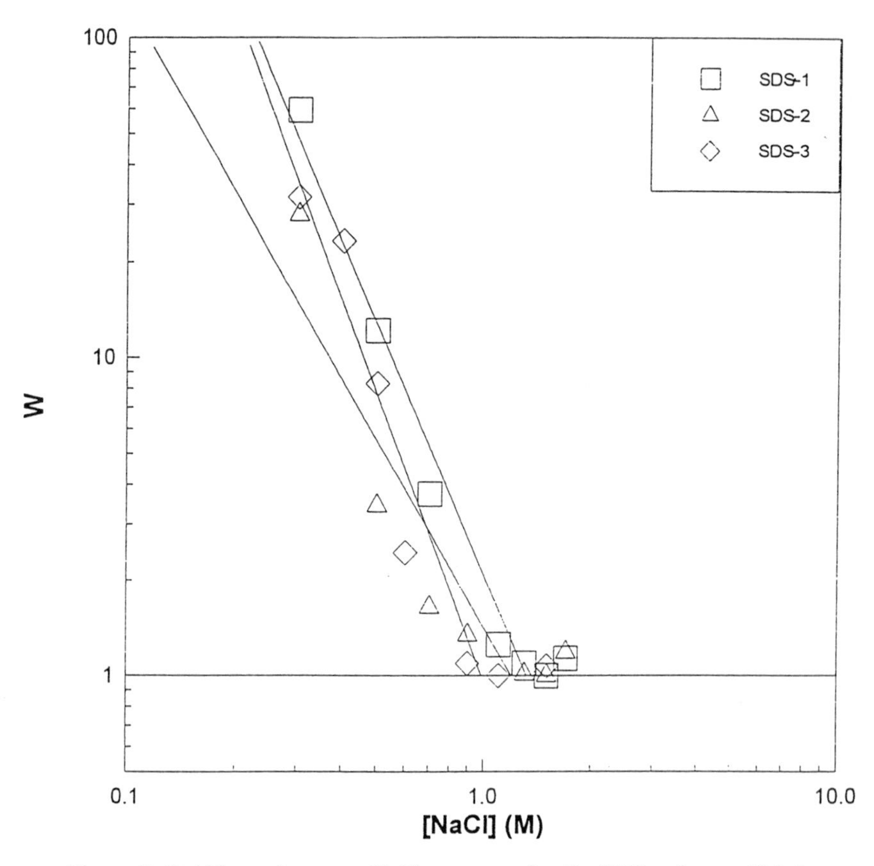

Figure 7. Stability ratio versus NaCl concentration for SDS series at pH 5-6.

parameters CCC and A also show similar trends with respect to pH. Thus the stability of the latex sample toward added salts increases with increasing pH.

As shown in Tables II and IV, at constant pH, the parameter ψ_δ increases with increasing $[S]_i$, which is attributed to the increased particle surface charge density with $[S]_i$. Consequently, the latex product containing a higher level of surfactant in the initial reactor charge should be more stable toward added salts and, indeed, it generally shows a higher value of CCC. The parameter A also increases with increasing $[S]_i$, which is consistent with the work of Patey (*21*). Considering two approaching particles possessing a relatively high particle surface charge density, electrostatic attraction force can be induced between the adsorbed anionic surfactant molecules on one particle surface and the counterions around another particle. This factor can greatly enhance the affinity between two approaching particles and, therefore, increase the apparent magnitude of A. No significant difference between the JS-2 and SDS stabilized latex products is observed in terms of chemical stability.

Finally, the zeta potential (ζ) data of the JS-2 or SDS stabilized particles at pH 5 as a function of [NaCl] are shown in Figure 8. The ζ data at pH 3 or pH 11 also show similar trends. There is no apparent correlations between ζ and those parameters such as type of surfactant, $[S]_i$, or pH. Therefore, the ζ data alone are insufficient to predict the relative stability of latex products toward added electrolytes. On the other hand, Figure 8 shows that the value of ζ decreases rapidly with increasing [NaCl]. This trend is due to the fact that the added counterion (Na^+) can compress the electric double layer and, consequently, reduce the zeta potential of the particles. The decreased ζ then can greatly reduce the potential energy barrier against coagulation. The measured zeta potential (ζ) has been generally assumed to be the same as the theoretical diffuse potential (ψ_δ ; ca. -10~ -20 mV in this work). In this study, this assumption is valid only when [NaCl] is in the range of 0.5-0.9 M (see Figure 7), which is consistent with the work of Bastos and de las Nieves (*12*).

Table IV. Critical Coagulation Concentration, Hamaker Constant, and Diffuse Potential Data Obtained from Experiments of Coagulation Kinetics (Reproduced with permission from reference 17).

	CCC (M)			A (10^{-21} J)			ψ_δ (-mV)		
Latex	pH 3	pH 5-6	pH 11	pH 3	pH 5-6	pH 11	pH 3	pH 5-6	pH 11
JS-2-1	1.17	1.14	2.07	1.95	2.85	3.03	15.92	19.16	23.06
JS-2-2	0.87	1.02	1.72	1.14	1.49	1.74	11.25	13.39	16.54
JS-2-3	0.78	1.04	-	0.49	0.95	-	7.13	10.70	-
SDS-1	0.92	1.32	1.66	1.89	2.80	3.01	14.70	19.72	21.74
SDS-2	0.83	1.19	1.57	0.94	1.08	1.53	10.09	11.89	15.17
SDS-3	0.72	0.98	1.51	0.90	1.07	1.35	9.52	11.22	14.08

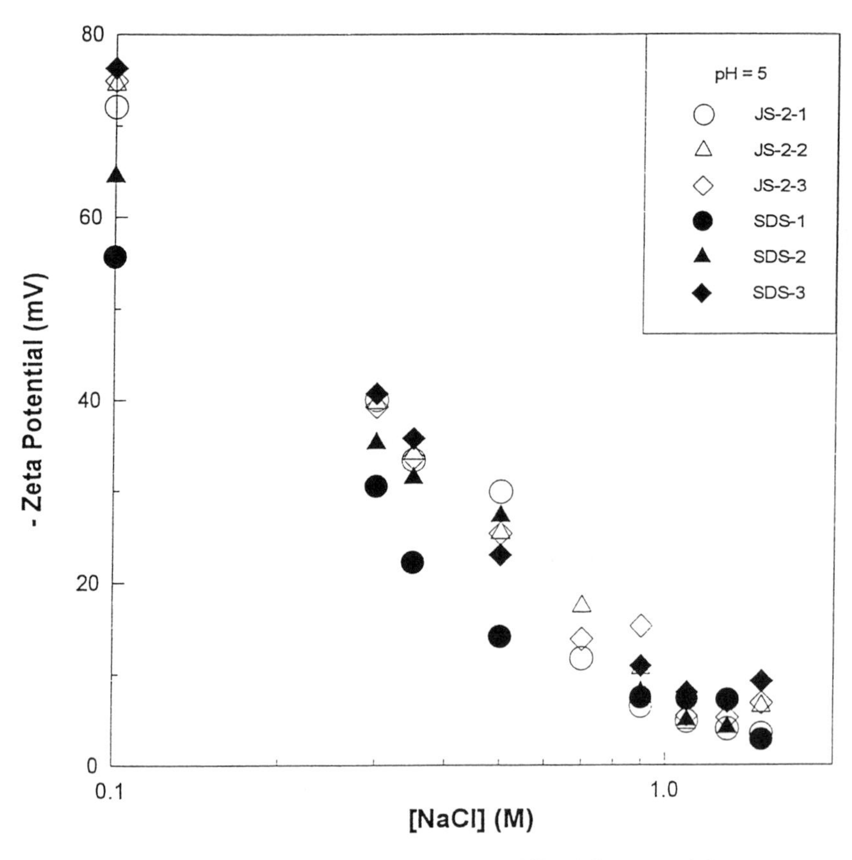

Figure 8. Zeta potential of JS-2 or SDS stabilized latex particles at pH 5 versus NaCl concentration.

Conclusions

In semibatch emulsion polymerization of butyl acrylate, the reactive surfactant sodium dodecyl allyl sulfosuccinate (JS-2) increases the particle surface charge density significantly and, thereby, reduces the particle size (d_p) of the latex product to as small as 70 nm. Such a small d_p has not been achieved by the competitive surfactant-free technique (normally $d_p > 300$ nm). The concentration of JS-2 in the initial reactor charge predominates during the particle nucleation period and it is the most important parameter in controlling the d_p of the latex product.

The JS-2 stabilized latex particles were characterized by the soap titration method and experiments of coagulation kinetics. The greater the amount of chemically incorporated JS-2 on the particle surface is, the larger is the saturated particle surface area occupied by one surfactant molecule (A_m) used in the soap titration experiment. Furthermore, the fraction of covalently bonded JS-2 that is buried inside the particles increases with increasing d_p. This is because the probability for JS-2 molecules to be buried inside the particles should be greater for the larger particles. This trend is in agreement with the above mentioned A_m data. The critical coagulation concentration data obtained from the experiments of coagulation kinetics show that the stability of the latex product toward added salts increases with increasing pH. The latex product containing a higher level of JS-2 in the initial reactor charge also shows a better chemical stability. At constant surfactant concentration, the JS-2 stabilized latex product shows worse chemical stability than its counterpart stabilized by the conventional surfactant sodium dodecyl sulfate. This result can be attributed to the fact that chemically incorporated JS-2 molecules can be buried inside the particles and, hence, reduce the chemical stability of the JS-2 stabilized latex.

Acknowledgment

The financial support from National Science Council, Taiwan, Republic of China (NSC84-2216-E011-024) is gratefully acknowledged.

Appendix

The following procedure outlines the method of estimating the fraction of chemically incorporated JS-2 that has been buried inside the P1 or P3 particles.

(1) Estimate PBA particle surface area covered by one molecule of sulfate end-group derived from the persulfate initiator (A_{sulf}) according to the following equations.

$$A_{sulf} = (a_t - a_{ads})/N_{sulf} \tag{A1}$$

$$a_t = 6w_p/(d_p \rho) \tag{A2}$$

$$a_{ads} = [(w_s - w_{sw})/MW_s]\, N_a \times (0.36\ nm^2/\text{molecule for PBA} \qquad (A3)$$

where the parameter a_t is total particle surface area in the latex sample, w_p is the weight of polymer in the latex sample, d_p is the particle size, ρ_p is the density of polymer, a_{ads} is the particle surface area covered by the adsorbed surfactant (JS-2) at the titration end-point, w_s is the weight of surfactant (JS-2) used at the end-point, w_{sw} is the weight of surfactant (JS-2) dissolved in water at the end-point that can be determined by the intercept of the X_s-vs.-X_p plot (see equation 1), and the parameter N_a is Avogadro's number. The parameter N_{sulf} in equation A1 is the number of sulfate end-groups on the particle surface that can be estimated with the knowledge of the amount of initiator used in the polymerization recipe, the initiator decomposition rate constant ($1.096 \times 10^{-4}\ s^{-1}$ at 80 °C), the initiator efficiency factor (assumed to be 0.6 in this work), and the reaction time.

(2) Estimate the weight of JS-2 chemically incorporated onto the P1 or P3 particle surface during polymerization (w_{chem}).

$$w_{chem} = (a_t - N_{sulf}A_{sulf} - a_{ads})MW_s/(0.36N_a) \qquad (A4)$$

Note that the parameters a_t, N_{sulf}, and a_{ads} in equation A4 represent those for the soap titration experiment dealing with the P1 or P3 particles in this step.

(3) Estimate the fraction of chemically incorporated JS-2 buried inside the P1 or P3 particles (f_{buried}).

$$w_{buried} = (w_{st} - w_{chem} - w_{unreact} \qquad (A5)$$

$$f_{buried} = w_{buried}/w_{st} \qquad (A6)$$

where w_{buried} is the weight of chemically incorporated JS-2 that has been buried inside the P1 or P3 particles and w_{st} is the total weight of JS-2 in the latex sample. The parameter $w_{unreact}$ in equation A5 is the weight of unreacted JS-2 that can be determined by measuring the surface tension in the dialysate and then performing interpolation of the surfactant concentration by Figure 2 shown in reference 6.

References

1. C. S. Chern and C. H. Lin, *Polym. J.*, **27**, 1094 (1995).
2. C. S. Chern and C. H. Lin, *Polym. J.*, **28**, 343 (1996).
3. M. B. Urquiola, V. L. Dimonie, E. D. Sudol, and M. S. El-Aasser, *J. Polym.*

Sci., Polym. Chem. Ed., **30**, 2619 (1992).
4. M. B. Urquiola, V. L. Dimonie, E. D. Sudol, and M. S. El-Aasser, *J. Polym. Sci., Polym. Chem. Ed.*, **30**, 2631 (1992).
5. M. B. Urquiola, E. D. Sudol, V. L. Dimonie, and M. S. El-Aasser, *J. Polym. Sci., Polym. Chem. Ed.*, **31**, 1403 (1993).
6. C. S. Chern and Y. C. Chen, *Polym. J.*, **28**, 627 (1996).
7. S. H. Maron, M. E. Elder, and I. N. Ulevitch, *J. Colloid Sci.*, **9**, 89 (1954).
8. T. R. Paxton, *J. Colloid Interface Sci.*, **31**, 19 (1969).
9. S. I. Ali, J. C. Steach, and R. L. Zollars, *Colloids Surf.*, **26**, 1 (1987).
10. S. L. Tsaur and R. M. Fitch, *J. Colloid Interface Sci.*, **115**, 463 (1987).
11. F. Carrique, J. Salcedo, M. A. Cabrerizo, F. Gonzalez-Caballero, and A. V. Delgado, *Acta Polymerica*, **42**, 261 (1991).
12. D. Bastos and F. J. de las Nieves, *Colloid Polym Sci.*, **272**, 592 (1994).
13. B. V. Deryagnin and L. D. Landau, *Acta Physicochim USSR*, **14**, 633 (1941).
14. E. J. W. Verwey and J. Th. G. Overbeek, *Theory of the Stability of Lyophobic Colloids*, Elsevier, New York, 1943.
15. J. E. Glass, In *Water Soluble Polymers: Beauty with Performance*, J. E. Glass,Ed., Advances in Chemistry Series **213**, American Chemical Society: Washington D.C., 1986, p. 391.
16. R. D. Jenkins, Ph.D. Dissertation in Chemical Engineering, Lehigh University, USA, 1990.
17. C. S. Chern and Y. C. Chen, *Colloid Polym. Sci.*, **275**, 124 (1997).
18. N. Fuchs, *Z. Phys.*, **89**, 736 (1934).
19. H. Reerink and J. Th. G. Overbeek, *Disc. Faraday Soc.*, **18**, 74 (1954).
20. X. Z. Kong, C. Pichot, and J. Guillot, *Colloid Polym. Sci.*, **265**, 791 (1987).
21. G. N. Patey, *J. Chem. Phys.*, **72**, 5763 (1980).

Fractionation Methods and Applications

Chapter 13

Thermal Field-Flow Fractionation of Colloidal Particles

Sun Joo Jeon and Martin E. Schimpf[1]

Department of Chemistry, Boise State University, Boise, ID 83725

Thermal field-flow fractionation (thermal FFF) was used to retain and separate colloidal particles with various surface compositions using acetonitrile (ACN) and water as carrier liquids. The particles included several sizes of unmodified polystyrene, as well as carboxylated and aminated polystyrene. In aqueous carriers, the effects of pH, ionic strength, and surfactant concentration were examined. Compared to ACN, differences in retention time with particle size are greater in aqueous carrier liquids containing sodium azide, and in general retention increases monotonically with azide concentration until a plateau value is reached. A notable exception in this trend is carboxylated polystyrene, whose retention passes through a minimum with increasing sodium azide concentration before increasing to a plateau value. Retention also decreases in sodium azide as the pH is decreased or the concentration of FL-70 surfactant is increased. Yet FL-70 is beneficial because it prevents flocculation of the particles and erratic results at high levels of retention. Without sodium azide, only the carboxylated particle is retained, and its retention increases with FL-70 concentration.

Historically, thermal field-flow fractionation (thermal FFF) was developed for the separation of dissolved macromolecules, primarily lipophilic polymers, while other FFF subtechniques were applied to particulate materials. For example, both flow FFF and sedimentation FFF have been used to separate particles in the submicron-size range and beyond (1,2). Flow FFF separates particles according to their size alone,

[1]Corresponding author.

while retention in sedimentation FFF depends on both particle size and density. Electrical FFF has been used to separate particles by size and electrophoretic mobility (3,4). Only recently was the applicability of thermal FFF to particle separations demonstrated (5).

The attractive feature of thermal FFF is that retention depends on the chemical composition of the particle as well as its size. The compositional dependence arises from thermal diffusion, which is the driving force responsible for retention in thermal FFF. Although the compositional dependence of thermal diffusion has been studied extensively in polymer solutions (6,7), it is not as well characterized for suspended particles. It is clear, however, that the thermal diffusion behavior of particles is quite different in aqueous and nonaqueous suspensions (5,8).

An interesting feature of the compositional dependence of thermal diffusion is that it appears to be dominated by interfacial phenomena, whether that interface consists of the free-draining region of a dissolved polymer (9) or the surface of a suspended particle (10,11). This feature makes thermal FFF a promising technique for separating materials by their surface composition. The work presented here focuses on the effect that slight modifications in surface composition have on the retention of suspended particles, as well as the effect of pH, ionic strength, and surfactant content in the carrier liquid.

Retention Theory

The reader is referred to the literature (12) for a detailed discussion of the mechanism and theory of thermal FFF retention. In short, the elution time (t_r) of a sample component is governed by the component's movement in response to a temperature gradient applied across the thin dimension of the thermal FFF channel. That movement, which is referred to as thermal diffusion or thermophoresis, results in the concentration of sample at the cold wall. The thickness of the concentrated zone determines a component's elution time because the downstream velocity varies with distance from the wall. Components are separated because they form compressed zones of different thickness. For highly retained samples, t_r is related to the component's ordinary (mass) diffusion coefficient (D), its thermal diffusion coefficient (D_T), and the temperature drop across the channel (ΔT):

$$\frac{t_r}{t^o} = \frac{D_T \Delta T}{6D} \qquad (1)$$

Here t^0 is the time required to elute a low molecular weight component, which is not affected by the temperature gradient. In this work, t^0 was determined by the elution time of sodium benzoate in water.

In the case of spherical particles, D can be related to particle diameter (d) by the Stokes-Einstein equation:

$$D = \frac{kT}{3\pi\eta d} \qquad (2)$$

where k is Boltzmann's constant, T is the absolute temperature in the region occupied by the sample zone, and η is the viscosity of the suspending medium. By combining eqs 1 and 2 we obtain

$$\frac{t_r}{t^o} = \frac{\pi\eta\Delta T}{2kT} dD_T \quad (3)$$

Although eq 3 is accurate only in the limit of high retention, it serves to illustrate the sample parameters that control retention in thermal FFF, namely the particle's diameter and thermophoretic mobility. The latter, which is quantified by D_T, depends on chemical interactions between the particle and carrier liquid. Consequently, retention time varies with both the size and chemical composition of the particle, as well as the composition of the carrier liquid.

Experimental

The thermal FFF instrument has been previously described (13); it is similar to the Model T-100 from FFFractionation Inc. (Salt Lake City, Utah), except that temperature control is achieved using a proportional counter built in-house. The controller is based on an operational amplifier functioning as an astable multivibrator, which governs the on and off times of the hot-wall heaters. To enhance stability, a thermistor is used to drive a feedback loop that alters the reference point of the multivibrator and hence the on and off times. The temperature stability using this controller is better than ± 0.1 K.

The channel has a tip-to-tip length of 46 cm, a breadth of 2.0 cm, and a thickness of 127 μm; the resulting void volume is 0.97 mL. The cold wall was 27 °C with acetonitrile (ACN) as the carrier liquid and ranged from 27-39 °C with aqueous carriers, depending on the value of ΔT. The higher cold wall temperature in the aqueous carriers is due to the high thermal conductivity of water, which results in a more efficient transfer of heat from the hot to cold wall.

Carrier liquids were prepared from either deionized water or spectrograde ACN obtained from Mallinckrodt Chemical Co. (St. Louis, MO). The liquids were delivered with a model 222B HPLC pump from Scientific Systems (State College, PA). A pulse dampener was used to reduce pump noise. Samples were injected with a Valco (Houston, TX) injection valve; sample loads were limited to 2 μg. Particles were detected by a model L-3000 UV detector from Hitachi, Ltd. (Tokyo, Japan) with an operating wavelength of 254 nm. The detector signal was recorded using a Houston Instruments (Austin, TX) chart recorder. The signal was also collected by an IBM-compatible personal computer using FFFractionation, Inc. software. Elution profiles were smoothed with a Savitzky-Golay algorithm.

Detailed information on the particles is summarized in Table I. They were obtained from either Bangs Laboratories (Carmel, IN) or Duke Scientific (Palo Alto, CA).

Results and Discussion

Figure 1 displays the elution profiles of several polystyrene particles of different sizes in both ACN and water, obtained with a ΔT of 40 K. In ACN, Shiundu et al. (8) demonstrated that retention is enhanced by the addition of tetrabutyl ammonium perchlorate (TBAP). For the elution profiles illustrated in Figure 1, we used a TBAP concentration of 0.1 mM. For the aqueous work, we added 3 mM NaN_3 and 0.1 wt-% FL-70 surfactant. Both these additives are common in particle separations by FFF, but as we illustrate below, retention in thermal FFF is affected by their concentration. The NaN_3 is used as a bacteriocide, while FL-70 is used to stabilize the particle suspensions. Although other surfactants have been used, FL-70 contains a mixture of cationic, anionic, and nonionic surfactants, and therefore works well for a wide range of particles.

Table I. Summary of Particles

No.	Particle [abbreviation]	Diameter (nm)	Surface Group	Surface Charge[a] (meq/g)	Supplier
1	polystyrene [PS]	91	-		Bangs
2	PS	135	-		Duke
3	PS	173	-		Duke
4	PS	197	-		Duke
5	PS	222	-		Duke
6	PS	261	-		Duke
7	PS	304	-		Duke
8	PS/acrylic acid [PS/COOH]	214	V-COOH[b]	86	Bangs
9	PS/aminobenzyl [PS/AB]	205	$Ar-NH_2$[b]	10	Bangs

[a] titrated by manufacturer
[b] functional group: Ar=aromatic; V=vinyl

There are significant differences among the retention of polystyrene particles in the two carrier liquids. These differences are summarized by plots of retention versus particle size in Figure 2. The smallest particle (no. 1) is retained about twice as long in ACN, but the dependence of retention on particle size is greater in the aqueous carrier. As a result, the largest particle (no. 7) has nearly equal retention in the two carriers. The dependence of retention on particle size is typically expressed by the size-based selectivity, defined as

$$S_d = \frac{d \log t_r}{d \log d} \quad (4)$$

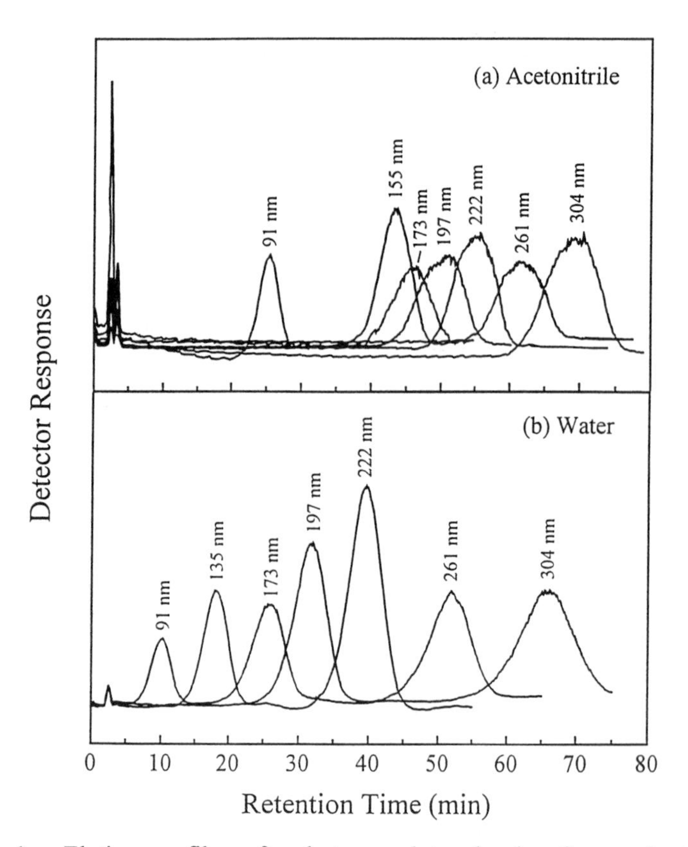

Figure 1. Elution profiles of polystyrene latex beads of several sizes in (a) acetonitrile containing 0.1 mM TBAP and (b) deionized water containing 0.1 wt-% FL-70 and 3 mM NaN_3; ΔT 40 K; T_C 37 °C; flow rate 0.4 mL/min.

From the data plotted in Figure 2, the values of S_d are 0.83 in ACN and 1.57 in water.

In a recent report (10), it was shown that small changes in surface composition can affect the retention of particles in aqueous carrier liquids. To explore this further, we compared the retention of polystyrene particles that have been modified with carboxylate and amine functionalities (nos. 8 and 9, respectively) to similarly-sized particles of underivatized polystyrene (no. 4). In ACN, retention is not significantly altered in the derivatized particles, as illustrated in Figure 3. Although the carboxylated particle is retained slightly longer than the other two, it has a larger diameter, and the increase is no greater than expected from the size dependence illustrated in Figure 2.

The situation is dramatically different in aqueous carriers, as illustrated in Figure 4. In 3 mM NaN_3, the underivatized and aminated particles are equally retained, while the carboxylated particle elutes in the void time. All the particles elute earlier than they do in ACN. As the concentration of NaN_3 is increased, the retention of all three particles increases, and slight differences appear between the aminated and underivatized particles. In 9 mM NaN_3, retention of the aminated and underivatized particles nearly reach that obtained in ACN, but the carboxylated particle continues to elute well ahead of the other two.

The effect of NaN_3 concentration can be explained by electrostatic effects. As the double layers of two surfaces of similar charge overlap, the surfaces repel one another. Electrostatic effects are particularly large in liquids with high dielectric constants, such as water (14). Electrostatic repulsion of particles from the wall force particles into the faster flowstreams located away from the wall, thereby decreasing retention. The thickness of the double layer is referred to as the Debye length and increases with the surface charge on the particle. FL-70, which is a mixture of cationic, anionic, and nonionic surfactants, makes an effective buffer having a pH of 8.5. At this pH, the highly charged carboxylated particle has a Debye length that is significantly larger than both the aminated and underivatized particles. As a result, its retention is the most affected by electrostatic effects. As the ionic strength is increased, the Debye length decreases, therefore retention of all three particles increases with ionic strength. However, the retention of the carboxylated particle is affected most, with t_r/t^o increasing 800% between 3 mM and 9 mM NaN_3, compared to 67-75% for the other particles. We note that the aminated particle contains aromatic amine groups, which are only weakly basic. Therefore, the aminated particle is not as highly charged at pH 8.5 as the carboxylated particle.

At high sample loads, electrostatic repulsion between particles will also decrease retention by causing expansion of the particle cloud. Particle-particle interactions generally manifest themselves as overloading effects, which decrease with sample loading. Although overloading effects are not generally significant with unmodified polystyrene unless several hundred micrograms of material is injected (15), overloading can be expected to occur at lower sample loads for carboxylated particles. In an effort to separate overloading effects from wall repulsion, we varied the sample loading of the carboxylated particle in 3 mM NaN_3 until it was barely detectable. Retention was still not observed at the lowest loading. Neither was retention affected by sample load in 9 mM NaN_3. Therefore, we believe electrostatic effects on retention are due to either particle-wall interactions or changes in thermophoresis rather than particle-particle interactions. This is significant because it means we can use ionic strength to manipulate retention without concern that retention

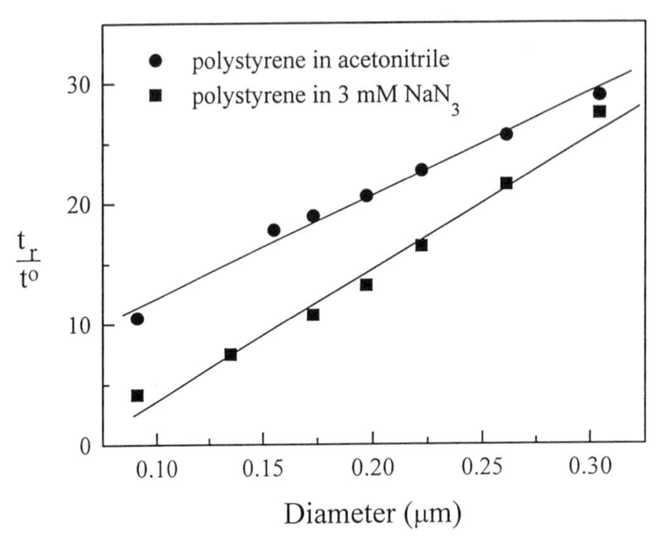

Figure 2. Plot of reduced retention time (t_r/t^o) versus particle diameter from elution profiles illustrated in Figure 1.

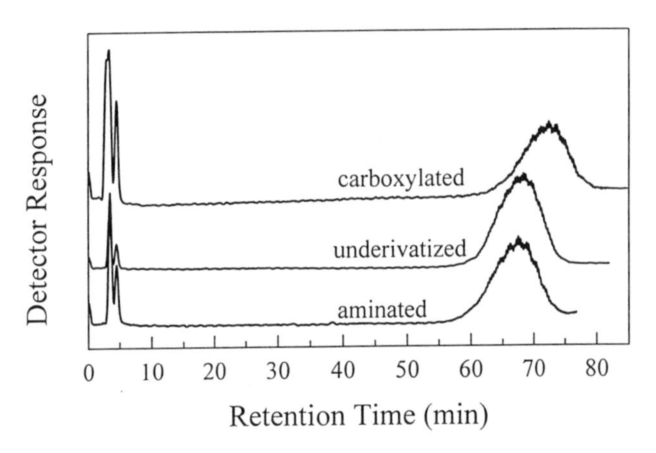

Figure 3. Elution profiles of underivatized, carboxylated, and aminated polystyrene of similar size in acetonitrile with 0.1 mM TBAP; ΔT 40 K; T_C 27 °C; flow rate 0.3 mL/min.

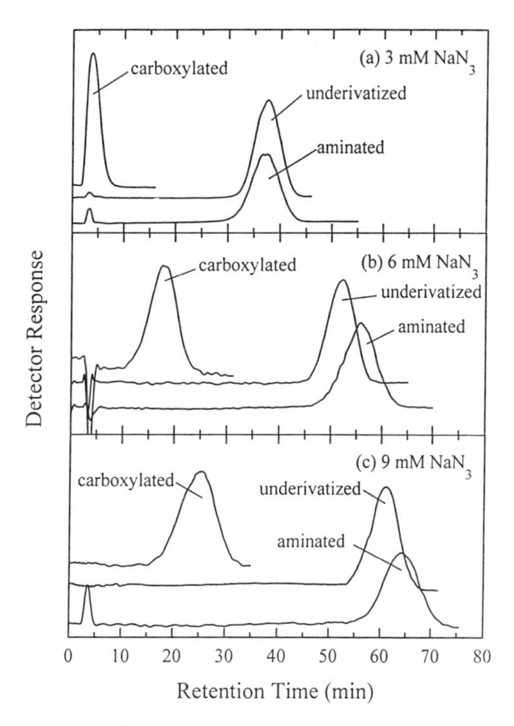

Figure 4. Elution profiles of underivatized, carboxylated, and aminated polystyrene of similar size in aqueous carrier with 0.1 wt-% FL-70 and (a) 3 mM NaN_3, (b) 6 mM NaN_3, and (c) 9 mM NaN_3; ΔT 40 K; T_C 37 °C; flow rate 0.3 mL/min.

will vary unpredictably due to variations in sample concentration from one analysis to the next.

In order to find a NaN_3 concentration where retention levels reach a plateau, we varied the ionic strength (I) over a wide range in a separate set of experiments. The results are displayed in Figure 5. These experiments were carried out in carrier liquids and without FL-70 surfactant. When FL-70 was used, it was used at a concentration of 0.1 wt-%, adding 1.1 mM to the ionic strength. We lowered ΔT to 30 K for this set of experiments because elution profiles became distorted at the high levels of retention achieved with increasing NaN_3 concentration using a ΔT of 40 K.

In the presence of FL-70, retention begins to level off as the ioic strength approaches 30 mM. Furthermore, the profiles parallel one another, with carboxylated polystyrene generally eluting first, followed by aminated and unmodified particles eluting at nearly the same time. A notable exception to this trend occurs in the absence of NaN_3. Here, the carboxylated particle is the only particle retained in 0.1 wt-% FL-70. This behavior was unexpected, but was reproduced with freshly-prepared and deionized carrier liquid.

The retention of each particle was higher in the absence of FL-70, again with the exception of the carboxylated particle in 0 mM NaN_3. However, the particles were more difficult to handle without FL-70. Suspending them in carrier liquid required sonication and the particles tended to flocculate over time. Above a NaN_3 concentration of 5 mM, retention times varied more and elution profiles became increasingly asymmetric. When t_r/t^o exceeded a value of 16, the particles failed to elute, therefore the concentration of NaN_3 where retention levels reached a plateau could not be found.

The effect of FL-70 was studied further by increasing its concentration to 0.5 wt-% in carrier liquids containing NaN_3 concentrations of 0 and 3 mM. The results are displayed in Figure 6. In 3 mM NaN_3, retention continues to decrease with increasing FL-70 concentration, even though the ionic strength increases from 3.0 mM (0 wt-% FL-70) to 8.5 mM (0.5 wt-% FL-70). The carboxylated particle was effected most by FL-70. In the absence of NaN_3, the carboxylated particle once again displays different behavior than the other two particles. Thus, neither the aminated nor the underivatized particle were retained, whereas the carboxylated particle is not only retained, but its retention increases with FL-70 concentration. This is opposite to its behavior in the presence of NaN_3.

The effect of FL-70 is difficult to understand. If surfactant molecules interact strongly with the particles, it is likely that the surface charge of the underivatized and aminated particles, which are not chemically charged at a neutral pH or higher, is increased; this would explain their decrease in retention. With the carboxylated particle, on the other hand, charged aliphatic amines in the surfactant mixture are likely to form ion pairs, thereby reducing the surface charge. This would explain the increased retention in 0 mM NaN_3 with FL-70 but not the opposite behavior in 3 mM NaN_3. However, it is possible that NaN_3 forms ion pairs with amines in the surfactant, thereby reducing the ionic strength while inhibiting the formation of ion pairs with carboxylate groups on the particle surface. We must also consider the effect of surfactant on thermophoresis, which may or may not be independent of electrostatic effects. Thus, FL-70 may alter the thermophoretic velocity of the particles whether or not adsorption to the surface occurs. In addition, the thermal diffusion of the surfactant

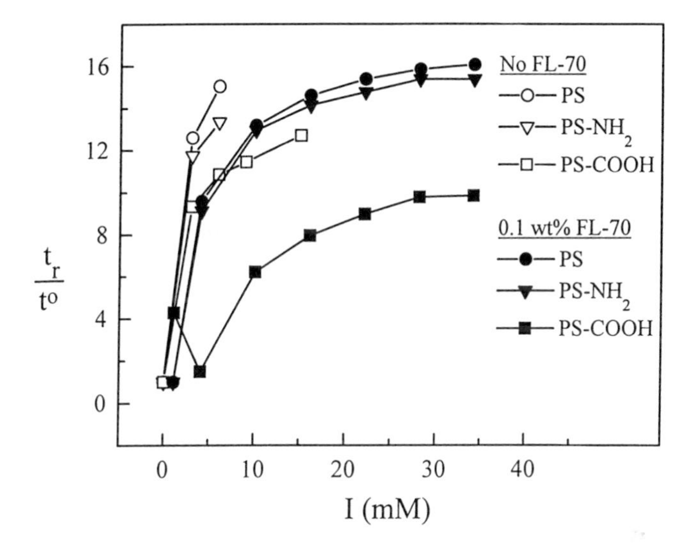

Figure 5. Reduced retention time versus NaN_3 concentration in aqueous carriers with and without FL-70; ΔT 30 K; T_C 27 °C; flow rate 0.4 mL/min

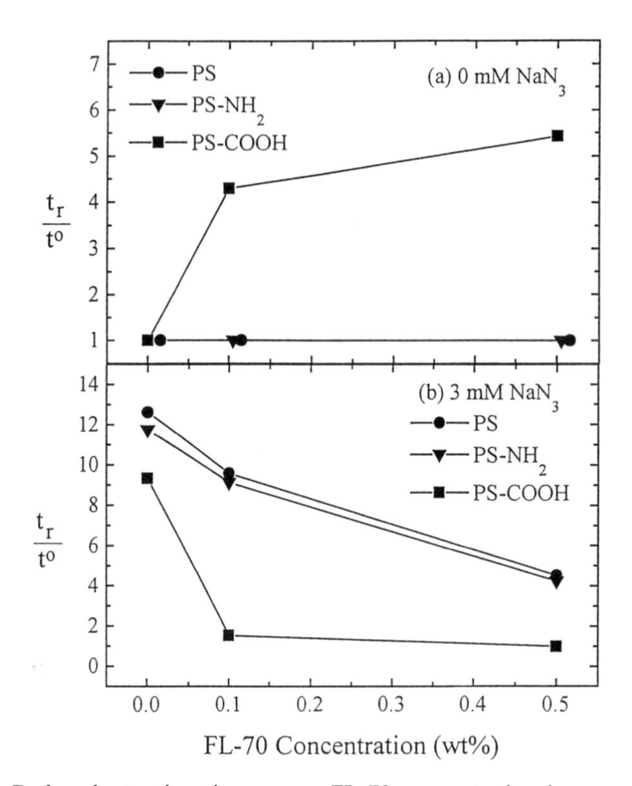

Figure 6. Reduced retention time versus FL-70 concentration in aqueous carrier liquids with (a) 3 mM NaN_3, and (b) without NaN_3; ΔT 30 K; T_C 27 °C; flow rate 0.4 mL/min.

itself may affect that of the particles, just as the thermal diffusion of polymers is altered in mixed carrier liquids (12). Further study is required to sort out these phenomena. Nevertheless, the effect of FL-70 on retention, whatever the mechanism, is not as strong as the effect of NaN_3.

Ion pairing also explains the increase in retention of the carboxylated particle in 0.1 wt-% FL-70 when the NaN_3 concentration is reduced from 3 mM to 0 mM. Thus, amines in FL-70 are more effective at forming ion pairs in the absence of NaN_3, and therefore retention is increased. But at NaN_3 concentrations above 3 mM, the increase in ionic strength has a more significant effect than ion pair formation. The occurrence of both phenomena produce a minimum value of t_r/t^o for the carboxylated particle at a NaN_3 concentration between 0 and 9 mM.

If electrostatic effects are primarily responsible for the differences in retention of the carboxylated particle, its retention should be increased by lowering the pH of the carrier liquid. Figure 7 displays elution profiles of the three particles in 10 mM phosphate buffer, adjusted to pH 4.7. Both the carboxylated and underivatized particles are well retained, with the former eluting slightly ahead of the latter. However, the particles are less retained than in 3 mM NaN_3, even though the ionic strength is significantly higher. The aminated particle, which is expected to be protonated at pH 4.7 (the pK_a of aniline is 4.9), elutes in the void time. This behavior is consistent with electrostatic effects playing an important role in particle retention. It remains unclear, however, how much of the electrostatic effect is manifested through electrostatic repulsion versus the effect of charge on thermal diffusion.

In an effort to look at pH effects independent of ionic strength, we measured retention in a series of carrier liquids containing 3 mM NaN_3 where the pH was adjusted by the addition of HCl. The ionic strength of these solutions vary from 3.0 mM at pH 6.8 to 3.1 mM at pH 4.0. The dependence of retention on pH is illustrated in Figure 8. As expected, the retention of the aminated particle was dramatically reduced by lowering the pH. However, retention of the carboxylated and underivatized particle was also reduced with pH, though to a lesser extent. It should be noted that in these unbuffered carrier liquids, the pH near the particle surface may be different than that in the bulk solution. Nevertheless, retention decreases with pH even though the ionic strength of the solution increases slightly. This behavior indicates that while pH can effect retention through changes in electrostatic potential, additional phenomena play a role in thermal diffusion and the associated retention of particles.

Conclusions

Thermal FFF is capable of separating particles by both size and composition. Unlike ACN, aqueous carriers can be use to separate similarly sized polystyrene colloids that are derivatized with different functional groups. Furthermore, the order of retention between carboxylated, aminated, and underivatized particles can be manipulated by additives to the aqueous carriers. As a result, the three types of particles can be distinguished by their differing levels of retention in 5 mM NaN_3.

The dramatic effects on retention of subtle changes in aqueous carrier liquids can be largely but not completely explained by electrostatic effects. Certainly the increase in retention with NaN_3 concentration is consistent with a significant electrostatic effect, as is the reversal in elution behavior of carboxylated and aminated

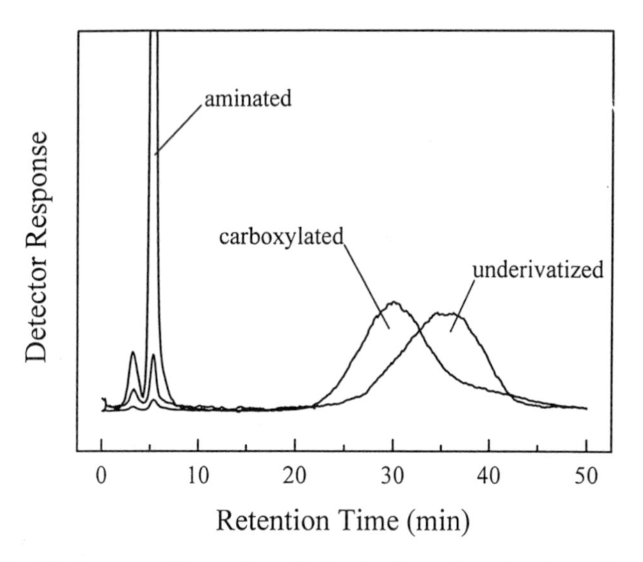

Figure 7. Elution profiles of underivatized, carboxylated, and aminated polystyrene of similar size in 10 mM phosphate buffer at pH 4.7; ΔT 45 K; T_C 39 °C; flow rate 0.2 mL/min.

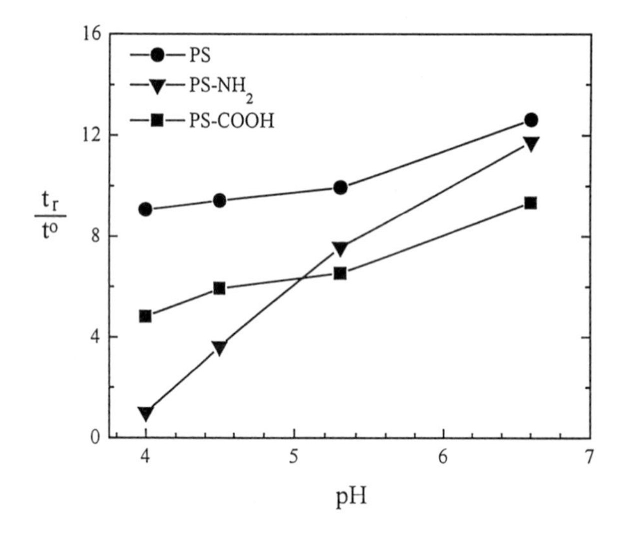

Figure 8. Elution profiles of underivatized, carboxylated, and aminated polystyrene of similar size in 10 mM phosphate buffer at pH 4.7; ΔT 45 K; T_C 39 °C; flow rate 0.2 mL/min.

particles when the pH is lowered from 8.5 to 4.7 in buffered carrier liquids. Even the distinct behavior of the carboxylated particle in solutions of NaN_3 and FL-70 can be explained by electrostatic effects if we consider the potential for amine components in the surfactant to adsorb onto the particle surface through ion pair formation. However, the decrease in retention of carboxylated polystyrene as the pH is lowered to 4.0 indicates that electrostatic effects are not the only factor in the retention of particles in aqueous carriers. More work is required to identify these other factors.

Electrostatic interactions make it difficult to differentiate the role played by thermal diffusion. In ACN, where electrostatic effects are expected to play a minor role, retention is not affected by the subtle differences in composition examined in this work. This is in contrast to reports (8) that differences in bulk composition affect particle retention in ACN. In aqueous carrier liquids, the effect of FL-70 may be due to adsorption of surfactant to the particle surface, changes in thermal diffusion, changes in electrostatic potential, or some combination of the three. The ability of surfactant adsorption to alter retention has great potential to expand the utility of thermal FFF through the judicious choice of surfactants.

The size-based selectivity differs significantly between aqueous and organic carrier liquids, being <1 in ACN and >1 in water. Retention theory predicts unit selectivity when D_T is independent of size, which is what we observe for dissolved homopolymers. The smaller selectivity in ACN indicates that D_T decreases with particle size, while the opposite dependence on size is indicated in water. Both observations are consistent with previous reports (8).

References

1. Ratanathanawongs, S. K.; Giddings, J. C. *Chromatography of Polymers: Characterization by SEC and FFF*; Provder, T., Ed.; ACS Symposium Series No. 521; American Chemical Society: Washington DC, 1993; Chapter 2.
2. Barman, B. N.; Giddings, J. C. *Chromatography of Polymers: Characterization by SEC and FFF*; Provder, T., Ed.; ACS Symposium Series No. 521; American Chemical Society: Washington DC, 1993; Chapter 3.
3. Caldwell, K. D.; Gau, Y. S. *Anal. Chem.* **1993**, *65,* 1764.
4. Schimpf, M. E.; Caldwell, K. D. *Am. Lab* **1995**, *27*, 64.
5. Liu, G.; Giddings, J. C. *Chromatographia* **1992**, *34,* 483.
6. Schimpf, M. E.; Giddings, J. C. *Macromolecules* **1987**, *20,* 1561.
7. Schimpf, M. E.; Giddings, J. C. *J. Polym. Sci.: Polym. Phys. Ed.* **1989**, *27*, 1317.
8. Shiundu, P. M.; Liu, G.; Giddings, J. C. *Anal. Chem.* **1995**, *67,* 2705.
9. Schimpf, M. E.; Giddings, J. C. *J. Polym. Sci.: Polym. Phys. Ed.* **1990**, *28,* 2673.
10. Shiundu, P. M.; Giddings, J. C. *J. Chromatogr. A* **1995**, *715,* 117.
11. Ratanathanawongs, S. K.; Shiundu, P. M.; Giddings, J. C. *Coll. Surf.* **1995**, *105*, 243.
12. Schimpf, M. E. *J. Chromatogr.* **1990**, *517,* 405.
13. Rue, C. A.; Schimpf, M. E. *Anal. Chem.* **1994**, *66,* 4054.
14. Adamson, A. In *Physical Chemistry of Surfaces,* 3rd ed.; Interscience: New York, 1976; pp. 197ff.
15. Hansen, M. E.; Giddings, J. C.; Beckett, R. *J. Colloid Interface Sci.* **1989**, *132,* 300.

Chapter 14

Charge and Hydrophobicity Fractionation of Colloidal-Size Polymers Using Electrical Field-Flow Fractionation and Liquid Chromatography

Saurabh A. Palkar[1], Robert E. Murphy[1], and Mark R. Schure[2]

[1]Analytical Research and [2]Theoretical Separation Science Laboratory, Rohm and Haas Company, Spring House, PA 19477

Electrical Field-Flow Fractionation and Reversed-Phase Liquid Chromatography are used to characterize polyelectrolytes with grafted hydrophobic ligands which are solution polymers but large enough to be considered in the colloidal particle size domain. The difficulties of fractionation by charge and by hydrophobicity are discussed in detail as are the conditions where charge and hydrophobicity are the key separation parameters.

There are many cases where water soluble polymer molecules have multiple types of groups, i.e. cationic, anionic, and hydrophobic groups may be present in the same molecule. These different groups may interact on intramolecular, intermolecular, or a mixed basis to form a wide variety of structures in solution. When these interactions are with surfaces, as is often the case with biopolymers, the interactions come in a wide variety of types. These range from relatively weak van der Waals interactions with little difference between the solution structure and adsorbed structure to large arrays of surface hydrogen bonded molecules where the solution structure shows little resemblance to that of the surface structure.

The fractionation of these types of molecules is difficult because separations based purely on size may miss some of the subtleties of the molecular structure. This aspect has been well recognized in the biological separations community where many molecules have an internal structure that may not be physically accessible under solvent conditions that resemble the biological environment.

In the case reported in this paper we discuss the characterization of polyelectrolytes with various hydrophobic groups bound to the polymer backbone in random intervals. These types of molecules have been reported previously in the literature (*1*,*2*) and are common industrial molecules used for a variety of purposes. The polyelectrolytes discussed here are of high molecular weight so that from a size perspective these molecules resemble colloids. Specifically, we will discuss the methodology whereby two solutes are characterized in terms of charge and hydrophobicity using Electrical Field-Flow Fractionation (Electrical FFF) and high performance reversed-phase liquid chromatography (RPLC).

EXPERIMENTAL

Samples. Two polycarboxylic acid polymers containing a variety of alkyl modified groups were synthesized and will be referred to as samples X_1 and X_2. Photon correlation spectrometry (PCS) using a BI-90 instrument (Brookhaven Instruments, Long Island, N. Y.) gave particle sizes of 120 nm and 100 nm for X_1 and X_2 respectively. Molecular weight analysis of these materials by aqueous Gel Permeation Chromatography (GPC) gave molecular weights of 400,000 and 1,000,000 for X_1 and X_2 respectively. Calibration here was done strictly by the use of acid polymer standards without hydrophobic grafts.

Electrical FFF. Characterization by charge is done using Electrical FFF because the colloidal size is similar for both samples. It has been demonstrated previously (*3*) that the product of particle size and charge control the retention characteristics of Electrical FFF. Larger particles and larger charge produce higher retention. Hence, when particle size is similar, elution time is determined by the particle charge.

The Electrical FFF instrument was obtained from Professor Karin Caldwell at the University of Utah and resembles the unit which was described in a recent publication detailing the new revised form of Electrical FFF (*4*). The fractionator consists of two graphite blocks and a Mylar spacer which comprise a channel of dimensions 178 μm thickness, 2 cm breadth, and 64 cm length. The solvent used for all Electrical FFF experiments is deionized water with no further buffering. The power supply provided typically 1 to 2 volts of direct current between 0.2 to 3.0 milliamps.

Reversed-Phase Liquid Chromatography. RPLC was carried out on a Perkin-Elmer (Norwalk, CT) system consisting of an ISS-100 autosampler (25 μl injection with 150 μl loop), Series 410 quaternary pump flowing at 0.5 ml/min, and solvent chamber (SEC-4). The solvents were purged and pressurized with helium prior to use. All solvents, except Milli-Q (Millipore Corporation, Milford, MA) water, were Baker Analyzed HPLC Reagent grade from J. T. Baker (Phillipsburg, NJ) and used without further purification.

Separations were carried out on a Supelcogel ODP-50 column (Supelco, Belfonte, PA), which contains 5 μm particles bonded with a C_{18} phase. The support material for the C_{18} phase is derived from polyvinyl alcohol and offers an extremely hydrophobic material. The column has a 4.00 mm internal diameter and is 15 cm long.

Detection was done with a Sedex 55 (Richard Scientific, Novato, CA) evaporative light scattering detector (ELSD) at 40 °C, 2.2 bar of nitrogen, and a gain setting of 8. Data was acquired at 1 data point per second with a Perkin-Elmer/Nelson 900 Series interface box and Perkin-Elmer Turbochrom software.

RESULTS

Electrical FFF. Fractograms of sample X_2 at different applied potentials are shown in Figure 1. Varying the applied potential, E, will also vary the current and hence retention when the wall electrodes are in the Faradaic region of the current vs. voltage curve. These experiments are very reproducible and well controlled when $E < 1.5$ V. However, when $1.5\text{V} < E < 2.0\text{V}$, the retention ratio R tends to limit at ≈ 0.1. Note that R is defined as the ratio t_0/t_r where t_0 is the elution time of an unretained low molecular weight marker and t_r is the channel retention time. For $E > 2.0$ V, the solute zones become irreversibly adsorbed, resulting in the loss of material on the column.

These results can be explained through the structure of the molecules themselves. As the zones become more compressed the repulsive nature of interchain interactions starts to dominate causing repulsion of the chains and no further retention increase (retention ratio R decrease). At some point as the potential is further increased, one of two likely scenarios can occur. First, a phase transition of the solute may occur in the vicinity of the electrode surface so that a strong hydrophobic interaction between the electrode surface and the hydrophobic chain elements is formed producing a very retained complex. Second, the solute chains may be electrochemically incorporated into the electrode surface through the oxidation process of the solute at the interface. This would form an irreversibly held complex and it is this scenario that we believe is happening in this case. Obviously, working at potentials less than 1.5 V is most desirable to minimize the zone compression and possible subsequent adsorption and/or reaction and yet cause retention to occur. This problem should not occur with hard, nondeformable, spherical colloidal material because the vast majority of the colloidal surface can not physically interact with the flat electrode. This is different than the case with flexible macromolecules which can deform easily and form surface complexes.

We show a comparison of X_1 and X_2 in Figure 2 at the same operating conditions. X_1 is more retained than X_2 in this figure and this is not surprising since X_1 is known to contain more acid than X_2. Since X_1 and X_2 have approximately the same colloidal size, this figure demonstrates that X_1 has more charge; this was suspected from the synthesis conditions. These experiments were conducted where the pK_a of the polyelectrolyte was below the pH of the deionized water, hence, most of the acid groups, but not necessarily all of the acid groups, are expected to be found in the dissociated state.

Initially both samples in this study were dialyzed using cartridge dialysis membranes for 12 hours prior to fractionation to eliminate any spurious salt effects that have been observed using the Electrical FFF technique. Further experimentation showed, however, that there was little difference in the results if these samples were run without dialysis. Also, experiments were conducted whereby we used larger sample sizes than the 1 μg injections shown in Figures 1 and 2. It was determined that the largest sample injection that produced analytically useful peaks was $\approx 10\mu$g.

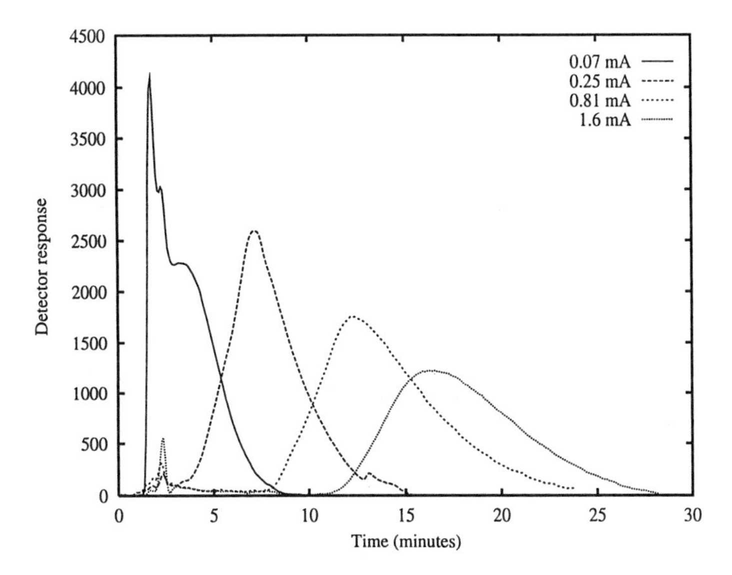

Figure 1. The Electrical FFF fractograms of various peaks obtained from sample X_2 at various field strengths, currents, and retentions. The voltages at the stated currents are: 1.04 V at 0.07 mA, 1.21 V at 0.25 mA, 1.40 V at 0.81 mA, and 1.60 V at 1.60 mA.

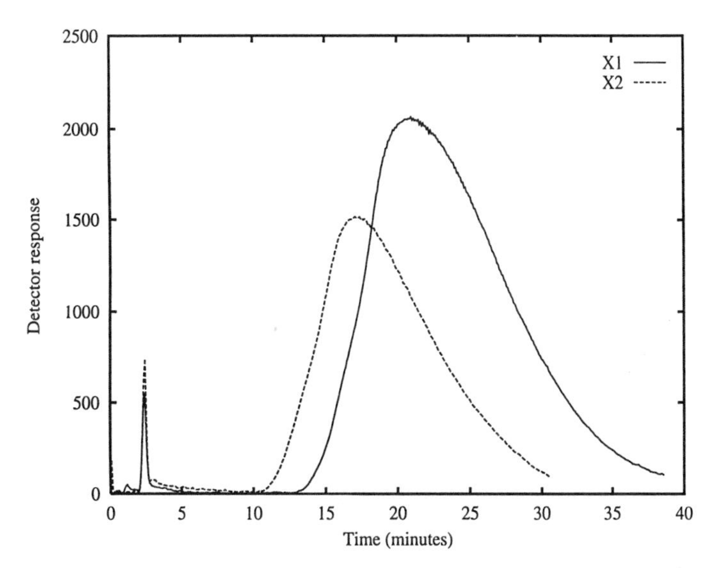

Figure 2. The comparison of Electrical FFF fractograms between sample X_1 and X_2 at 1.50 volts and 1.00 milliamp.

This was too low to use the Electrical FFF technique with liquid chromatography in a hyphenated mode where a sequential loading of the chromatograph with a finite plug of solute can be done to provide a very fast and selective analysis. The use of LC/LC is now increasing because its use has been demonstrated, there is commercially available software and hardware for this purpose, and because the dilution factor is not as large as with FFF. These techniques could be coupled if the ELSD detection limit was a factor of ≈ 200 smaller.

Liquid Chromatography. Taking into account the structure of the polyelectrolyte one can rationalize through simple chemical arguments that the analysis should be conducted at high pH so that the acid is completely dissociated and the chain completely extended by repulsion amongst the charge centers. In this way the full interaction of the hydrophobe can be experienced with the surface C_{18} chains of the retentive phase. At low pH it is expected that the polyelectrolyte will "ball up" and the hydrophobic portion of the molecule will be shielded from C_{18} interaction because of the attraction of the protonated acid groups for their neighbors. In this regard, we initially have tried to keep the solvent pH in the mildly basic region to promote the exposure of the hydrophobe to the C_{18} surface and to attempt separations based purely on hydrophobic mechanisms.

Prior to liquid chromatographic analysis, the samples were filtered and then subsequently diluted to different pH. The visual analysis of the sample under the different pH conditions resulting from the polymer and solvent buffer mixture is given in Table I.

Table I. Sample Appearance

Solvent	Sample pH	Visual Condition
Water	4.3	cloudy
5 mM Ammonia buffer	6.5	clear
50 mM Ammonia buffer	9.1	clear

For the pure water case although the solution was cloudy it was nonetheless filterable. The clear samples at pH 6.5 and pH 9.1 are highly viscous, however, the addition of small amounts of tetrahydrofuran (THF) reduces the viscosity in both cases. These results can be explained as follows. With an increase in pH, the polymer chains expand from the spherical random coil state to one of being rod-like so as to minimize charge repulsion. The rod-like expanded chains are much longer than the sphere diameter which promotes a higher probability of intermolecular interaction leading to higher viscosity. From an optical point of view, the higher pH state causes the colloid-like globular material, which is approximately on the length scale of visible light, to be transformed into an expanded polymeric rod where the length scales of the rod are no longer near the visible region.

The results of RPLC at the three different injection conditions given in Table I are shown in Figure 3 for sample X_1. In these experiments solvent A was 5 mM ammonium acetate/ammonium hydroxide buffer at pH 9.50 and solvent B was 5 mM ammonium acetate/ammonium hydroxide in methanol. The solvent mixture was varied from 40/60 A/B to 0/100 over 20 minutes and was then held steady at the 0/100 composition for 5 minutes before returning to the 40/60 A/B composition. The column was equilibrated for 30 minutes at the 40/60 A/B composition prior to the next analysis. As can be seen from Figure 3, there is a big difference between these three experiments. Zone definition is poor and although not shown, the speed of the gradient affects the position of the zones.

Further investigation suggests that polymer aggregation takes place when the solvent concentration is approximately 100% methanol. This aids in explaining the results shown in Figure 3 where some of the jagged edges of the peaks may be due to the formation of particulate material.

The solvent program is changed for the chromatograms shown in Figure 4 and includes the following solvent composition: solvent A was 10 mM ammonium acetate at pH 6.50, solvent B was water, and solvent C was methanol. The solvent program here consists of a linear program beginning with the composition 10/20/70 (A/B/C) and ending with 10/0/90. This gradient was developed over twenty minutes, whereupon the column was washed for 5 minutes with mobile phase of the end composition prior to a return to its initial conditions. After a 30 minute equilibration period the column was ready for a subsequent analysis. This results in chromatograms which are approximately 55 minutes long, as shown in these figures.

As shown in Figure 4, the results are different for samples X_1 and X_2. This information is useful in differentiating the two samples. Reinjection of fractionated solute material collected over a few minutes shows sharp boundaries at the beginning and end of the zone. However, no individual peaks appear from this myriad of different structures. Hence, the column method is indicative of the distribution of structures but the resolution is not high enough to resolve individual components. However, it is not clear here whether hydrophobic interactions, charge, or a combination of these two effects is controlling the retention. Since Electrical FFF demonstrates that X_1 is more charged, a retention mechanism based on charge would suggest that X_1 be more soluble in water and hence less retained. This is the case shown in Figure 4 but this is ambiguous because it is not clear to what extent the hydrophobic portions of the molecule dominate or contribute as compared to the charged portion.

One would postulate that under these conditions most of the polymer chain is charged. In addition, the presence of a low dielectric constant organic solvent like methanol tends to increase the interaction among charged sites. However, the presence of methanol decreases the number of dissociated protons on the acid groups because they are not dissociated in methanol to the same extent as the dissociation which occurs in water. Hence, higher methanol content

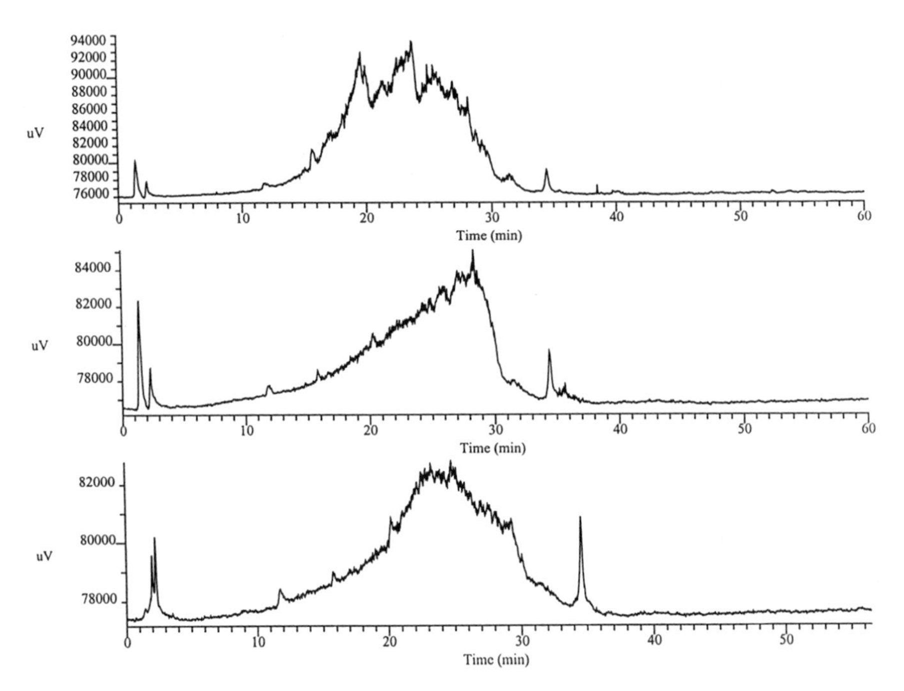

Figure 3. The RPLC analysis of sample X_1. The solvent program is described in the text. The various injection conditions correspond to those in Table I; top: pH 4.3, middle: pH 6.5, and bottom: pH 9.1.

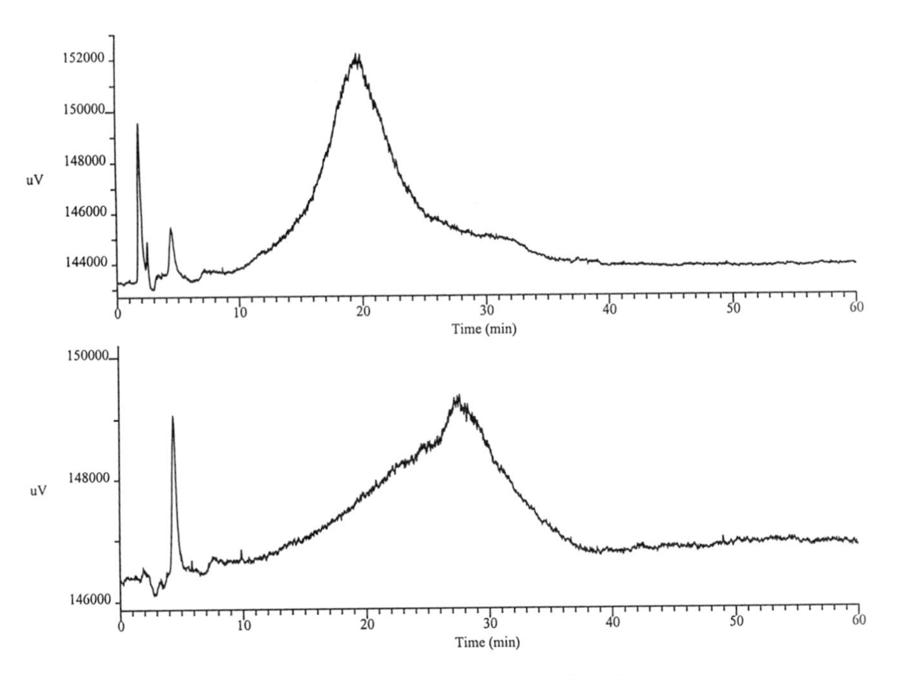

Figure 4. The RPLC analysis of sample X_1 and X_2. The solvent program is described in the text. The injection is made in an equivolume mixture of 10 millimolar aqueous ammonium acetate and THF.

causes the polyelectrolyte chain to be more in the random coil state than in the extended chain conformation suggesting that the hydrophobic groups may be more accessible at lower methanol concentrations. This physical picture is consistent with the onset of aggregation in pure methanol due to the loss of solvation of the acid groups and additional hydrogen bonding that can occur through interchain associations. However, the scenario of higher methanol concentration causing lower retention is also consistent with simple solvation models for free molecules.

Figure 5 shows what appears to be a very high resolution separation for the components of samples X_1 and X_2. The conditions for these chromatograms are the same as those in Figure 4, however, the injection step is done in pure water. These chromatograms appear to show high resolution and fine structure, however, the chromatograms are irreproducible. This may be due to a kinetic relaxation effect where the solute molecule is initially globular in the pure water state. In this state an extensive network of intramolecular and intermolecular hydrogen bonds may exist because most but not all acid groups are dissociated and hence able to form hydrogen bonds. In this condition, the polyelectrolyte is expected to be extensively physically crosslinked and penetration of this network by small molecules may be somewhat slow. The diffusion of methanol into this system is hindered because diffusion pathways are small and water must be displaced to allow methanol penetration. These processes most likely take place in minutes and may not be at complete equilibrium for hours and possibly even days. The cause of this apparent high resolution is probably due to the different structures that evolve through the solvent induced transformation of the solute structure. Since these structures probably have metastable regions with respect to their transformations it is reasonable, for lack of more direct proof, to suspect that some distinct structures may evolve during this transformation. Also, if these structures are micellar in nature, which is quite possible for polymers this large even at the very small concentrations that are found in the column, the aggregation number may change with the evolving solute structures resulting in different peaks. Hence, the slow transformation in methanol may cause both the irreproducibility and the "high resolution" that have been found when the column solvent conditions are different than the pure water environment used for sample dissolution here. This suggests that pure water should be used as the solvent, however, polymer retention is extremely high under pure water conditions and the methanol is necessary so that analysis time is finite.

It should be noted that acetonitrile, a very common organic solvent modifier and one that we have utilized previously for polymer separations, caused infinite retention of the X_1 and X_2 samples when substituted for methanol in these studies. This further suggests that the separations are mediated by some form of structure change of the polymer and not by hydrophobic solvation of the hydrophobe segments.

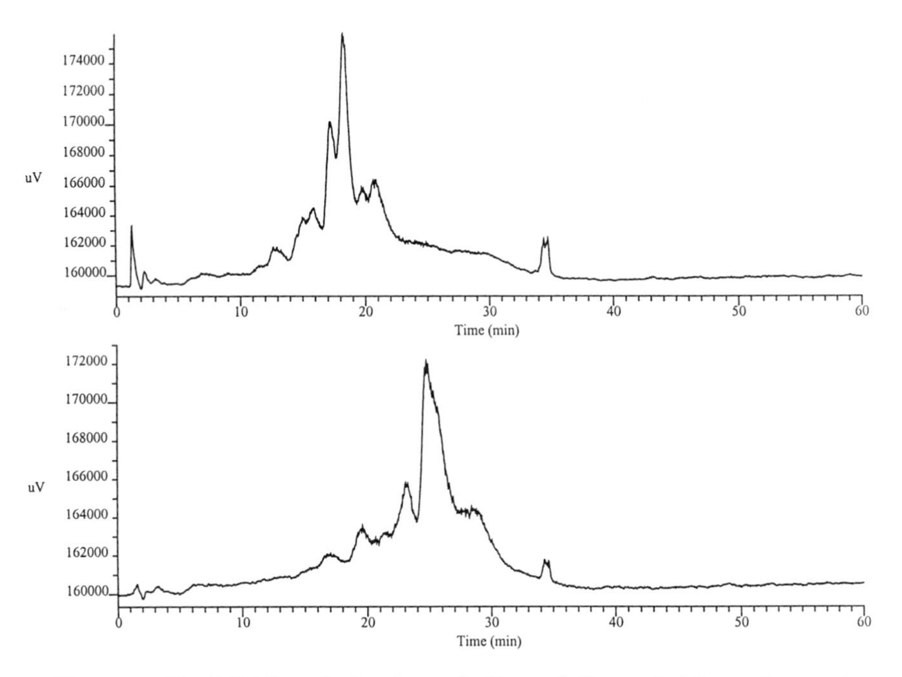

Figure 5. The RPLC analysis of sample X_1 and X_2, as in Figure 4, except that the injection is made in pure water.

DISCUSSION

As the results have shown, the solvent environment is critical to the success of the RPLC method. Situations like these are well known to occur in polymer chromatography because the solvent composition may be near a phase boundary of the polymer in a particular solvent system. We use small pore stationary phases in this work so that only a small amount of surface area is present; this practice has been noted previously (*5*) and allows for reasonably large retention for the large polyelectrolytes described here. Much like the case in the RPLC separation of proteins (*6*), large changes in retention are noted to occur over small changes in solvent composition.

Throughout this work we have noted that aggregation may occur if the gradient is produced faster than a 1% per minute change in solvent composition. This kinetic effect seems to predominate wherever polyelectrolytes are present in environments which are not high in salt. Hence, it may be envisioned that slow changes in polymer conformation, relative to the chromatographic time scale, may be taking place when the solvent composition is rapidly changed.

We have attempted to utilize the tools of Electrical FFF and RPLC to probe the charge and hydrophobicity of synthetic polymers. In that regard, it appears that increases in retention time of the various components eluting from the RPLC experiments are correlated with increasing hydrophobic character. However, as explained in this chapter, there are many ambiguous interpretations of the retention mechanism that are possible here. However, as an operational tool, RPLC, when combined with mass spectrometry, IR, and NMR, functions as a useful analysis tool. When calibration of RPLC is made with standard compounds and the spectroscopic tools are used to understand this calibration, we expect a further stage of knowledge to be forthcoming in understanding the chromatographic results which are illustrated here. Work is currently being done in our laboratory in this direction to further clarify the mechanisms of RPLC for synthetic polymers of the type described here.

LITERATURE CITED

1) Lockhead, R. Y. in *Polymers as Rheology Modifiers* Schulz, D. N. and Glass, J. E. editors, ACS Symposium Series *462*, Am. Chem. Soc., **1991**, Washington, D. C., p. 101-120.

2) Morishima, Y. in *Multidimensional Spectroscopy of Polymers, Vibrational, NMR and Fluorescence Techniques* Urban, M. W. and Provder T. editors, ACS Symposium Series *598*, Am. Chem. Soc., **1995**, Washington, D. C., p. 490-516.

3) Caldwell, K. D. *Anal. Chem.*, **1988**, *60*, 959A-971A.

4) Caldwell, K. D., Gao, Y. S. *Anal. Chem.*, **1993**, *65*, 1764-1770.

Chapter 15

High-Resolution Particle Size Characterization for Quality Control

Jose Gabriel DosRamos

Matec Applied Sciences, 56 Hudson Street, Northborough, MA 01532

A new particle size instrument capable of providing high-resolution measurements in a Quality Control role is presented. This instrument is based on the technique known as Capillary Hydrodynamic Fractionation (CHDF). Sample analysis automation, fractionation column temperature control, and disposable fractionation column design were incorporated into this instrument. The effect of eluant temperature on Taylor dispersion of particles in capillary flow was quantified. Qualitative agreement with theory was obtained. Particle average velocity was found to decrease with increasing eluant temperature. By-product particle size populations were detected in a colloidal silica sample.

Particle size characterization is a critical step in the manufacturing and production of colloids. Properties of colloids such as film gloss, packing density, rheology, and Chemical Mechanical Polishing efficiency are affected by particle size and particle size distribution (PSD). Traditional Quality Control methods for particle size characterization have only provided average particle size and standard deviation information, but not a detailed particle size distribution [1]. Today's more stringent colloid performance requirements have made it increasingly necessary to measure not only the average particle size, but also detailed, high-resolution, particle size distributions of colloids.

High-resolution measurement of particle size distribution allows identification of multiple particle size populations within a given sample. Multiple particle size populations may be found in colloids where only one population is expected. This may occur, for example, in seeded emulsion polymerization of polymer latexes. Secondary nucleation may result in generation of undesirable populations with small particle size. These small particles have a large total surface area capable of adsorbing most of the stabilizers in the latex resulting in catastrophic particle coagulation.

There are many Particle Characterization techniques [2]. Traditional particle size characterization techniques include Photon Correlation Spectroscopy (PCS), Fraunhoffer Diffraction, and Turbidimetry. These are ensemble (non-fractionation) techniques and as a result provide mainly average particle size information. They have the advantage of being fast, and easy to operate. More detailed particle size distribution data is provided by fractionation techniques.

Fractionation methods include Capillary Hydrodynamic Fractionation (CHDF), Sedimentation Field Flow Fractionation (SFFF), and Disc Centrifugation. These methods offer high-resolution particle size distribution measurement. These methods also have the potential to measure true particle size distributions. Ensemble techniques need to assume a shape to the PSD, often Gaussian. Those colloids which do not exhibit the assumed PSD shape will be inaccurately characterized by ensemble techniques. On the other hand, fractionation methods are usually more difficult to operate, and slower.

SFFF consists of separating particulate or macromolecular materials based on their mass, and size [3]. A spinning ribbon-like channel is used to separate the particles, which are carried by a flowing liquid in the channel. Particle size is measured from the elution time of the particles if the particle density is known. Disc Centrifugation is based on separating the particles by mass and size in a fast-rotating disc [4]. The particles are injected at the center of the disc and travel outwards due to the centrifugal forces in the disc. Particle size is measured from the retention time in the disc along with the particle density.

CHDF particle size measurement is performed by fractionating the particles according to size in a narrow-bore open capillary tube where an eluant continuously flows in laminar flow [5]. The particles are separated as they flow in the capillary and exit in order of decreasing size. Particle size is measured from the elution time of the particles from the pre-calibrated capillary. CHDF has the following advantages over other fractionation techniques: (i) results are independent of particle density; (ii) analysis time is under ten minutes, independent of the sample's PSD [13]; and (iii) operation is simple.

CHDF's short analysis time is achieved by using fractionation-capillary dimensions (length and diameter) such that molecular species (such as salts) exit the capillary within ten minutes of injection; molecular species have near-zero particle size and therefore exit the capillary last; actual particles, regardless of the sample's PSD, elute earlier.

The ideal Quality Control particle size analyzer would combine the speed, and ease of use of a non-fractionation technique with the high-resolution data obtained from a fractionation technique. This paper describes a new High-Resolution Quality Control instrument based on the Capillary Hydrodynamic Fractionation technique. The traditional CHDF instrument has been optimized for use as a Quality-Control instrument including automated (unattended) particle size analysis, and temperature control of the fractionation capillary.

Background

Several investigators have attempted to separate particles by flow in capillaries [6, 14-19]. A discussion of their results is found in several references [6, 7]. Silebi and DosRamos [6] were the first to perform an analytical separation of submicrometer colloidal particles by flow through small-bore open capillary tubes. This method is called Capillary Hydrodynamic Fractionation (CHDF).

Figure 1 shows a diagram of a traditional CHDF setup. CHDF high-resolution particle size analysis consists of injecting through an HPLC injection valve a given colloidal sample into a narrow-bore capillary where an eluant continuously flows. An HPLC pump is used to deliver the eluant, at constant flow rate, through the injection valve, the fractionation capillary, and a UV-detector. The UV-detector is normally operated at 220 nm, or 254 nm, and is used to detect the particles as they exit the fractionation capillary. The UV signal intensity is utilized to quantify the different particle size populations in the sample through the use of the particle extinction cross section data [7].

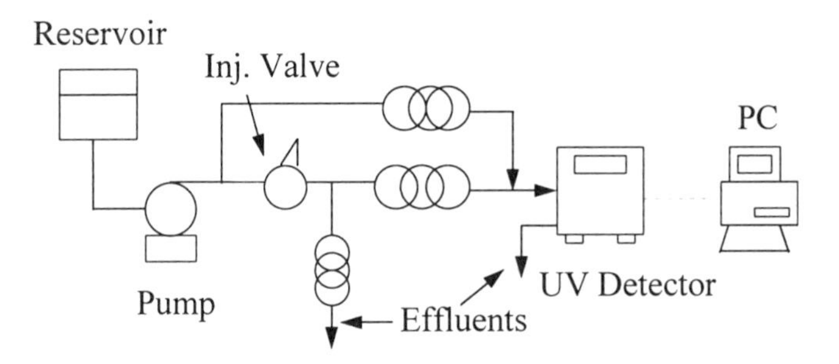

Figure 1. Schematic diagram of a traditional CHDF setup.

CHDF particle fractionation occurs as the particles sample the different eluant streamlines in the capillary. As shown in Figure 2, large particles are excluded from the slow-moving fluid near the capillary wall. Small particles approach the wall to a closer distance, equal to the particle radius. As a result of this size exclusion effect, large particles exit the capillary ahead of smaller particles.

Other factors affect the flow of particles in a microcapillary [7]. At low eluant electrolyte concentration, an electric double layer repulsion force exists between the capillary and the particles. This force increases as the eluant electrolyte concentration decreases. A fluid inertial force is exerted on the particles by the eluant pushing them toward an equilibrium position between the capillary axis and the capillary wall. This force, also known as the "Lift Force", increases with increasing eluant average velocity, increasing particle size, and decreasing capillary inner diameter. Non-ionic surfactants present in the eluant can adsorb on the capillary wall thereby creating a steric repulsion between the capillary and the particles [8].

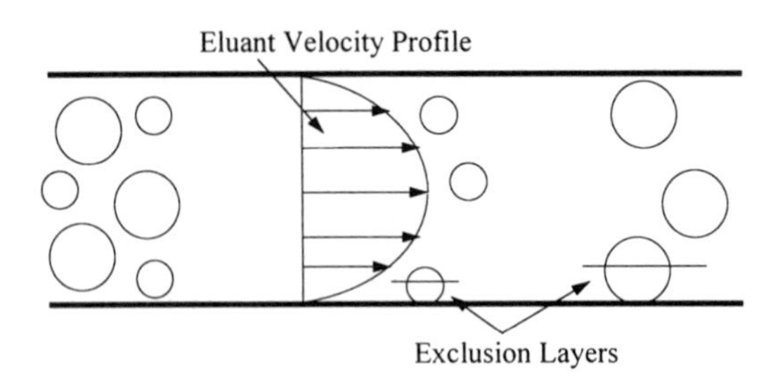

Figure 2. Particle flow fractionation in a CHDF capillary.

Particle Size Measurement by CHDF

A calibration curve of the fractionation capillary consists of a plot of Calibration Standards Particle Size against their Separation Factor, Rf, which is defined as follows [9]:

$$Rf = tm/tp \qquad (1)$$

where tm is the elution time for a molecular species such as sodium benzoate, and tp is the elution time for a calibration standard. A number of standards (usually about 8) covering the whole submicron range are injected into the fractionation cartridge. The separation factor for each standard is recorded and the calibration curve is generated. Rf is calculated for unknown samples, and the particle size is obtained from the calibration curve. The particle's extinction cross section curve is taken into account in order to quantify the different particle size populations in the sample. Figure 3 shows a CHDF fractogram for a blend of three monodisperse Polystyrene standards injected into the CHDF capillary at time zero. Large particles exit first. High-resolution is shown as seen from the baseline separation.

Quality-Control CHDF Requirements

A suitable Quality-Control particle size analyzer should have the following characteristics:

(i) easy, automated, operation.
(ii) fast analysis time.
(iii) 24-hour a day reliability.

(iv) low maintenance.
(v) accurate, and reproducible data.

This work aimed at developing a CHDF instrument with the following features: (i) automated particle size analysis by use of a commercial HPLC auto-sampler which would allow analysis of up to 150 samples unattended; (ii) temperature control of the fractionation capillary (see below, temperature studies); (iii) low-cost, disposable fractionation column; and (iv) in-line monitoring of the eluant pH, conductivity, and temperature.

Temperature Effects on Particle Fractionation by CHDF

Ambient temperature is expected to fluctuate noticeably in production plants. The effect of ambient temperature on CHDF fractionation performance was therefore studied. The CHDF capillaries from a commercial CHDF instrument (Matec Applied Sciences, 56 Hudson St., Northborough MA 01532) were immersed in a constant-temperature water bath. The water bath temperature was varied. Calibration standards were injected and their separation factors along with the pump pressure were recorded.

Figure 4 shows overlaid fractionation data for a blend of three monodisperse Polystyrene calibration standards with mean average particle sizes of 804 nm, 305 nm, and 44 nm. These were injected into the fractionation capillary at temperatures of 20, 30, and 45 degrees Celsius. As the temperature increases, the particle elution time increases. This temperature effect is more noticeable for larger particles. The elution time increases most noticeably for the 804 nm standard. The 305 nm standard is less retarded as the temperature increases than the 804 nm standard. The changing temperature least affects the 44 nm particles.

Figure 5 presents the measured separation factor for three calibration standards of sizes 620 nm, 350 nm, and 60 nm. Rf noticeably decreases with increasing temperature for large particles. The 60 nm particles show little effect.

The calculated particle size for the three standards is presented in Figure 6. These drop for larger particles as the capillary temperature increases. The change is more pronounced for the 620 nm particles because the slope of the calibration curve (particle size versus Rf) is larger in the 620 nm region. These results point to the need for temperature control in CHDF fractionation.

Figure 7 presents the fractogram half-height width (HHW) as a function of capillary temperature for the same three standards. The HHW measures the degree of axial dispersion that the particles undergo in the capillary. Larger values of HHW indicate that the particles are spreading further in the axial direction in the capillary, thereby producing wider fractograms. The 620 nm particles show a small increase in HHW as the temperature increases. The 305 nm particles are little affected by the

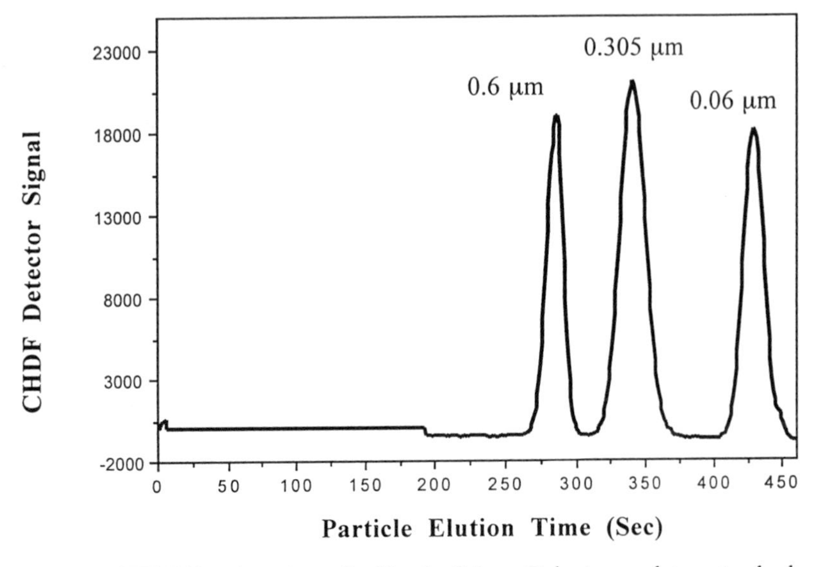

Figure 3. CHDF Fractionation of a blend of three Polystyrene latex standards. High-resolution fractionation is evident from baseline separation.

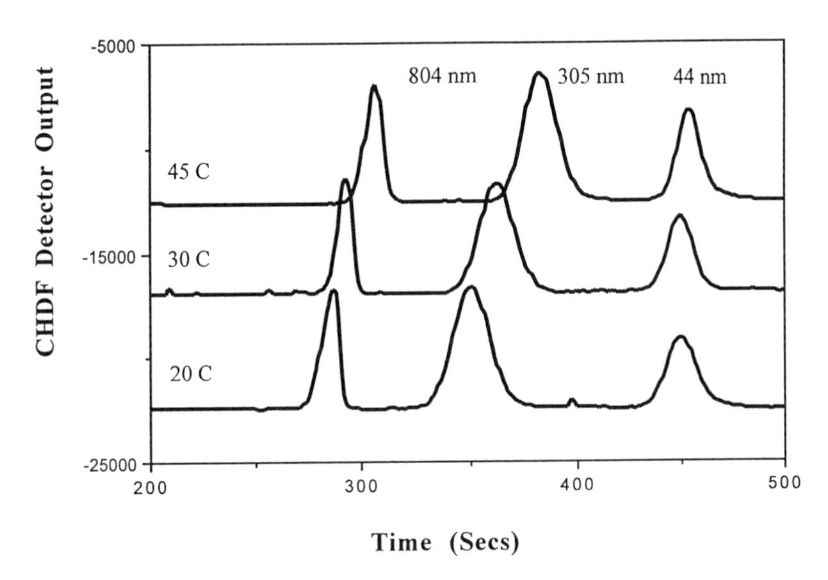

Figure 4. Ambient temperature effect on CHDF fractionation. Large particles are retarded as temperature increases.

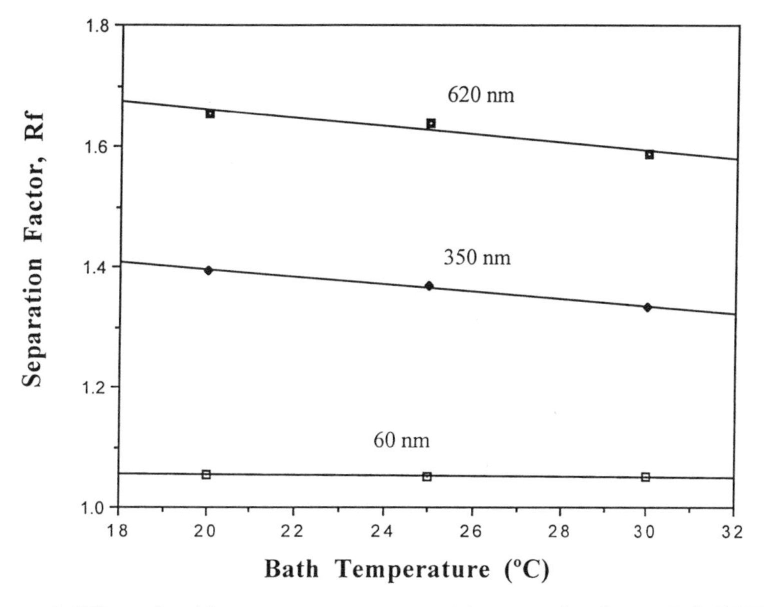

Figure 5. Effect of ambient temperature on particle separation factor, Rf. CHDF capillaries are immersed in a water bath.

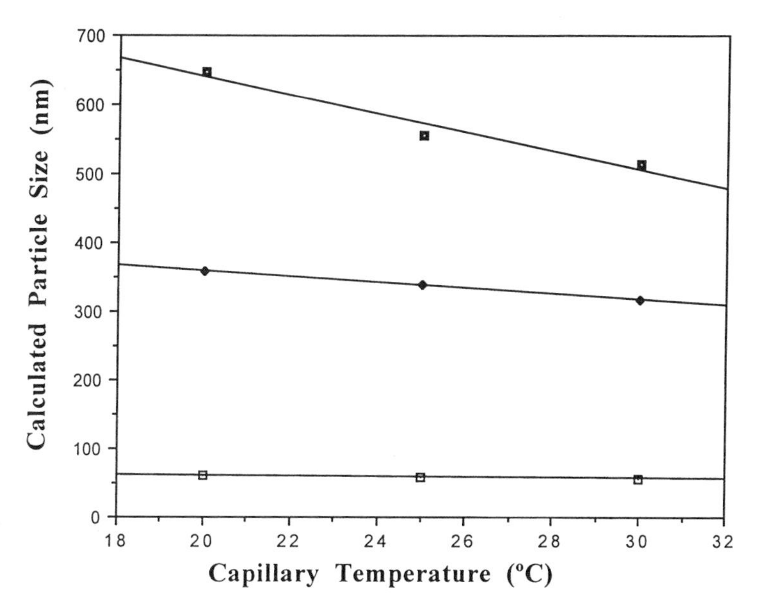

Figure 6. CHDF particle size results for a capillary immersed in a varying-temperature water bath.

temperature. The 60 nm particles show a decrease in HHW with increasing temperature.

Figure 8 contains data relating the pressure drop in the capillaries against capillary temperature. Also shown is the marker elution time against temperature. The pressure drop decay from 5400 PSI to 4350 PSI is due to the reduction in eluant viscosity with temperature (2% per degree C). The change in marker elution time is relatively small (4 seconds in a span of 10 degrees C). This change might be due to small variations in capillary dimensions due to capillary thermal expansion.

This marker data indicates that the eluant average velocity in the capillary remains unchanged as the capillary temperature is varied. Therefore, the above-presented changes in particle elution time are not due to a change in eluant average velocity with temperature in the capillary.

Temperature-Controlled CHDF Fractionation Capillary

The above results demonstrated that is imperative to provide temperature control of the fractionation capillary in order to ensure reproducible particle sizing data, independent of ambient temperature.

The capillary column configuration was modified in order to allow for temperature control facilities. A column heater element was mounted on a cylindrical heat sink component. The capillary column was mounted on the heat sink element in order to ensure efficient heat transfer between the heat sink and the capillary column.

Figure 9 presents particle size data from a temperature-controlled CHDF capillary column. The ambient temperature was varied from 65 to 75 degrees F while the capillary column was kept at constant temperature. It can be seen that the measured particle size is insensitive to ambient temperature fluctuations. High reproducibility from these high-resolution measurements is therefore achieved.

Automated Particle Size Analysis

In order to maximize operator-independent data, as well as, minimize personnel requirements, the manual HPLC injector was replaced by an HPLC auto-sampler. Up to 150 samples can be analyzed unattended by use of the HPLC auto-sampler.

Figure 10 shows four consecutive repeat injections of a Chemical Mechanical Polishing (CMP) silica slurry sample (Cabot Corp., Tuscola IL) analyzed on a CHDF instrument fitted with temperature control and auto-sampler operation. These CMP slurries are used for polishing semiconductor wafers during production of high-speed data processors. Highly planar wafer surfaces are required in this process. The reproducibility of the measurement is acceptable for Quality Control operation.

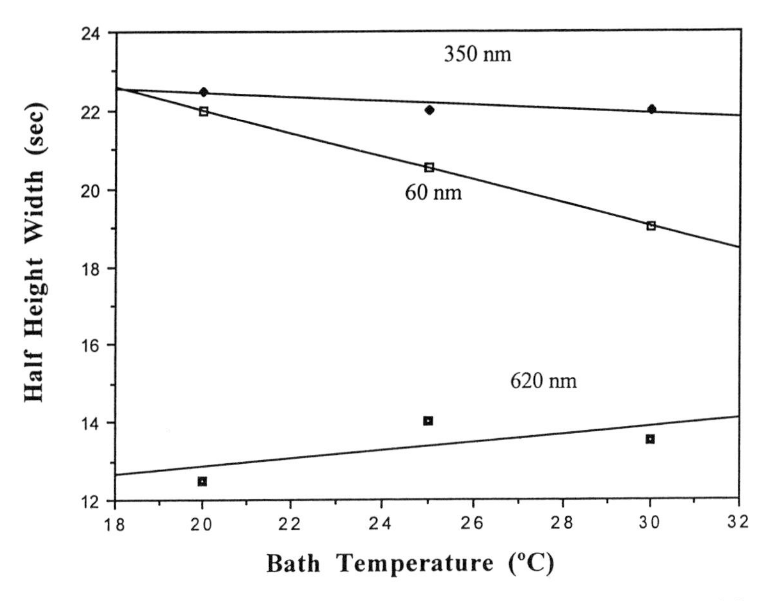

Figure 7. Eluant temperature effect on particle axial dispersion measured from fractogram half-height width.

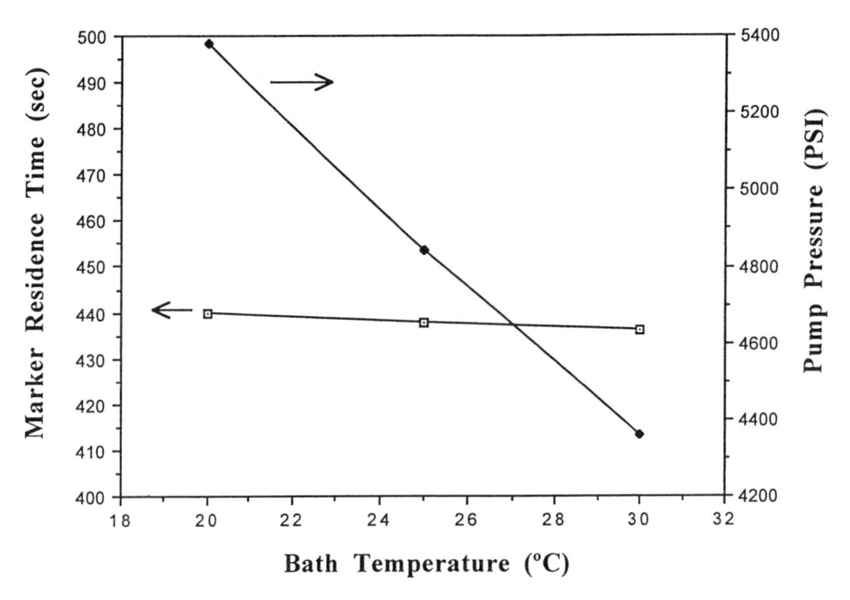

Figure 8. Eluant temperature effect on capillary pressure drop, and marker species (sodium benzoate) residence time.

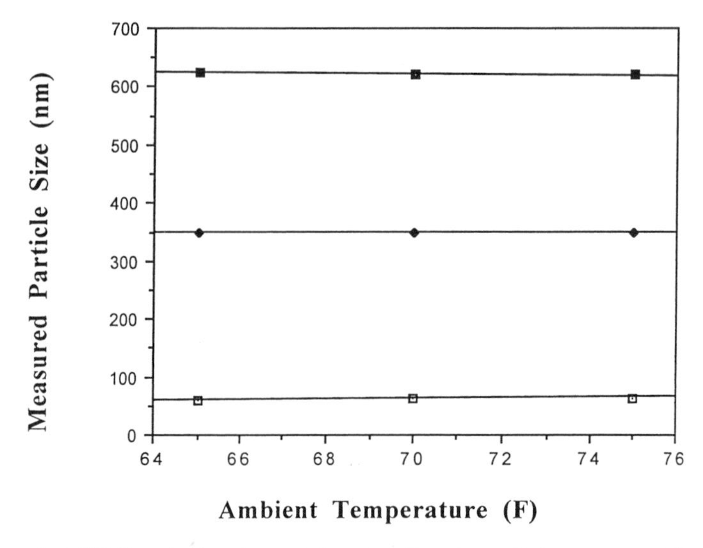

Figure 9. CHDF calculated particle size for three Polystyrene calibration standards. Fractionation capillaries kept in a constant-temperature water bath as the ambient temperature is varied.

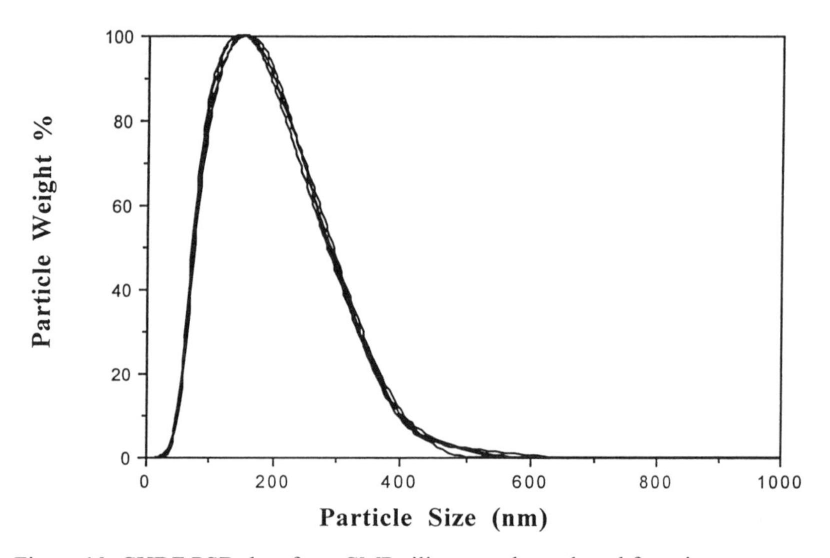

Figure 10. CHDF PSD data for a CMP silica sample analyzed four times.

Figure 11 presents CHDF high-resolution particle size distribution data for a colloidal silica standard (Nissan Chemical, Japan). This standard has a nominal average particle size of 300 nm. CHDF results indicate that most of the particles have a particle size close to 300 nm; however, 150 nm, and 450 nm particles are also detected. These particles are byproducts of the production process. The manufacturer has taken steps in order to eliminated these unwanted particles.

In-Line monitoring of Eluant pH, Temperature, and Conductivity

It has been reported that CHDF particle fractionation is affected by the eluant composition, i.e. ionic strength, pH, and anionic surfactant concentration [6, 8]. The separation factor decreases as the eluant ionic strength increases. As a result, smaller than expected particle size values are measured by CHDF if the eluant ionic strength is raised by either eluant contamination or degradation. This drop in separation factor with ionic strength is due to the fact that the electric double layer repulsion between the particles and the capillary surface decreases with increasing eluant ionic strength. This allows the particles to approach the capillary more closely while sampling slower-moving eluant streamlines. The particle average velocity decreases as a result. Non-ionic surfactant concentration also plays a role. A maximum in the separation factor is obtained at about the critical micelle concentration (cmc) of the non-ionic surfactant in the eluant.

As a result of the role that the eluant composition plays in CHDF fractionation, in-line eluant pH, conductivity, and temperature probes were installed on the CHDF setup. Further details of this implementation are given below.

Particle Sizing Instrument for Quality Control

The results above show that temperature control of the fractionation capillary is required for suitable performance by CHDF in a QC environment. Also, the eluant pH, conductivity, and temperature should be monitored. Figure 12 describes relevant features of the QC CHDF instrument.

The manual HPLC injection valve has been replaced by an auto-sampler, similar to those used for HPLC. This auto-sampler can be fully programmed from the CHDF computer. Samples can be analyzed unattended. Overnight operation is possible. Operation becomes operator independent.

pH, conductivity, and temperature sensors are located downstream from the UV-detector. This location was selected rather than upstream from the pump in order to eliminate the possibility of introduction of air bubbles into the system through the sensors. The presence of particles in this stream does not alter these readings since their concentration is low.

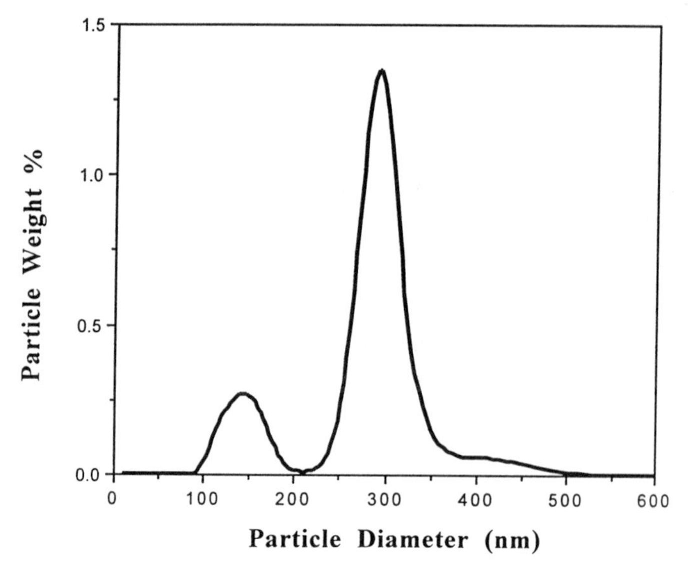

Figure 11. CHDF PSD data for 300 nm silica standard. Unexpected populations are detected.

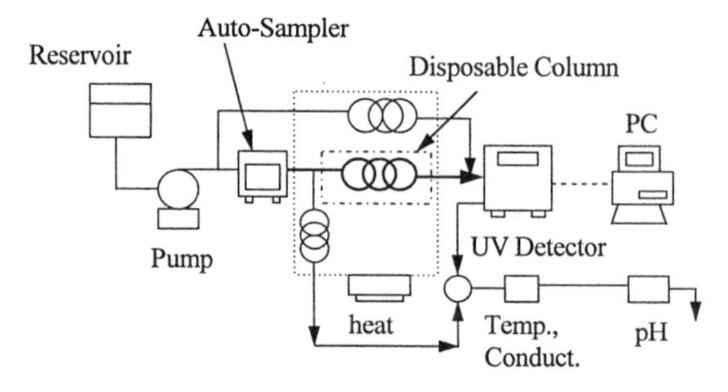

Figure 12. Schematic diagram of a Quality Control CHDF instrument. Sample analysis automation, temperature control, and disposable column format are highlighted.

All three capillaries have temperature control. The fractionation capillary is user-replaceable and disposable for quick recovery in case of capillary failure.

Discussion

Temperature Effect on Rf

The decrease in Rf values with temperature can be attributed to the increase of the particle's Brownian motion with temperature. As the Brownian motion increases, large particles start to overcome the Lift Force. The Lift Force tends to keep the particles in fast eluant streamlines near the center of the capillary. As these large particles are able to sample slower streamlines, their elution time increases. Small particles such as the 60 nm standard are subject to weak lift forces and therefore show no temperature effect on Rf.

The Lift Force, F_H, is defined as follows [7]:

$$F_H = 6 \Pi R_p \mu v_{pr} (1 + \kappa K_A) \quad (2)$$

where κ is the ratio of particle radius, R_p, to capillary inner radius, R_o; v_{pr} is the particle's radial velocity in the capillary; μ is the eluant viscosity; K_A is defined as follows:

$$K_A = \frac{9}{8(1 - \frac{r}{R_o})} \quad (3)$$

where r is the particle's radial position. The radial migration velocity of a freely rotating particle in a general bounded flow is given by [10, 11]:

$$vpr = \frac{6\pi Rp}{\nu}[v_{ps}{}^2 h(\frac{r}{R_o}) - v_{ps} v_m g(\frac{r}{R_o}) + \frac{5}{9} v_m{}^2 \kappa^2 (f_1(\frac{r}{R_o}) + f_2(\frac{r}{R_o}))] \quad (4)$$

where ν is the kinematic viscosity of the fluid. The functions h, g, f_1, and f_2 include volume integrals containing the Green's function of the Stokes equations and are not evaluated explicitly.

Silebi and DosRamos [5] used the expressions above to produce a mathematical model of particle fractionation in microcapillaries. The effects of colloidal and inertial forces, as well as, wall retardation, and size exclusion phenomena were studied. It was determined that the inertial forces become dominant as the eluant average velocity reaches a critical minimum value for given particle size. Small particles required unattainable high velocities in order to be influenced by the lift force.

Temperature Effect on Axial Dispersion

The reduction in axial dispersion with increasing temperature for small particles can be attributed to a reduction of Taylor Dispersion. For species not subject to Fluid Inertial Forces, an increase in Diffusivity results in a decrease in Axial Dispersion [7]. The Taylor-Aris modified axial dispersion coefficient in capillaries is shown below [12]:

$$D_{DM} = D_{\infty} + \frac{v_m^2 R_o^2}{48 D_{\infty}} (1 - \frac{R_p}{R_o})^6 \qquad (5)$$

where D_{DM} is the total axial Diffusivity undergone by particles of size R_p, and Diffusivity D_{∞} in a capillary of radius R_O; v_m^2 is the eluant kinematic viscosity. The axial Diffusivity for a molecular species can be calculated by setting R_p equal to zero. This expression does not take into account fluid inertial, electric double layer, or steric forces on the flowing species. This predicts that axial dispersion decreases for increasing particle Diffusivity.

Figure 7 shows that the 60 nm particles follow the behavior predicted by equation 5, i.e., axial dispersion decreases as the species Diffusivity (temperature) increases. The 620 nm particles are subject to a large Fluid Inertial Force. Higher temperatures increase the Brownian motion of these particles, which then overcome the Lift Force and sample slower velocity eluant streamlines. This results in a larger degree of axial dispersion. The 350 nm particles do not show a change in axial dispersion. The reason for this is that the decrease in Lift Force effect is offset by a reduction in axial dispersion due to higher Brownian Diffusivity.

Conclusions

A new high-resolution particle size analyzer has been developed. This instrument, based on Capillary Hydrodynamic Fractionation (CHDF), is capable of performing in a Quality-Control environment. Ambient temperature dependence of CHDF data had to be addressed. The disadvantages of ensemble techniques are overcome while providing detailed, true particle size distribution data. Undesirable particle size populations can be identified readily.

Predictions by Taylor Dispersion theory of axial dispersion dependence on eluant temperature were qualitatively demonstrated. A decrease of particle average velocity with increasing eluant temperature was shown as predicted by Silebi and DosRamos [6].

References

1. Weiner, B., Preprints: *Particle Size Analysis 1991,* Royal Society of Chemistry, **1991**.

2. Barth, H. G., and Flippen, R. B., *Anal. Chem.*, 67, 257R-272R, **1995**.

3. Giddins, J. C., Kaldwell, K. D., and Myers, M. N., *Macromolecules,* **9**, 106, **1976.**

4. Devon, M. J., Provder, T., and Rudin, A., in *Particle Size Distribution II: Assessment and Characterization,* ACS Symposium series 472, p. 279, **1991**.

5. Silebi, C. A., and DosRamos, J. G., *AIChE J.*, 35, 165, 1989.

6. Silebi, C. A., and DosRamos, J. G., *Journal Colloid Interface Sci.*, 130, 1, **1989**.

7. DosRamos, J. G., Ph.D. dissertation, Lehigh University, **1988**.

8. Venkatesan J., DosRamos, J. G., and Silebi C. A., in *Particle Size Distribution II: Assessment and Characterization,* ACS Symposium series 472, p. 279, **1991**.

9. DosRamos, J. G., and Silebi, C. A., *Journal Colloid Interface Sci.*, 135, 1, **1990**.

10. Cox, R. G., and Brenner, H., *Chem. Eng. Sci.,* 23, 147, **1968**.

11. Ishii, K., and Hasimoto, H., *J. Phys. Soc. Jpn.*, 48, 6, 2144, **1980**.

12. DiMarzio, E. A., and Guttman, C. M., *Macromolecules*, 3, 131, **1970**.

13. DosRamos, J. G., and Silebi, C. A., *Polymer International*, 30, 445, **1993**.

14. Krebs, V. K. F., and Wunderlich, W., *Angew. Makromol. Chem.,* 20, 203, **1971.**

15. Small, H. J., *J. Colloid Interface Sci.,* 48, 147, **1974.**

16. Mullins, M. E., and Orr, C., *Int. J. Multiphase Flow,* 5, 79, **1979.**

17. Noel, R. J., Gooding, K. M., Regnier, F. E., Ball, D. M., Orr, C., and Mullins, M. E., *J. Chromatogr.,* 166, 373, **1978.**

18. Brough, A. W. J., Hillman, D. E., and Perry, R. W., *J. Chromatogr.,* 208, 175, **1981.**

19. De Jaeger, N. C., Trappers, J. L., and Lardon, P., *Part. Charact.*, 3, 187, **1986**.

Chapter 16

Application of Capillary Hydrodynamic Fractionation To Study the Breakup of Aggregates by Sonication and Shear

Mehdi Durali and Cesar A. Silebi

Emulsion Polymers Institute and Department of Chemical Engineering, Lehigh University, Bethlehem, PA 18015–4732

Breakup of aggregates formed by coagulation with electrolyte and flocculation with long chain associative polymers were studied using sonication and shear forces. Size distribution of aggregates was analyzed using capillary hydrodynamic fractionation (CHDF) technique. Sonication of aggregates formed by flocculation released more than 90% of the primary particles, while only 40% of the aggregates formed by coagulation with electrolyte broke up into single particles, suggesting a stronger interparticle force among the particles in the latter case. Calculating fractal dimension of aggregates formed at electrolyte concentrations of below and above critical coagulation concentration (ccc) revealed the two regimes of diffusion limited aggregation and cluster-cluster collision. Shearing such aggregates increased the value of fractal dimension, indicating the formation of more compact structures.

Despite the practical importance of the processes involved in floc strength and breakup, there is very little fundamental understanding of the factors affecting the strength of aggregates or their mode of breakup under stress. Lacking proper quantitative methods, most of the experimental investigations in this area are of an empirical nature. The number of particle-particle contacts and the extent of interparticle forces are the controlling factors in determining floc strength and breakup. Aggregates can be converted into single particles quite easily if there is a weak interaction among the particles. Generally speaking, the extent of flocculation depends on particle concentration [1]; in dilute solutions only single particles exist, but beyond a critical concentration, an equilibrium exists between the aggregates and single particles. Aggregates might be broken up by lowering the electrolyte concentration [2] so that the repulsion potential among the particles increases, causing breakup. In a similar manner, the flocculation of sterically stabilized particles can be reversed by altering the solution conditions.

In practical processes involved in handling colloidal dispersions some shear forces are present. Under such conditions, an equilibrium state exists where some aggregates grow by collision while others break up by fluid stresses, resulting in a constant floc size [3-5]. At higher shear rates, the equilibrium aggregate size decreases, so, a criterion for floc strength is apparent: at a given shear rate, the stronger the aggregate, the larger the floc size.

In fundamental work by Powell and Mason [6], breakup of flocs under various forms of laminar flow fields was studied. The authors found that extensional flow was much more effective than rotational shear in the breakup of model aggregates.

The fractal structure of the aggregates plays an important role in determining floc strength and breakup. In a compact structure (higher fractal dimension), the number of particle-particle contacts in the aggregates are greater; therefore, these aggregates should be stronger than those having lower fractal dimensions. On the other hand, empirical studies on floc strength generally consider the size of flocs under given shear rates as a measure of floc strength. For a specific mass of particles, the size of aggregates will depend on the fractal dimension. Thus, interpretation of such measurements would not be straightforward. It is also interesting to note that low collision efficiency refers to more compact floc structure leading to lower interparticle attraction and hence lower floc strength. It is seen that any attempt on combining the particle-particle interaction with the fractal behavior of the aggregates in determining floc strength and breakup would be of great value.

The theoretical study of floc breakup starts by assuming the floc as an isolated body subjected to simple shear flow. The governing hydrodynamic equations are solved to determine the velocity and stress field in the vicinity of the floc. By assuming a model for the internal structure of the flocs, the stresses within the floc are calculated. Then the maximum floc size which can exist at a certain shear rate is determined by comparing these stresses with the cohesive strength among the particles in the floc. One of the critical steps of this analysis is the assumption of floc shape and structure. Bagster et al. modeled the floc as a solid impermeable sphere [7]. Later this model was modified by Adler and Mills [8] who modeled the floc as a uniformly porous sphere. Experimental investigation of Sonntag and Russel [9] indicated a fractal structure for the flocs. Using small angle light scattering, they followed the variation in average floc size with shear rate, and found that the average floc size decreases with increasing the shear rate demonstrating a power dependence. A value of 2.48 obtained for fractal dimension indicated a nonuniform structure, contradicting the uniform structure assumed in the previous models.

Study of floc breakup reported in the literature is basically focused on the bulk properties of aggregates. Almost nothing has been done to study the effect of fluid stresses on the breakup of small aggregates. Probably the only published article in this regard is the work of Barman et al. [10] on the effect of ultrasonification on the breakup of small aggregates fractionated by SFFF. They found that sonicating a coagulated sample more than 20 min disrupted all the larger aggregates, converting them to original singlet particles.

This work presents the experimental results obtained on the breakup of aggregates. Aggregates formed by coagulation of latex particles with different concentrations of electrolyte and those created by flocculation of particles with long chain associative polymers were studied. In the first part, breakup of aggregates by ultrasonication will be discussed, and then the effect of shear forces on the floc breakup will be evaluated.

EXPERIMENTAL

Materials and Methods. Monodisperse polystyrene particles used in this study were provided by Dow Chemical Company. The average particle size of these standard particles were 91, 190, and 234 nm, respectively. They were cleaned using serum replacement technique [11] to remove left over monomer, initiator, surfactant, and other low molecular weight impurities. Sodium chloride (Fisher Scientific) was implied as the electrolyte and used as received. Two nonionic surfactants were used to stop the coagulation reaction which were Triton N101, polyoxy ethylene (10) nonyl phenol ether (HLB 13.4, Union Carbide), and Brij 35, polyoxy ethylene (23) lauryl ether (HLB 16.9, ICI Americas Inc). Both surfactants were used as received. Flocculation of particles was performed using an associative polymer from a series of model polymers provided by Union Carbide Corporation. This polymer had a linear urethane backbone with two hydrophobic end groups, each containing 16 carbon atoms. The overall molecular weight of this polymer was 64,000. Such a polymer forms associative clusters in aqueous solutions and also adsorbs on the surface of polystyrene latex particles. Under certain conditions this phenomena could lead to bridging flocculation. The interaction of polystyrene latex particles with model associative polymers has been reported in previously [12].

Breakup of aggregates was accomplished by both ultrasonication and shear forces. For the ultrasonication experiment, test tubes containing aggregated dispersions were exposed to sonication forces in a sonication bath capable of generating 1200 watts at 28 kc for different periods of time to break the aggregates. Breakup by shear was performed in a Bohlin Rheologi concentric cylinder rheometer where aggregates were subjected to shear forces for different periods of time. A maximum shear rate of 1460 sec^{-1} was available with this instrument for the specified geometry. Special procedures had to be developed to collect samples from the rheometer for particle size analysis. The critical point was that sheared samples must be protected against recoagulation; several methods for accomplishing this task were tried. The best experimental approach was to shear the coagulated samples for a certain period of time and, while the instrument was still running, to add the nonionic surfactant to the solution and continue the shearing process for 10 seconds to provide complete mixing between the particles and nonionic surfactant. The particle size distribution of broken aggregates was analyzed using capillary hydrodynamic fractionation (CHDF), in which the fractionation of different size species occurred according to their difference in size [13,14], details of the method are also given in other papers of this symposium book. Mixture of particles with diameters between 0.03 and up to 2 mm can be analyzed with our CHDF unit.

RESULTS AND DISCUSSION

Breakup by Sonication

Breakup of Aggregates Formed by Coagulation. Two different size polystyrene latex particles (91 nm and 234 nm) were used in this study. The 91 nm latex was coagulated at two electrolyte concentrations of 0.1 M NaCl (6 min coagulation) and 0.2

M NaCl (3 min coagulation). A 0.5% solution of a nonionic surfactant, Triton N101, was used to stop the coagulation. The critical coagulation concentration (ccc) for this latex was 0.15 M NaCl, allowing us to examine the breakup of aggregates formed at slow and rapid Brownian coagulation. Such a variation in electrolyte concentration provides the opportunity to study the effect of ultrasonication on the floc strength. The aggregates formed with 0.1 M NaCl were expected to be broken in a shorter sonication period because of weaker interparticle attraction. In Figure 1 we compare the breakup of aggregates formed at two electrolyte concentrations of 0.1 and 0.2 M NaCl. In this Figure, the normalized number of singlet particles (N_1/N_0) is plotted vs the sonication time. The normalized number of singlet particles is given by the ratio of the signal intensity (I_1), in the fractogram of the coagulated sample, due to the singlets in the coagulated sample and the signal intensity (I_0) of the fractogram of the original (uncoagulated) dispersion. As predicted, aggregates formed with 0.1 M NaCl are disrupted before the other one indicating a weaker interparticle attraction. It is also interesting to note that although the flocs formed with 0.2 M NaCl were stronger than the other case (0.1 M NaCl), in both cases sonication for 900 sec was enough to reform 95 percent of the original singlet particles.

In another experiment, a 234 nm polystyrene latex particle was coagulated with 0.4 M NaCl (ccc = 0.33 M) for 2 days. The high ionic strength and long coagulation time were chosen to ensure the formation of large and strong aggregates. A 0.5 wt.% solution of a nonionic surfactant, Brij 35, was added to stop the coagulation. The addition of nonionic surfactant to the coagulated sample also prevented the aggregates from reforming during or following sonication. The same amount of coagulated sample was added to several vials and each vial was sonicated for a certain period of time. As before, the aggregates were fractionated using the CHDF apparatus and the floc size was analyzed by the particle size distribution program. Figures 2 shows the evolution in the breakup of the coagulated sample. The number distribution of the aggregates was normalized with respect to the number density of the singlets and hence, the peak corresponding to the singlets always reaches the value of one. The sonication process can be monitored by following the variation in number distribution of the other aggregates. The numbers on the PSD curves indicate the number of particles in each aggregate. As observed, the number density of doublets increased with increasing sonication time, reached a maximum at about 8 minutes of sonication, and decreased at longer sonication times. The increase in the concentration of doublets early in the sonication process is caused by the breakup of larger aggregates, which results in an increase in the number of doublets. As the sonication proceeds, almost all of the aggregates break apart to form single particles. This situation is clearly demonstrated in Figure 3. The number of singlets and doublets are plotted against the sonication time and both of the curves pass through a maximum. The position of the maximum for the singlets appear at a longer sonication time as compared to the position for the doublets which reflects the breakup of the doublets back into single particles. Sonicating the sample beyond 11 minutes had a reverse effect where the sample started to reaggregate.

This behavior might be due to the extensive heat generated in the sample at long sonication times. With an increase in temperature, the nonionic surfactant covering the particles can start to desorb creating bare surfaces on the particles which could result in coagulation upon collision.

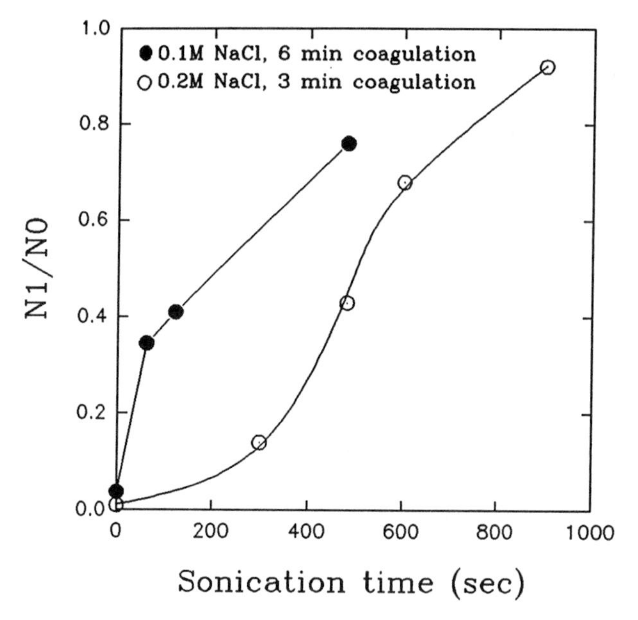

Figure 1. Normalized singlet number vs sonication time for flocs formed at two electrolyte concentrations of 0.1 and 0.2 N NaCl.

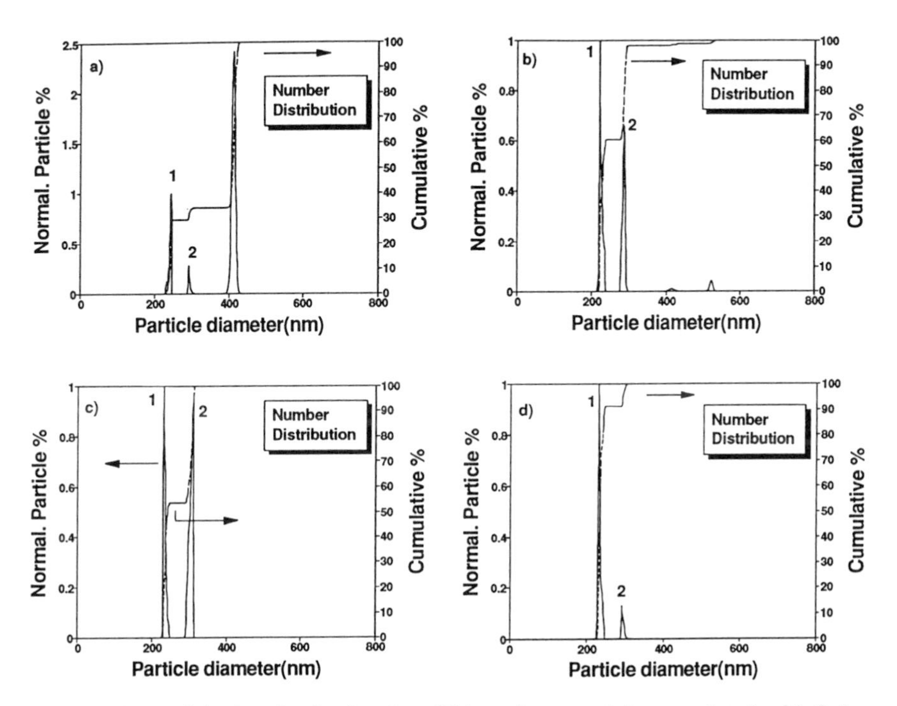

Figure 2. Particle size distribution for a 234 nm latex particle coagulated with 0.4 M NaCl for 2 days: a) no sonication, b) 5 min sonication, c) 8 min sonication, and d) 20 min sonication.

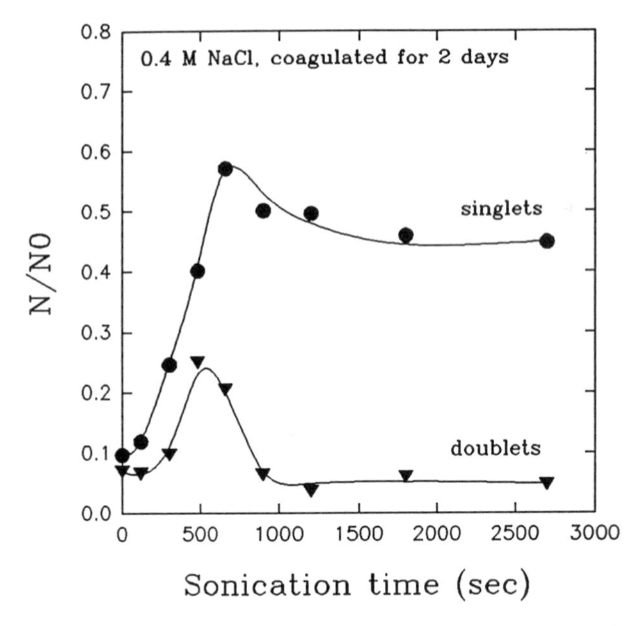

Figure 3. Normalized number of singlets and doublets as a function of sonication time. N_0 is the original number of singlet particles.

Breakup of Aggregates Formed by Flocculation. Flocculation of latex particles with model associative polymers was discussed previously [14]. It was shown that aggregates can be formed either through particle bridging or by cooperative networking of free polymer in the solution and the polymer adsorbed on the surface of the particles. In order to study the effect of ultrasonication on the breakup of aggregates formed by associative polymers, a sample of a 190 nm polystyrene latex flocculated with a hexadecyl terminated associative polymer (Mw=67,000) for more than one year was sonicated for different periods of time. The aggregates were fractionated using CHDF to analyze the aggregate size distribution. Figure 4 presents the normalized number of singlet particles as a function of sonication time. The number of singlet particles increases with sonication time reaching a value of N_1/N_0=0.95 at about 900 sec of sonication indicating an almost complete breakup of the flocs formed by associative polymers. Comparing Figures 3 and 4 indicates that the flocs formed with electrolyte are stronger than the ones formed with associative polymers (60% breakup for the former case compared to 98% for the later) which is probably due to a closer interparticle distance in flocs formed with electrolyte.

Breakup of Aggregates by Shear Forces

As stated before, the effect of shear forces on the breakup of aggregates was studied using a Bohlin Rheologi rheometer equipped with concentric cylinder geometry. A maximum shear rate of 1460 sec^{-1} was available with this instrument for the specified geometry. Samples were sheared for an interval of 90 sec at constant shear rate. Sampling from the rheometer required great care, since sheared samples must be protected against recoagulation. Several methods of preventing recoagulation were used. In one method, a certain amount of Brij 35 was added to the unsheared sample prior to shearing so that this surfactant would cover the fresh particle surfaces formed when the aggregate broke up and prevent them from reaggregation. The CHDF analysis of the sheared samples, however, proved that this technique was unsuccessful (the concentration of singlets decreased with increasing shear time) probably because the surfactant adsorbed in multilayers on the surface of the aggregates prior to shearing and so was unavailable to cover the new surfaces. In another method, the coagulated samples were sheared for certain time intervals, then the instrument was stopped and a solution of surfactant (Brij 35) was injected into the sheared sample to protect the particles from reagglomeration. Again, the CHDF results showed that this was not successful due to the shorter time scale of reaggregation compared to the restabilization process. The optimum technique was to shear the coagulated samples for a certain period of time and, while the instrument was still running, to add the nonionic surfactant to the solution and continue the shearing process for 10 seconds to provide complete mixing between the particles and nonionic surfactant. Shear breakup of aggregates formed from the 234 nm monodisperse polystyrene latex was studied at two electrolyte concentrations of 0.2 and 0.5 M NaCl to check the effect of interparticle forces on the process of breakup. The coagulation process was allowed to proceed for more than a day to ensure reaching equilibrium in the coagulation. The coagulated samples were sheared in the Bohlin rheometer at different shear rates, but for a constant shearing time. After each shearing interval, the sample was protected against reaggregation according to the procedure explained above and the particle size distribution was analyzed by CHDF.

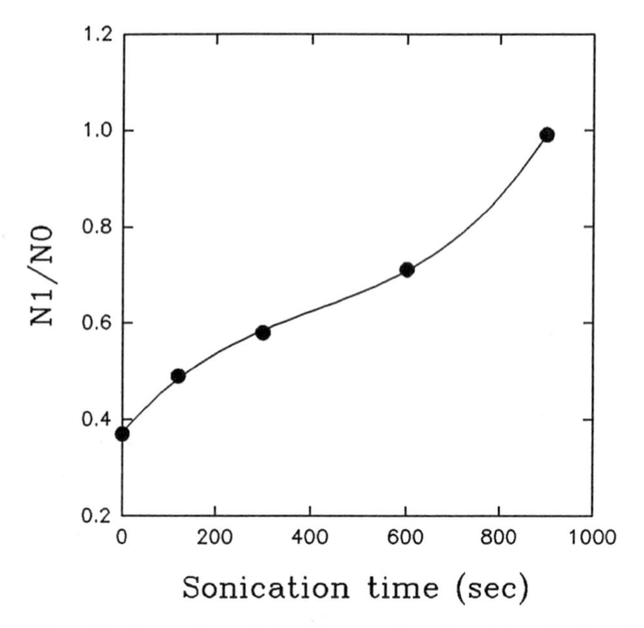

Figure 4. Normalized number of singlets as a function of sonication time for a 10.0% solid polystyrene particle (190 nm) flocculated with 1.0% hexadecyl terminated associative polymer (M_w=67,000) for 1 year.

Figures 5 and 6 illustrate the evolution of breakup for the sample coagulated with 0.2 M NaCl. In both number distribution analysis (Figure 5) and weight distribution analysis (Figure 6), three cases are presented: no shear, shearing at 580 sec^{-1}, and shearing at 1460 sec^{-1}. As the breakup proceeds, the intensity of the peaks assigned to the larger aggregates decreased and the peaks for the smaller ones increased. The sharp increase in the intensity of the singlet peaks is specially notable. However, it should be mentioned that the complete breakup of aggregates into singlets was not possible in the range of shear rates available. This can be observed from the right shoulder of the singlet peak for the 1460 sec^{-1} case in Figure 5, which is more apparent for the weight distribution presented in Figure 6.

The variation in number and volume average aggregate diameters with shear rate for the 0.2 M NaCl case is shown in Figure 7. The aggregate diameter decreases with increasing shear rate and reaches a minimum value of 270 nm number average and 340 weight average diameter at 1460 sec^{-1}. This again shows that complete breakup of the aggregates was not possible in the range of shear rates available. Figure 8 presents the variation of the normalized number concentration of singlets with shear rate for the aggregates formed by 0.2 and 0.5 M NaCl. For the 0.2 M NaCl case there is a linear relationship between the shear rate and the increase in the number of singlets, which means that the process of breakup starts at very low shear rates. In the case of 0.5 M NaCl, a critical shear rate (~ 800 sec^{-1}) is observed below which the number of singlets is unchanged. This is an indication of greater interparticle attraction between particles for the 0.5 M NaCl case which creates a resistive force against shear breakup.

Another way of studying the interparticle forces holding the aggregates together is by following the variation in the average number of singlets per aggregate as a function of shear rate. Figure 9 presents this situation for flocs formed by 0.2 and 0.5 M NaCl. The number of primary particles in the aggregate decreases as the shear rate increases and reaches a minimum at 1460 sec^{-1}. The aggregates formed with 0.5 M NaCl show a larger number of singlets at each shear rate indicating a higher attraction force among the particles forming the flocs.

Breakup of aggregates can also be monitored by following the variation in the viscosity of coagulated sample as a function of shear rate. Figure 10 presents this measurement for the 234 nm polystyrene latex coagulated with 0.2 M NaCl for 1 day. Unlike the discrete shearing method used for the previous samples, in this experiment the sample was sheared continuously and the viscosity was monitored simultaneously. Results are presented for the coagulated sample and the uncoagulated sample both at the same solids concentration. At very low shear, the uncoagulated sample shows a viscosity of 1.3 cp (centipoise) while the coagulated sample has a viscosity of 500 cp despite having the same solids content. This very large difference in viscosity is due to the formation of large aggregates which are very porous and entrap large amounts of liquid among the interstices between the particles forming them. Consequently, the apparent volume fraction of solids increases several times, and therefore, the viscosity of the coagulated sample increases drastically. The sharp decrease in viscosity with an increase in shear rate is also a consequence of the porous nature of the large aggregates. These structures are relatively loose and consist of several smaller, stronger flocs. Upon shearing, these aggregates start to break and release the entrapped liquid; therefore, the apparent or effective volume fraction of solids decreases as the shear rate increases.

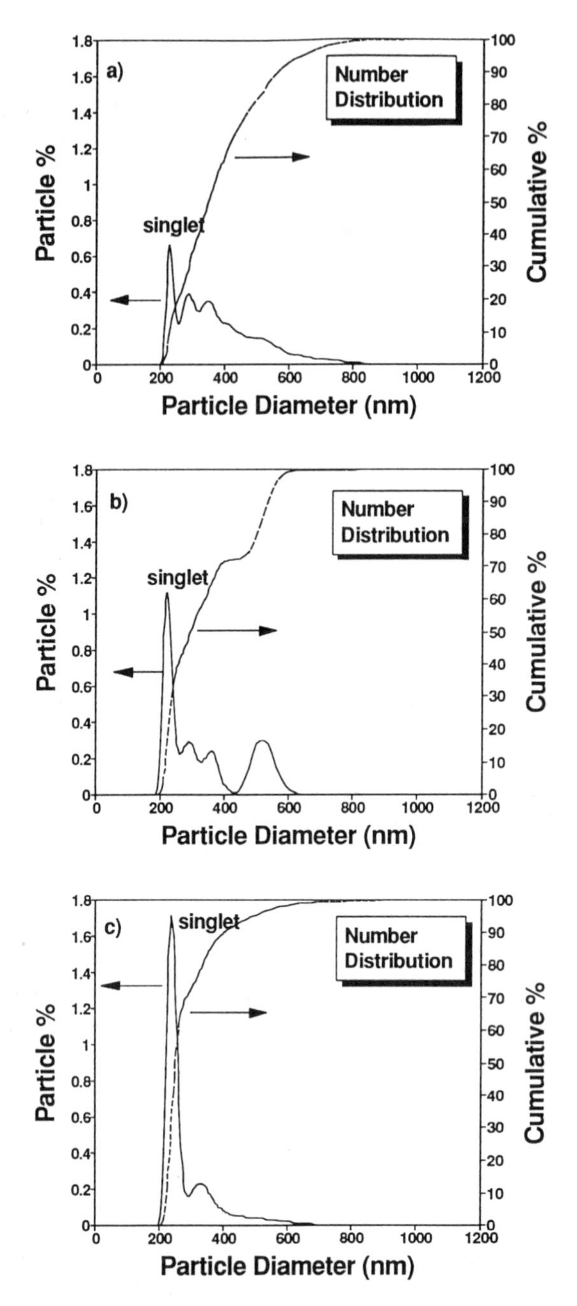

Figure 5. Particle number distribution for a 234 nm polystyrene latex coagulated with 0.2 M NaCl for 1 day. a) zero shear rate, b) 581 sec^{-1}, c) 1460 sec^{-1}.

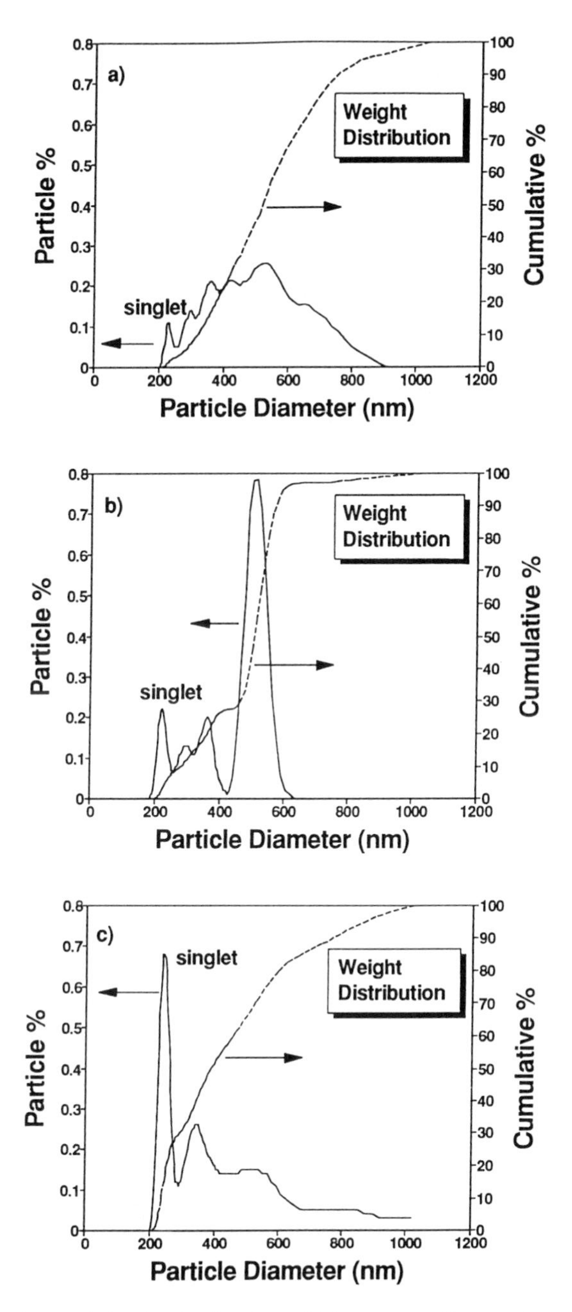

Figure 6. Particle weight distribution for a 234 nm monodisperse latex coagulated with 0.2 M NaCl for 1 day. a) zero shear rate, b) 581 sec^{-1}, c) 1460 sec^{-1}.

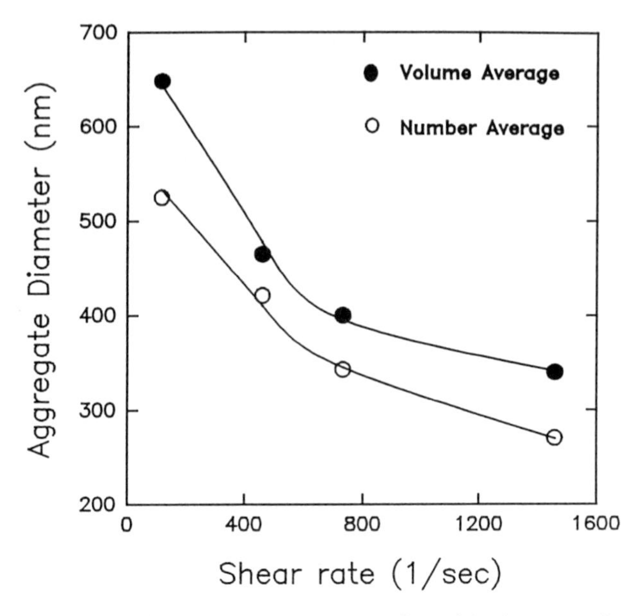

Figure 7. Variation of the average aggregate size with shear rate for the same sample described in figure 6.

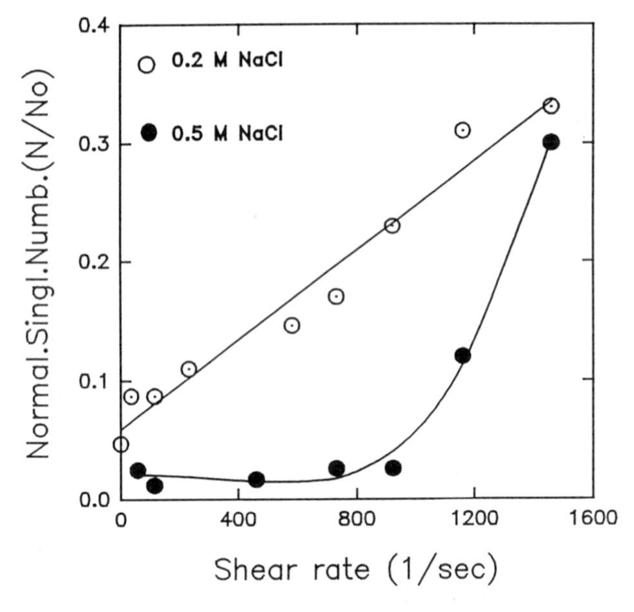

Figure 8. Variation of the normalized singlet number with shear rate for samples coagulated with 0.2 and 0.5 M NaCl.

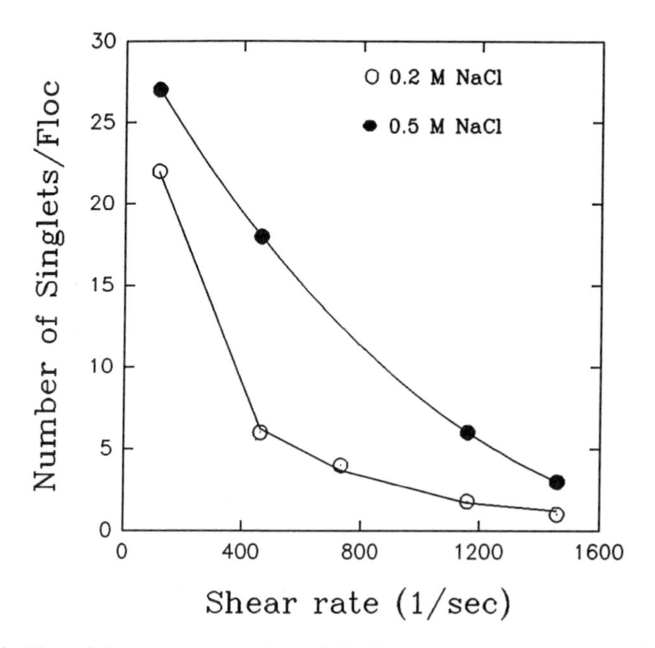

Figure 9. Plot of the average number of singlets per aggregate versus shear rate for a 234 nm polystyrene latex coagulated with 0.2 and 0.5 M NaCl for 1 day.

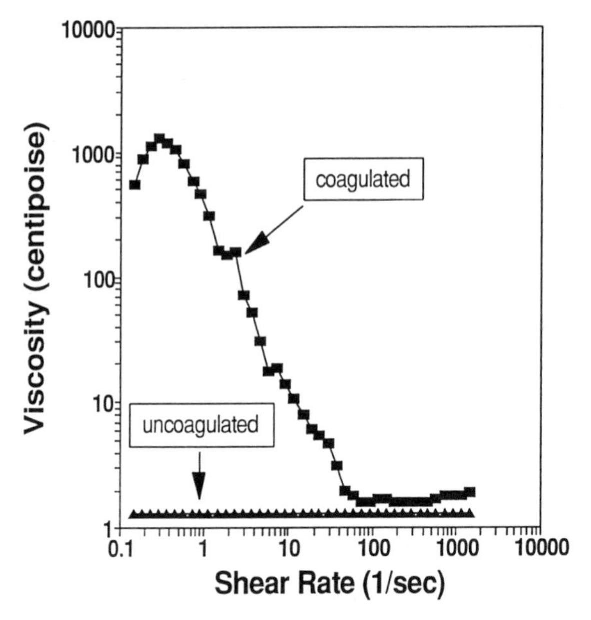

Figure 10. Viscosity versus shear rate for a 234 nm polystyrene latex coagulated with 0.2 M NaCl for one day. The viscosity of the uncoagulated sample is shown for comparison.

Thus, the viscosity of the sample drops significantly. At higher shear rates, the viscosity reaches a plateau which should not be mistaken for the complete breakup of the aggregates into single particles. This situation can be better observed in Figure 8 where the variation in the number of singlets as a function of shear rate is demonstrated for the same sample (0.2 M NaCl case). It is observed that at the highest shear rate available (1460 sec^{-1}) it was only possible to recover 35% of the single particles. So, this indicates that the plateau value in the viscosity curve corresponds to small aggregates which have a compact structure and do not entrap any liquid; therefore, the volume fraction of solids remains unchanged and the viscosity stays constant close to the value for the uncoagulated sample.

The effect of shearing time at a constant shear rate is shown in Figure 11 for the sample undergone the shear profile presented in Figure 10. The viscosity of the uncoagulated sample is shown for comparison. It is observed that the viscosity of the coagulated sample is unchanged over a period of 5 min. at the highest shear rate available suggesting that smaller aggregates are held together strongly and are not affected by the shearing time.

In order to compare the efficiency of sonication and shear breakup techniques, the coagulated sample in Figure 11 which was sheared at 1460 sec^{-1} for 5 min, was sonified for 10 min in the sonication bath described before to check if aggregates could be broken into smaller species. Figures 12 and 13 are the CHDF fractograms (raw data) for the unsonified and sonified samples, respectively. The signal intensity is plotted against particle elution factor which is the ratio of the particle elution time (t_p) to the marker elution time (t_m). The singlet peak is marked in both figures. Sonicating the sample for 15 min increased the intensity of the singlet peak several times which is a clear indication that in the range of shear rates available with our shearing device, sonication was a more efficient tool for breakup of aggregates.

Analysis: Fractal Dimension

Industrial aggregation processes usually involve particle clusters including several hundreds of original particles with unlimited number of configurations. Fractionating such aggregates into their individual components is practically impossible and provides no valuable information. In practice, it is necessary to evaluate the aggregates in terms of a universal value which is fairly simple to determine while providing useful information. Aggregates are now identified as *fractal* bodies [15,16] meaning that they do not have integer dimensions. For solid, three-dimensional objects such as a sphere, the mass of the object correlates with its characteristic dimension (diameter) to the third power. Therefore, if log mass is plotted versus log size, a straight line with the slope of three will be obtained. Creating these kind of plots for aggregates (usually number of original particles versus aggregate diameter) provides lower slopes with non-integer values. The slope of such a line is known as *fractal dimension*, d_f. In three dimensional space, d_f may have values from 1 to 3 where lower value represents linear aggregates while upper one demonstrates aggregates with uniform density (or porosity). Generally, fractal dimensions of intermediate value are obtained. As d_f decreases, a more open structure is found. Earlier computer simulations modeled aggregation process [16] as the random addition of single particles to growing clusters. This approach is now called diffusion-limited aggregation (DLA) and gives a fractal dimension

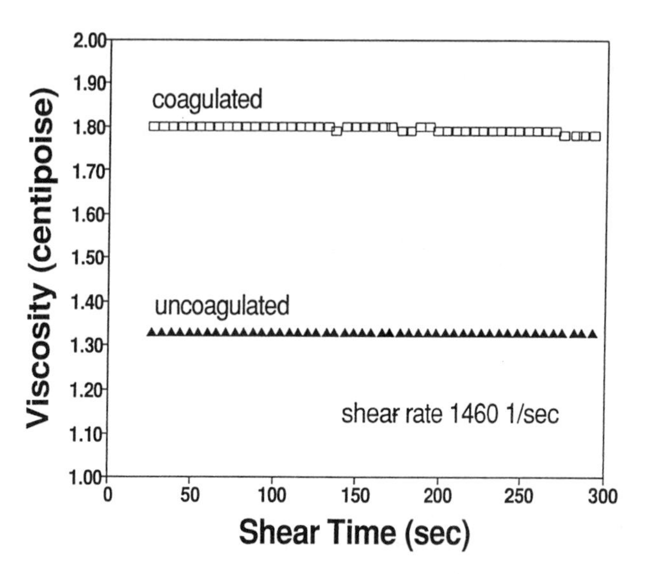

Figure 11. Viscosity vs shearing time at 1460 sec^{-1} constant shear rate for a sample coagulated with 0.2 M NaCl for 1 day and sheared according to the procedure of figure 10.

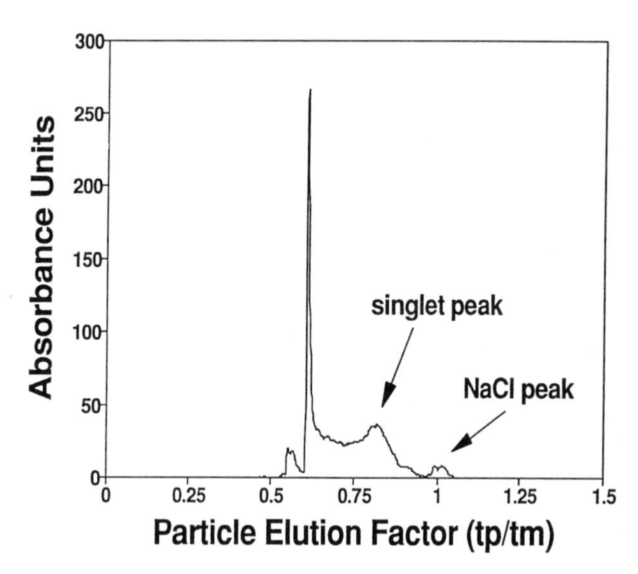

Figure 12. CHDF fractogram for a 234 nm polystyrene latex coagulated with 0.2 M NaCl for 1 day and then sheared at 1460 sec^{-1} for 5 min.

of about 2.75 suggesting a fairly compact aggregate structure. In practice, this model is observed for the low ionic strength aggregation where some degree of repulsion still exists between the particles and the slow rate of coagulation allows the single particles to diffuse in some way into the existing aggregates before making any contact. Another simulation technique assumes cluster-cluster aggregation which is more likely to happen in most of practical aggregation processes. This model gives a relatively lower value of d_f ± 1.75, representing a more open structure. This case is encountered for coagulation at high electrolyte concentration where the coagulation process is so fast that the particles do not have time to reorient themselves, and floc-floc collisions are more possible. Figure 14 illustrates these concepts schematically.

Fractal dimension of colloidal systems has been the focus of several experimental studies. Weitz et al. [18] found a value of 1.7 for fractal dimension of aggregated colloidal gold supporting the cluster-cluster simulation approach. For rapid aggregation of colloidal silica, Aubert and Cannell [19] obtained d_f values of 1.75 ± 0.05. A study by Tence et al. [20] on aggregation of spherical iron particles led to fractal dimensions of 1.88 ± 0.09.

In order to calculate the fractal dimension, the average number of primary particles per aggregate is plotted versus the volume average aggregate size. In CHDF operation, the total number of aggregates per unit volume is given by the ratio of the number of singlets to the cumulative percentage of singlets,

$$N_T = \frac{N_1}{(cumulative\%)_1} \tag{1}$$

cumulative percentage of each species is calculated by the CHDF analysis software. Here, N_T is the total number of aggregates per unit volume and N_1 is the number of singlets per unit volume. The average number of particles per aggregate can be determined as

$$< N >= \frac{N_0}{N_T} \tag{2}$$

where <N> is the average number of primary particles per aggregate and N_0 is the original number of particles per unit volume. The fractal dimension of aggregated polystyrene latex particles were determined in this study for two cases of slow and rapid Brownian coagulation. Figure 15 is a plot of <N> versus DV (volume average aggregate size) for the aggregates formed with 0.2 and 0.33 M NaCl for different coagulation times. For coagulation with 0.2 M NaCl (slow coagulation), fractal dimension happened to have a value of 2.79 which is in close agreement with diffusion-limited aggregation predictions. For rapid coagulation (0.33 M NaCl), a value of 2.32 was observed which approximates the cluster-cluster aggregation model.

Figure 16 shows the fractal dimension of aggregates formed with a 234 nm polystyrene latex particle at two electrolyte concentrations 0.2 and 0.5 M NaCl (ccc = 0.33 M), and sheared using the Bohlin rheometer. The idea is that by shearing the coagulated sample at different shear rates, the total number of particles per aggregate

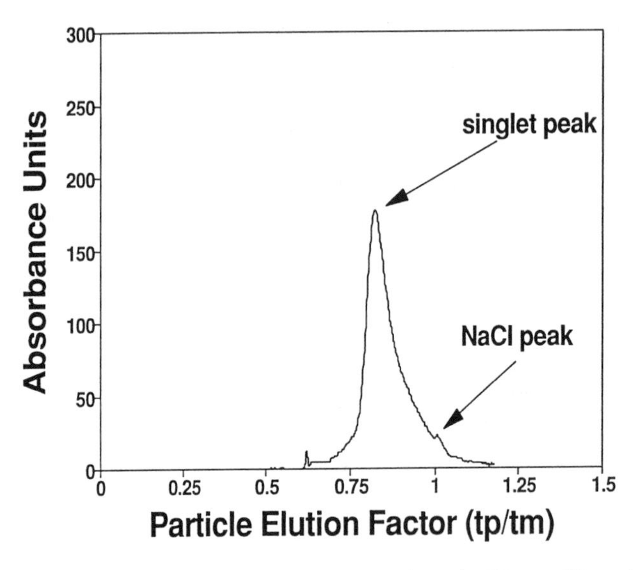

Figure 13. CHDF fractogram for the same sample as in Figure 12 sonicated for 10 min in a sonication bath.

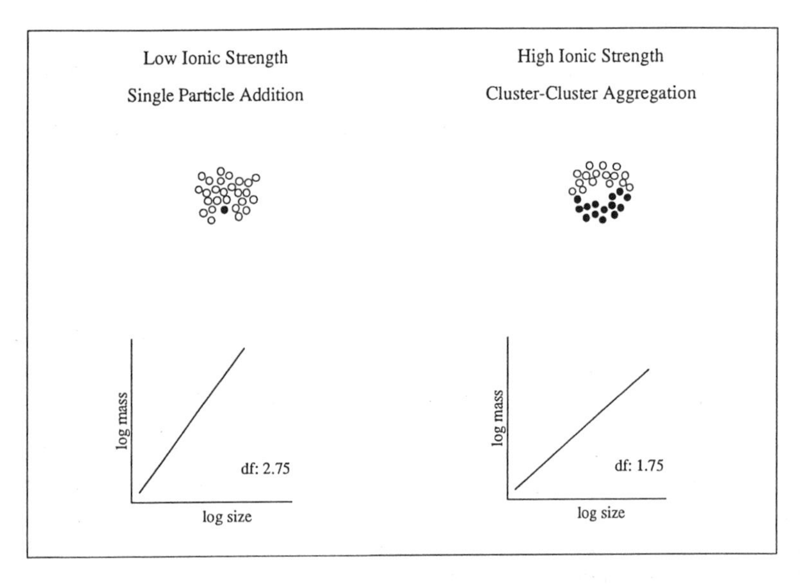

Figure 14. Schematic representation of fractal dimension at low and high ionic strength [15].

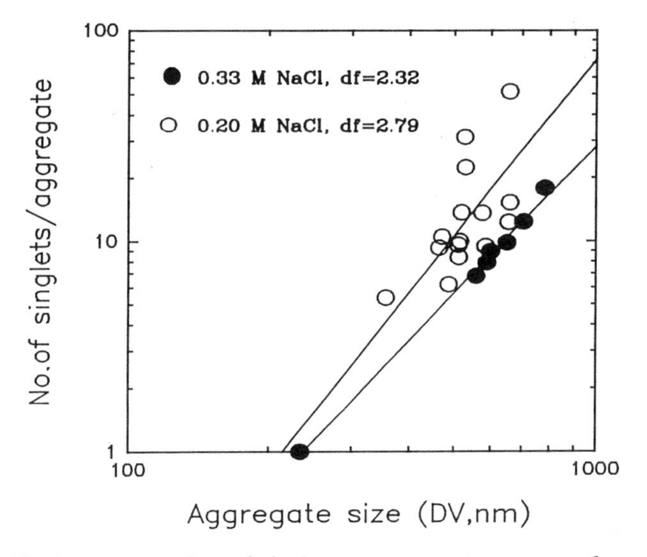

Figure 15. Average number of singlets per aggregate versus volume average aggregate diameter for two electrolyte concentrations of 0.2 and 0.33 M NaCl.

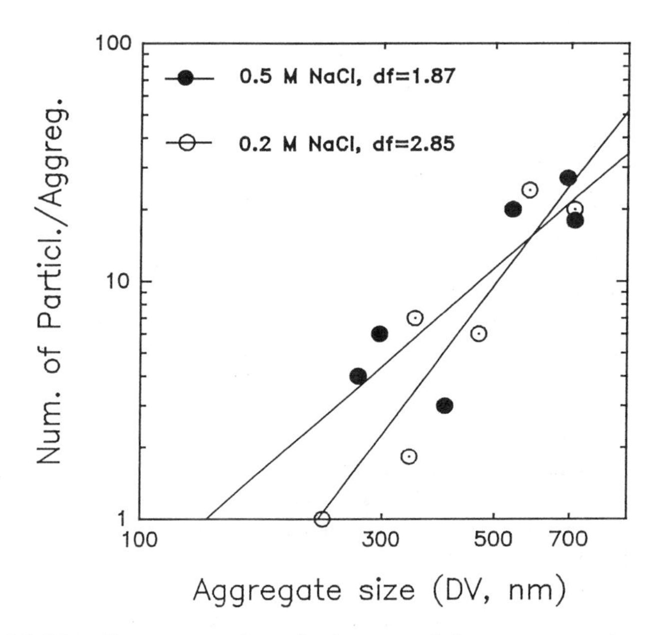

Figure 16. Plot of average number of primary particles per aggregate vs volume average aggregate diameter for a 234 nm polystyrene latex coagulated with 0.2 and 0.5 M NaCl for 1 day and then sheared using the Bohlin rheometer.

would decrease which this would possibly appear as an increase in the value of the fractal dimension. A value of 2.85 was obtained for the 0.2 M NaCl case which shows an increase compared to the value found for the unsheared case (d_f=2.79). For the rapid coagulation (0.5 M NaCl, sheared) a value of 1.87 is observed which in comparison to the previous case (0.33 M NaCl, unsheared, d_f=2.32) shows a decrease in fractal dimension. This is an indication of the formation of looser aggregates. It was difficult to evaluate the effect of shear rate in this case because the same concentration of electrolyte has not been used in the two experiments (0.33 M versus 0.5 M NaCl).

SUMMARY AND CONCLUSIONS

Sonication breakup of aggregates formed by electrolyte was studied for two different particle sizes. The particle size distribution of the samples was analyzed by CHDF. The 91 nm polystyrene particles coagulated at two electrolyte concentrations of 0.1 and 0.2 M NaCl, were sonicated for different times and it was found that the aggregates formed with 0.2 M NaCl were stronger than the ones formed with 0.1 M NaCl due to a weaker repulsion potential between the particles in the aggregate. Sonicating the aggregates formed from 234 nm particles (coagulated with 0.2 M NaCl) revealed that larger aggregates brokeup into singlets and doublets with up to 20 min of sonication. Longer sonication times has a reverse effect and reaggregated the particles into clusters due to the excessive heat generated in the sample by sonication which enhanced the Brownian motion of the particles and increased their collision frequency. Sonicating a 190 nm polystyrene particle flocculated with a hexadecyl hydrophobe associative polymer showed that the aggregates could be converted more than 90% into single particles after 15 min sonication.

Shear breakup of 234 nm polystyrene particles coagulated with 0.2 and 0.5 M NaCl revealed that in the range of shear rates available, only 35 percent of aggregates could be converted into primary particles. The flocs formed with 0.2 M NaCl showed a continuous breakup while the ones formed with 0.5 M NaCl displayed a yield stress below which aggregates remained intact. Following the viscosity of the aggregated dispersion showed a significant increase in zero shear viscosity of the aggregated sample as compared to single particle dispersion. This was probably due to the open structure of aggregates which held a considerable amount of liquid in the interstices between the particles and increased the apparent volume fraction of the solids in the system. Increasing the shear rate decreased the viscosity of the coagulated sample drastically as a result of breakup of the larger aggregates and releasing the entrapped liquid which decreased the volume fraction of the particles. At higher shear rates, the viscosity profile reached a plateau indicating the existence of smaller aggregates which formed a more compact structure and did not entrap any liquid.

The fractal dimension of the aggregates formed by electrolyte addition was studied for both slow and rapid Brownian coagulation. A value of 2.79 was obtained for the slow coagulation indicating the formation of a fairly compact structure. For rapid coagulation a fractal dimension of 2.32 was obtained suggesting cluster-cluster collision and the formation of a looser structure. Shearing the aggregates formed by slow coagulation increased the value of the fractal dimension indicating the breakup of aggregates and formation of more compact structures.

The applicability of CHDF to the separation and characterization of aggregated colloidal clusters based on differences in their size has been demonstrated. Some demonstrated applications of this technique include the following: (a) observing changes of the population of latex aggregates formed by charge neutralization (using electrolyte) and by particle bridging (using associative polymers), (b) tracking the breakup of aggregates, formed by charge neutralization and by particle bridging, by sonication and shear forces, (c) studying the breakup of aggregates to singlets and doublets as a function of shearing and sonication time, and (d) tracking the fractal dimension of aggregates formed by charge neutralization and particle bridging after being sheared or sonicated for various periods of time.

Literature Cited

1. Long, J. A.; Osmond, D. W. J., and Vincent, B. *J. Colloid Interface Sci.* **1973**, 42, 545.
2. Lessard, R. R., and Zieminski, S. A. *Ind. Eng. Chem. Fund.* **1971**, 10, 260.
3. Michaels, A. S., and Bolger, J. C. *Ind. Eng. Chem. Fund.* **1962**, 1, 153.
4. Hunter, R. J., and Frayne, J. *J. Colloid Interface Sci.* **1980**, 76, 10.
5. Fair, G. M., and Gammell, R. S. *J. Colloid Sci.* **1964**, 19, 360.
6. Powell, R. L., and Mason, S. G. *AIChE J.* **1982**, 28, 286.
7. Bagster, D. F., and Tomi, D. *Chem. Eng. Sci.* **1974**, 29, 1773.
8. Adler, P. M., and Mills, P. M. *J. Rheol.* **1979**, 23, 25.
9. Sonntag, R. C., and Russel, W. R. *J. Colloid Interface Sci.* **1986**, 113, 2.
10. Barman, B. N., Giddings, J. C. *PSME* **1990**, 62, 186.
11. Ahmed, S.M., El-Aasser, M.S., Pauli, G.H., Poehlin, G.H., and Vanderhoff, J.W. *J. Colloid Interface Sci.* **1980**, 73, 388.
12. Jenkins, R.D., Durali, M., Silebi, C.A., and El-Aasser, M.S. *J. Colloid Interface Sci.* **1989**, 133, 302.
13. DosRamos, J. G., and Silebi, C. A. *J. Colloid Interface Sci* **1990**, 135, 165.
14. DosRamos, J. G., and Silebi, C. A., In Particle Size Distribution II; Provder, T., Ed., ACS Symposium Series 472, Washington, DC, 1991; pp. 292-307.
15. Gregory, J. *Critical. Rev. in Environment. Cont.* **1989**, 19, 185.
16. Meakin, P. *Adv. Colloid Interface Sci.* **1988**, 28, 249.
17. Vold, M.J. *J. Colloid Sci.* **1963**, 18, 684.
18. Weitz, D.A., and Oliveria, M. *Phys. Rev. Lett.* **1984**, 52, 1433.
19. Aubert, C., and Cannell, D.S. *Phys. Rev. Lett.*,**1986**, 56, 738.
20. Tence, M., Chevalier, J.P., and Jullien, R. *J. Phys.* **1986**, 47, 1989.

Chapter 17

Electrokinetic Lift Effects on the Separation Factor in Capillary Hydrodynamic Fractionation

A. D. Hollingsworth and Cesar A. Silebi

Department of Chemical Engineering and Emulsion Polymers Institute, Lehigh University, Bethlehem, PA 18015–4732

Above an eluant ionic strength of 10^{-4} M (typical of normal CHDF operation), the normalized average residence time of monodisperse polystyrene nanospheres (average diameter < 0.36 μm) being pumped under laminar flow conditions is independent of the flow rate. At lower ionic strength, however, a significant flow rate dependence has been observed using the same latexes. This behavior is inconsistent with an inertial lift force mechanism and supports the hypothesis that the anomalous particle elution was electrokinetically induced. The incorporation of a *total* electrokinetic lift force expression into our mathematical model has improved the prediction of the separation factor (a relative particle average velocity), indicating that the shear flow contribution to the lift force is significant in the CHDF simulation. Particle separation experiments have demonstrated that the use of very low conductivity eluants in CHDF provides better resolution of colloidal particles than that presently obtained using higher electrolyte concentrations.

Capillary hydrodynamic fractionation (CHDF) is an established analytical method for particle size characterization in the submicron size range (*1-3*). In this hydrodynamic flow fractionation process, colloidal dispersions are pumped through an open-bore microcapillary tube prior to optical detection, eluting in order of decreasing particle diameter. The fractionation step results in much higher resolution than would otherwise be possible in determining particle size distributions, particularly if the sample is multimodal or has a broad size distribution. A mathematical model which was developed from first principles without the use of any adjustable parameters is capable of predicting both the particle average velocity as well as the degree of axial dispersion in CHDF (*4,5*). At very low eluant ionic strength (*i.e.*, less than 10^{-4} M), however, the phenomenological coefficients corresponding to monodisperse polystyrene latexes could not be predicted quantitatively (*6*). This result was observed when nonionic surfactants were dissolved in the eluant. A recent study revealed that the separation factor (a parameter equal to the ratio of the mean displacement velocities of particle and eluant) depended strongly on eluant flow rate when the fluid conductivity was less than 10 μS/cm (*7*). The greatest deviation between the experimental separation factors and the predicted values corresponded to the *lowest* electrolyte concentration (using a surfactant-free eluant), thus precluding an inertial lift force mechanism (*8*), as well as the combined effect of steric and osmotic repulsive interactions (*9*). When an electrokinetic lift force expression was incorporated into the CHDF model, these data showed good qualitative agreement with the theoretical results (*7*).

Electrokinetic lift is defined as the transverse electroviscous force acting on a charged particle moving relative to another surface in an electrolyte solution (*10*). For translational particle motion, the electrokinetic lift of colloidal particles may be due to an induced streaming potential in the particle-wall gap (*10-12*) or the asymmetry of the electric field around the charged spheres caused by the presence of the wall (*13*). In both cases, a nonzero Maxwell stress tensor gives rise to an additional force acting on the surfaces which is proportional to the square of the ratio of translational particle velocity to the fluid conductivity. All of these theories are restricted to the special case of sliding motion and only two of them (*12,13*) are valid for the case under consideration, *i.e.*, particle radius and separation distance from the wall of comparable magnitudes. To more fully account for the particle hydrodynamics in CHDF, the electrokinetic lift force expression must also include the effects of particle rotation and shear flow, both of which strengthen the magnitude of the theoretical force. Bike and Prieve (*14*) have determined the electrokinetic lift on a charged sphere, freely rotating and translating in linear shear flow along a plane wall. Because the electrical stress is a nonlinear function of the electric field, the total lift force, F_{ek}, is comprised of six components including three cross terms, *i.e.*, F_{ek}^{jk} (j, k = translation, rotation, or linear shear flow). For large separations, translation and shear flow over a stationary sphere each contribute about 25% of the total lift, while the coupling between them accounts for the balance of the total force; the rotational contribution (and its negative coupling with translation) are important only in the lubrication limit. This numerical analysis assumed that the zeta potentials of the sphere and plate were equal and that the double layer thickness was sufficiently small that the space charge density would not be perturbed by the flow. As was the case with the previous theories, the assumption of thin, equilibrium double layers is probably not valid for the lowest conductivity conditions in CHDF.

The theory of electrokinetic lifting was proposed to explain the anomalous migration of colloidal particles away from one wall of a slit-like flow cell apparatus (*15*). Bike *et al.* (*16*) conducted a similar series of experiments using 5 and 10 μm diameter polystyrene latex spheres in solutions of varying viscosity and conductivity. They observed the same phenomenon which supports the hypothesis that the flow rate dependent separation distances were due to an electrokinetic lift force acting between the particles and the wall. Despite the qualitative agreement, the predicted lift force is two or more orders of magnitude smaller than the one deduced experimentally (*14*). This discrepancy has been attributed to the presence of non-equilibrium electrical double layers and emphasizes the need to account for ion diffusion within the double layers. Using a colloidal particle scattering apparatus (microcollider), Wu *et al.* (*17*) have studied the electrokinetic lift force acting on a 5 μm diameter latex particle moving close to a glass surface in a wall shear flow. As was the case in the flow cell experiment, these investigators found that the equilibrium particle-wall separation distances increased with increasing shear rate and decreased with increasing glycerol solution conductivity. Two recently developed electrokinetic lift force theories (*18,19*) were applied to the interpretation of this data. The agreement between experimental and theoretical results was reasonably good, with the differences never exceeding a factor of two. Unlike any of the previous analyses, the new theories include the solution of the ion conservation (convective-diffusion) equation with the appropriate boundary conditions. In this more rigorous approach, the main contribution to the lift force comes from a nonzero normal hydrodynamic stress in the particle-wall gap caused by the tangential flow of ions in the double layer which results from the streaming potential built up outside of the double layer. Since both are based on the lubrication approximation, neither theory is applicable to the analysis of CHDF experiments. It is not yet known how the hydrodynamic stress will compare with the Maxwell stress when the particle radius and separation distances are of comparable magnitudes.

In the first part of this paper, separation factors corresponding to injections of monodisperse polystyrene latexes into a CHDF system are compared with simulated results. In the absence of a non-linear theory accounting for the polarization of thick double layers at arbitrary particle-wall separation distances, the Bike-Prieve result (*14*) was incorporated into our mathematical model in an attempt to improve the prediction of the separation factors at low eluant conductivity conditions. In this analysis, we have also assumed that the particle's small size relative to the capillary internal diameter will not invalidate the use of a force expression developed for linear shear flow conditions. Here, the ratio of particle to tube radii ranged from 0.004 to 0.04. The second section presents the results of some preliminary particle separation experiments. An important implication of an electrokinetic lift force in CHDF is that it may significantly reduce axial dispersion or instrumental band broadening phenomena. These undesirable effects are a serious obstacle to obtaining the particle size distribution, and are the main cause of imperfect resolution in CHDF and in other fractionation methods.

Experimental

The CHDF system used in this study is an experimental unit built in our laboratory and has been described previously (*7*). Briefly, eluant was pumped from a reservoir using a positive displacement type metering pump equipped with a pulse dampener, into a low dispersion injector, a fused silica capillary tube, and a multi-wavelength detector with a 15 μl fluid cell. The average internal diameter of the capillary tubes was calculated hydrodynamically using the Hagen-Poiseuille law which is justified by the low Reynolds number conditions in CHDF. The capillary tube length was determined from the minimum residence time required for full development of the radial concentration profile. In order to minimize dead volume effects associated with the low flow rates used, sample splitting at the entrance and eluant make-up at the exit of the capillary were used. A constant flow rate pump was used to deliver the make-up stream of distilled de-ionized (DDI) water to the detector cell. The detector output (fractogram) was monitored at various wavelengths in the UV range and interfaced with a microcomputer and chart recorder for data acquisition. The detector wavelength was chosen such that the strongest peak would be obtained at a minimum noise to signal ratio; monitoring several wavelengths was necessary for the turbidity ratio analysis. A commercial CHDF instrument (model CHDF-1100, Matec Applied Sciences) was also used to perform some of the particle separation experiments.

Eluant solutions of a non-ionic, water-soluble surfactant belonging to the polyethyleneoxide/alkyl ether class of compounds (BRIJ 35 SP, ICI Americas, Inc.) were used at concentrations above the critical micelle concentration. To remove the residual ionic contamination from this surfactant, a concentrated solution was treated by a mixed-bed ion exchange resin. The ionic strength was then adjusted with the addition of a small quantity of NaCl to the diluted solution. Sodium chloride and sodium lauryl sulfate (SLS) solutions were also used as eluants. The water used in this study was distilled and then cleaned by passing it through ion exchange columns and particulate filters. All eluants were filtered through a 0.22 μm membrane and an in-line filter installed upstream of the injection valve. For the low conductivity eluants, ionic strength was calculated from the measured specific conductance and pH, assuming the electrolyte to be composed of H^+, OH^-, Na^+, and Cl^- ions. A series of uniform polystyrene (PS) latexes manufactured by the Dow Chemical Company were used in this investigation. All latex standards were cleaned using an ion-exchange procedure to remove excess surfactant and ionic contaminants. Experiments were conducted at the lowest possible particle concentrations to avoid the influence of particle-particle interaction on the separation factor. Samples were sonicated before their injection in order to break up any aggregates that might have formed during the dilution with a nonionic surfactant solution.

Particle separation was quantified experimentally in terms of the separation factor, R_f, determined as the ratio of elution times associated with the marker and particle peaks. As in previous studies, sodium benzoate was used as the marker species. The marker and the monodisperse standards were injected separately into the CHDF system to prevent flocculation of the latexes. For a symmetrical fractogram (detector response), the average residence time of the particles corresponds to the peak maximum. In CHDF, the separation factor is always greater than unity, indicating that on an average basis, the particles travel through the microcapillary tube faster than the eluant. The maximum value of the separation factor is 2 which occurs when the particles are traveling exclusively at the centerline position in the tube. This situation never occurs in practice due to the retarding effect of the capillary wall on the particle velocity.

CHDF Model

When the radial distribution of particle concentration is fully developed, the average axial particle velocity is given by:

$$\langle v_{pz} \rangle = \frac{\int_0^{1-\lambda} v_{pz}(\beta,\lambda)\exp[-E(\beta,\lambda)]\beta\, d\beta}{\int_0^{1-\lambda} \exp[-E(\beta,\lambda)]\beta\, d\beta} \tag{1}$$

where

$$E(\beta,\lambda) = \frac{1}{kT}\left[\Phi(\beta,\lambda) - R_o \int_0^{\beta} \frac{v_{pr}(\beta',\lambda)}{M_\perp(\beta',\lambda)} d\beta' - R_o \int_0^{\beta} F_{ek}(\beta',\lambda)\, d\beta'\right] \tag{2}$$

represents the total interaction potential normalized with respect to thermal energy kT. The dimensionless parameters $\lambda = R_p/R_o$ and $\beta = r/R_o$ are the ratio of the particle to tube radii and the dimensionless radial position, respectively; the functions $\Phi(\beta,\lambda)$ and $F_{ek}(\beta,\lambda)$ represent the colloidal interaction potential and the electrokinetic lift force. The integral terms on the right-hand side of equation 2 are the inertial and electrokinetic "potentials", neither one of which is a true potential since the corresponding forces are non-conservative. Here, $v_{pr}(\beta,\lambda)$ is the radial velocity, describing the hydrodynamic response of the particle to the inertial force exerted by the fluid, and $M_\perp(\beta,\lambda)$ is the radial component of the hydrodynamic mobility dyadic; β' is a dummy variable of integration. The initial, or reference, point is the lower limit of each of these integrals which is taken as the tube axis. At this radial position, both the inertial and total electrokinetic forces are zero. Analytical expressions for the colloidal interaction potential, the local particle velocity in the axial direction, $v_{pz}(\beta,\lambda)$, and the inertial force, $F_h(\beta,\lambda) = v_{pr}/M_\perp$, have been published (*4,7*). The total electrokinetic lift force in cylindrical coordinates is given in Appendix I. Equation 1 is valid for $D_\infty t/R_o^2 >> 1$, a criteria which was satisfied for each of the CHDF experiments presented below. Here, the diffusivity of the sphere center in an unbounded fluid, D_∞, is given by the familiar Stokes-Einstein relation, and t is time (*e.g.*, the average residence time of the solute in the tube). Note that the hydrodynamic lift force (which the fluid exerts on particles suspended in laminar flow through a straight tube) causes the particles to migrate away from both the tube axis and the tube wall to a particular radial position. The electrokinetic lift force acting on the particles is always directed away from the wall towards the centerline of the tube.

The particle average velocity was calculated by weighting the local particle velocity with the concentration at that radial position (given by the probability density function, $\exp[-E(\beta,\lambda)]$) and integrating over the cross section of the tube accessible to the particle. The upper integration limit in equation 1 represents the difference of the tube and particle radii, accounting for the size exclusion layer adjacent to the wall. In order to calculate $\langle v_{pz} \rangle$ numerically, the tube radius was divided into 4 x 10^3 equidistant points and integrated using Simpson's Rule; the inertial and electrokinetic potentials were determined by applying a five-point quadrature formula at each integration point. The theoretical separation factor was then obtained by dividing the eluant mean velocity, v_m, into the previous result. Experimentally determined parameters (*e.g.*, eluant ionic strength, v_m, and D_p, the particle average diameter) together with specific material parameters were used in all of the simulations presented here.

Results and Discussion

Comparison of Theory and Experiment. Separation factors corresponding to a nonionic surfactant solution are shown in Figure 1 for five polystyrene standards ranging in size from 0.091 to 0.357 μm diameter. Salt (6.25 x 10^{-4} M NaCl) was added to the surfactant solution to increase the fluid conductivity to 3.0 μS/cm. Under these conditions, fluid inertia was insignificant in CHDF (*4,7*) and electrostatic repulsion between the particles and the tube wall could not explain the large deviation between the data and the theoretical result (curve C). The incorporation of the translational variant of electrokinetic lift (*12*) into our model increased the simulated particle velocities; however, the results underestimated the data for all particle sizes (curve B). When equations I2 and I3a,b were used to predict the separation factor-particle diameter behavior (curve A), quantitative agreement with the data was achieved.

Figure 2 shows the separation factors measured using an eluant of lower conductivity. In this case, salt was not added to the purified nonionic surfactant solution, reducing the estimated ionic strength by a factor of 10/3. As shown, all of the theoretical results (curves A-C) underestimated the data for each particle size. Under the lower conductivity condition, we expect that the assumption of thin, unperturbed electrical double layers is violated more severely than in the previous set of experiments. The use of the linear theory in the CHDF model is therefore questionable; nonetheless, the qualitative agreement with our data suggests that the rapid particle elution was due to an electrokinetic lift effect. A more general theory accounting for the polarization of thick double layers in Poiseuille flow would be appropriate for this case.

Figure 3 shows the effect of eluant ionic strength on the separation factor at three flow rate ranges using a 25.6 μm I.D. capillary tube and the 0.234 μm standard. This tube diameter is 2.5 to 3 times *larger* than those used in standard practice, which allowed a wider range of flow rates than was possible with the 9.0 μm I.D. tube. Curve D represents the theoretical result when an electrokinetic lift force expression was not included in the R_f simulation, whereas curves A-C correspond to the incorporation of equations I2 and I3a,b into the model. As shown, the measured separation factors were quite sensitive to flow rate below an ionic strength of 10^{-5} M, despite the rather small particle to tube diameter ratio of 0.009. Although R_f exhibited a strong flow dependence using the 4 x 10^{-6} M eluant, it remained nearly constant for all flow rates at an ionic strength of 10^{-3} M. Electrostatic repulsion together with fluid inertia could not explain this unusual behavior; furthermore, the highest separation factor observed ($R_f \approx 1.8$) indicated that the non-inertial lift force

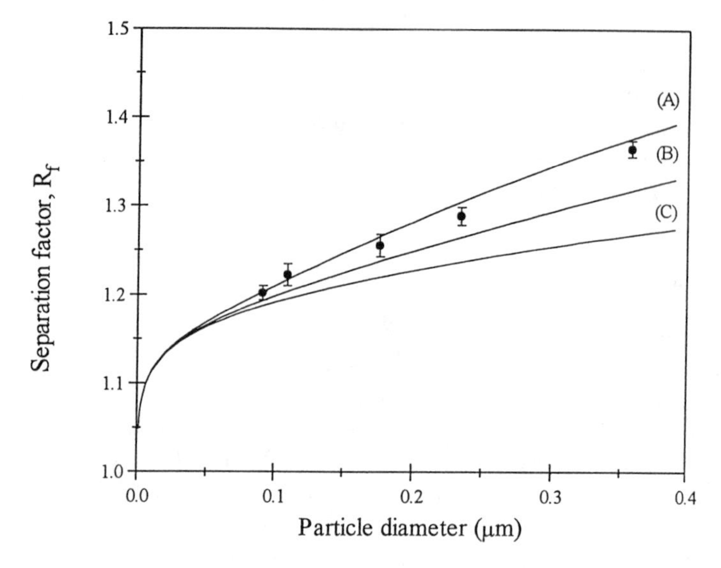

Figure 1: Separation factor–particle diameter behavior for 0.01% (w/w) BRIJ 35 SP eluant. Solid lines, R_f simulation: *total* F_{ek} (A); translational variant of F_{ek} (B); F_{ek} not included (C). Eluant average velocity, 2.1 cm/s; estimated ionic strength, 2.3 x 10^{-5} M; K = 3.0 μS/cm; R_o = 4.5 μm; ζ = –60 mV.

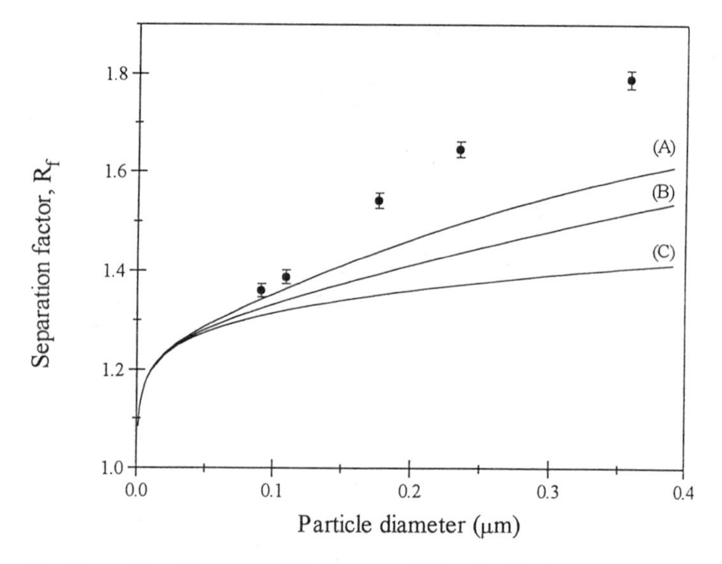

Figure 2: Separation factor-particle diameter behavior for 0.02% (w/w) BRIJ 35 SP eluant. Solid lines, R_f simulation: *total* F_{ek} (A); translational variant of F_{ek} (B); F_{ek} not included (C). Eluant average velocity, 1.6 cm/s; estimated ionic strength, 7.0 x 10^{-6} M; K = 1.0 μS/cm; R_o = 4.5 μm; ζ = –60 mV.

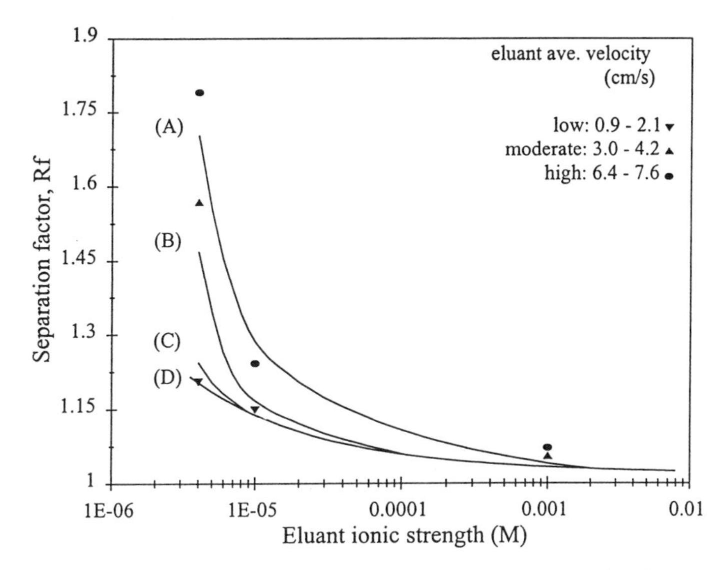

Figure 3: Comparison between predicted and measured separation factor-eluant ionic strength behavior for three flow rate ranges. Solid lines, R_f simulation: v_m = 7.6 cm/s (A); v_m = 3.0 cm/s (B); v_m = 0.9 cm/s (C); F_{ek} not included (D); R_o = 12.8 μm.

resulted in particles migrating laterally to a position about 0.3 tube radii from the axis. This radial position is far from the equilibrium point of 0.6 radii ("tubular pinch effect") observed for neutrally buoyant spheres (*8*). An electrokinetic lift effect is entirely consistent with these trends; thus, for large flow rates and low eluant conductivities, the separation factor was larger than that which was observed for smaller v_m/K ratios. The analysis of turbidity ratios verified that all fractogram signals, including those corresponding to the surfactant-free eluant, represented single particles and not agglomerated species. In all of these experiments, the ion Peclet number was $O(10^{-1})$ which is not too far from $Pe_{ion} << 1$ where the theoretical result given by equations I1 and I2 applies.

The effect of flow rate (expressed as eluant mean velocity) on the separation factor is shown in Figure 4 for four latex standards ranging from 0.109 to 0.357 μm in diameter. A 10^{-5} M sodium chloride solution (K = 1.6 μS/cm, pH = 6.0) was pumped through the 25.6 μm I.D. capillary tube at four flow rates. As shown, the eluant flow rate begins to affect the value of the separation factor at velocities above about 1 cm/s in this system. At the lowest flow rate, the model (solid lines) could predict the results without the use of an electrokinetic lift force expression for all but the largest particle size. Qualitative agreement with the experimental results was observed for the higher flow rates using equations I2 and I3a,b in the R_f simulation; however, all of the theoretical results overestimated the data at the highest flow rate. We should be able to improve the fit for large v_m by using a more general electrokinetic lift force expression in the CHDF model. When the electrokinetic lift

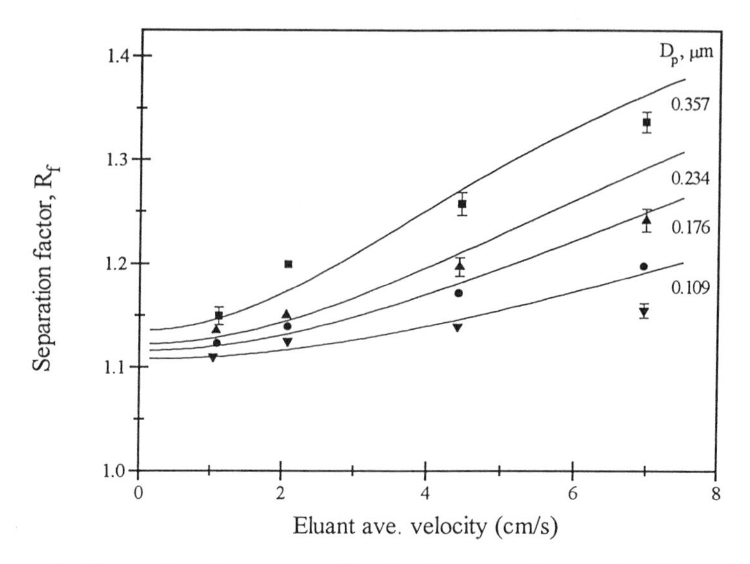

Figure 4: Comparison between predicted and measured separation factor-eluant mean velocity behavior for four particle sizes. Eluant, NaCl solution; estimated ionic strength, 10^{-5} M; $K = 1.6$ μS/cm; $R_o = 12.8$ μm.

force was omitted from the model, the theoretical separation factors were constant for all flow rates (*i.e.*, $R_f = 1.104$, 1.113, 1.119, and 1.131 for the 0.109, 0.176, 0.234, and 0.357 μm particles, respectively).

The strong zeta potential dependency indicated by equation 12 requires the accurate determination of ζ_p and ζ_w. The particle zeta potential for each latex was estimated from measurements of the electrophoretic mobilities of the polystyrene latex standards in the presence of 10^{-5} M SLS surfactant in a 25 μm internal diameter capillary tube (*20*). Ottewill's numerical tables (*21*), which take into account the electrophoretic retardation and double-layer relaxation effects (*22*), were used to determine the particle zeta potentials presented in Table I. This approach is quite valid for 1-1 electrolyte systems and zeta potentials less than 100 mV.

Table I. Particle Zeta Potentials Calculated Using 10^{-5} M SLS and a 25 μm I.D. Capillary Tube

Latex Designation	*Diameter (μm)*	*Zeta Potential (mV)*
LS-1044-E	0.109	−64.0
LS-1045-E	0.176	−49.0
LS-1047-E	0.234	−42.3
LS-1010-E	0.357	−36.8

Electroosmotic velocities were also measured for various SLS and NaCl concentrations using a 25 μm I.D. capillary (*20*). From these values, we estimated the zeta potential of the fused silica wall to be −90 mV at an eluant ionic strength of 10^{-5} M and applied electrical field strength of 286 V/cm. In this case, ζ_w was calculated using the classical Smoluchowski-Helmholtz equation which is valid for large κR_o. Here, κ^{-1} is the Debye length or characteristic double-layer thickness. Note that the calculation of F_{ek} was based on the case of equal zeta potentials, so we have assumed an average value for this parameter.

The next set of experiments involved a latex standard which was large enough to give rise to significant hydrodynamic effects in CHDF. The critical eluant velocity for this size was 0.8 cm/s as determined by the product of the particle Reynolds and Peclet numbers (*4*). Above this value, the simulated separation factor (in the absence of electrokinetic lift) changed appreciably with v_m. Figure 5 shows the effect of flow rate on R_f for the 0.794 μm diameter particles (λ = 0.031) at a relatively high eluant ionic strength. A 1.5 x 10^{-3} M SLS solution (K = 103 μS/cm) was pumped through the 25.6 μm I.D. capillary tube at velocities ranging from 0.9 to 7.2 cm/s. As shown, the separation factor increased from 1.10 to a plateau value of about 1.23 as the eluant flow rate was raised. The solid lines represent the theoretical result for three situations: both F_{ek} and F_h included in the model (curve A); only F_h included (curve B); and only F_{ek} included (curve C). At the high eluant conductivity, equations I2 and I3a,b did not influence the simulation (curves A and B represent the same result), while curve C shows a constant separation factor of 1.07. These results were consistent with the inertial lift force mechanism.

The experiments were repeated after replacing the surfactant solution with DDI water (K = 0.66 μS/cm). Figure 6 shows that the experimental R_f increased sharply when v_m exceeded 1 cm/s; at higher velocities, the separation factors remained nearly constant ($R_f \approx$ 1.85). This limiting value probably reflects the particles' near centerline positions in the capillary tube, as well as significant particle-particle interactions. Under these conditions, the CHDF model shows that an inertial force would offset the effect of electrokinetic lift, tending to push the particles towards the capillary wall (note the negative slope of curve B, and that curve C is above curve A). The data indicates that $F_{ek} >> F_h$ at the higher flow rates in this set of experiments. All of the models (curves A-C) underpredicted the data presented in Figure 6, although they showed the right qualitative features.

As noted above, the latex cleaning was carried out to remove excess surfactant and ionic contaminants from the PS standards used here. This procedure ensured that the fluid conductivity of injected dispersion was similar to that of the eluant being pumped through the capillary tube. It was suspected that the introduction of uncleaned dispersions into the capillary would result in smaller separation factors. A simple calculation was made to estimate the retarding effect of the initially high conductivity environment in which the particles would travel. Figure 7 shows the predicted value of R_f as a function of particle diameter for two eluant conductivities. Curves A and B correspond to K = 1.0 μS/cm (estimated ionic strength, 0.01 mM) and K = 560 μS/cm (estimated ionic strength, 10.0 mM), respectively. The higher conductivity was chosen to represent that of a diluted,

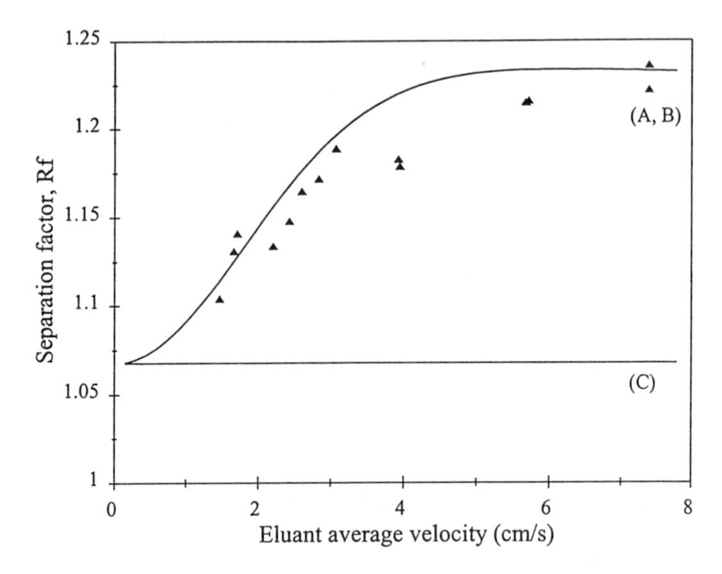

Figure 5: Comparison between predicted and measured separation factor-eluant mean velocity behavior for a 0.794 μm diameter PS standard. Solid lines: F_{ek} and F_h included in the simulation (A); F_{ek} not included (B); F_h not included (C). Eluant, 1.5 x 10^{-3} M SLS; $K = 103.2$ μS/cm; $R_o = 12.8$ μm; $\zeta = -60$ mV.

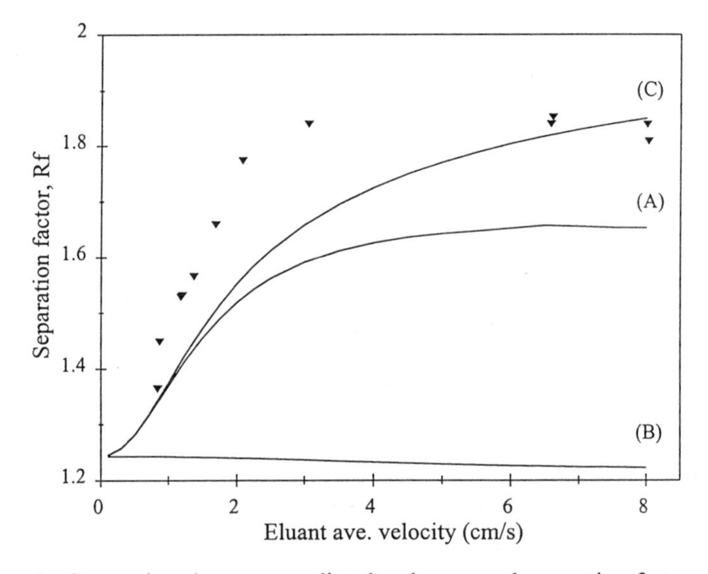

Figure 6: Comparison between predicted and measured separation factor-eluant mean velocity behavior for a 0.794 μm diameter PS standard. Solid lines: F_{ek} and F_h included in the simulation (A); F_{ek} not included (B); F_h not included (C). Eluant, DDI water; estimated ionic strength, 4 x 10^{-6} M; $K = 0.66$ μS/cm; $R_o =$ 12.8 μm; $\zeta = -60$ mV.

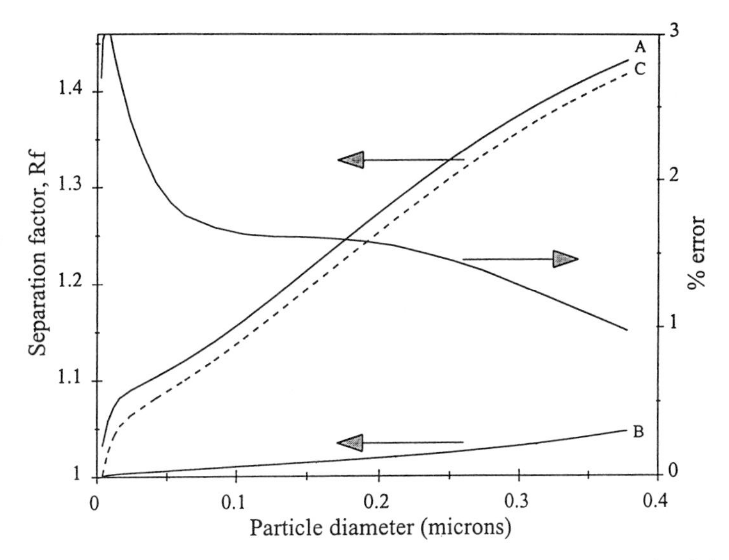

Figure 7: Theoretical calculation showing the effect of sample conductivity on the separation factor as a function of particle diameter. Eluant conductivity: 1.0 μS/cm (A); 560 μS/cm (B). Dashed line (C) based on equations II1 and II2 with δl = 2.8 cm. Capillary length, 1220 cm; eluant average velocity, 5 cm/s; R_o = 12.8 μm; $\zeta = -60$ mV.

uncleaned sample. The intermediate (dashed) curve is the result expected when a slug of finite length (2.8 cm) is introduced via the 5 μl sample loop using a flow rate of 0.53 ml/min. The analytical expression used here is derived in Appendix II. As shown in Figure 7, the deviation between the low conductivity result (Curve A) and that obtained using equations II1 and II2 was typically 1 to 2% in the particle size range of interest, suggesting that sample cleaning in a practical analysis may not be necessary. Here, the relative error was expressed as $100(1 - R_f/R_f^L)$. The calculation also implies that the use of long capillary tubes may help to ensure reproducible CHDF operation at very low eluant conductivity. An absolute particle sizing technique would eliminate any possible problems associated with a calibration-based procedure (*6*).

Particle Separation Experiments. Table II summarizes nine separation experiments conducted at various eluant conductivities using synthetic mixtures of the PS latex standards. Three of the particle separations were obtained using a commercial CHDF instrument.

The fractograms presented in Figure 8 correspond to the experiments involving the larger diameter microcapillary. The vertical scale labeled *Detector signal* is common to all data, with the three fractograms stacked for clarity. At the highest eluant conductivity, typical of that specified in standard practice, no peak separation was achieved (fractogram A), although the presence of two populations is suggested by the signal asymmetry. When the experiment was repeated using the 1.4

Table II. Summary of Particle Separation Experiments

Fractogram Designation	*Particle Diameters, 50:50 mixture (μm)*	*Particle Diameter Ratio*	R_o *(μm)*	*K (μS/cm)*
A	0.357-0.091	3.9	12.8	13.4
B				1.4
C				0.66
A	0.176-0.109	1.6	*	14.2
B			4.8	1.7
A	0.234-0.176	1.3	*	14.2
B			4.8	1.7
A	0.357-0.234	1.5	*	14.2
B			4.8	1.7

*Matec CHDF-1100

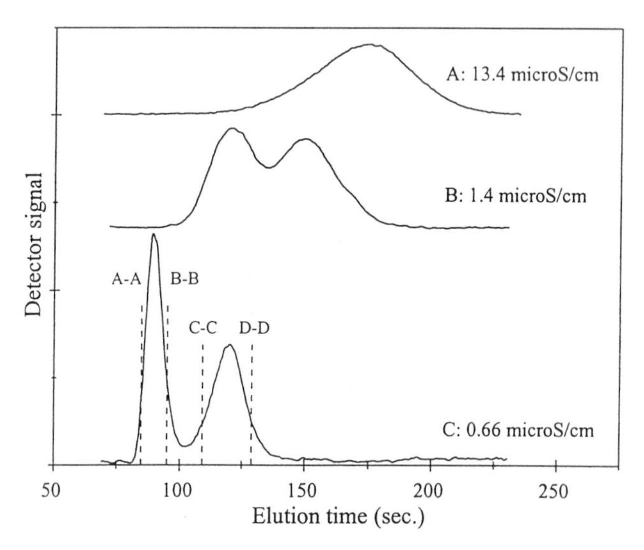

Figure 8: Detector response-elution time data for a 50:50 mixture of PS particles (D_p: 0.357–0.091 μm). Eluant: 2.0 x 10^{-4} M SLS (A); NaCl solution (B); DDI water (C). Eluant average velocity, 7.3 cm/s; R_o = 12.8 μm. vertical lines A-A, B-B, *etc.*, denote beginning/end of relatively constant turbidity ratio.

μS/cm eluant, peaks corresponding to each particle size were observed, but the signals were not fully resolved (fractogram B). Here, the difference between the peak elution times was 30 seconds. Using the DDI water eluant, the bimodal mixture was resolved almost completely within 140 seconds (fractogram C); the peak elution time difference was the same as that observed in fractogram B. Note that this tube internal diameter was 2.5 to 3 times *larger* than that of typical commercial instruments which are capable of completely resolving the same mixture of particles within 6 minutes using a 14 μS/cm eluant. However, the peak width corresponding to the 0.357 μm particle size was actually much narrower in the larger bore tube indicating significant particle focusing at the lower eluant conductivity.

The turbidity ratio-elution time data presented in Figure 9 corresponds to fractogram C. The two solid horizontal lines represent the theoretical turbidity ratios at wavelengths of 220 and 254 nm for particle diameters of 0.091 and 0.357 μm as determined using Mie theory (*6*). As shown, the portions of the fractogram occurring between the vertical lines denoted A-A and B-B, as well as between C-C and D-D, correspond to periods of relatively constant turbidity ratio. For elution times from 87 to 97 seconds, an average turbidity ratio of 0.688 was recorded, as compared with the theoretical result of 0.672 for the larger particle size. An average turbidity ratio of 6.14 compares well with the theoretical value of 6.21 for the 0.091 μm diameter particles (112 to 131 seconds). This analysis verifies the separation of the two particle populations; no evidence of particle agglomeration was observed in any of the experiments.

The results of the previous experiments suggest that greater separation efficiencies may be realized by pumping very low conductivity eluant through a smaller diameter microcapillary tube. Figures 10-12 present the fractograms obtained using a 9.6 μm I.D. tube with a 1.7 μS/cm eluant (fractogram B), and the commercial CHDF system (fractogram A). Within 325 seconds, partial resolution of the 0.176-0.109 μm and 0.357-0.234 μm mixtures was obtained using the 14.2 μS/cm eluant. Nearly complete resolution of both mixtures was achieved within 175 and 150 seconds, respectively, using the nonionic surfactant solution. The 0.234-0.176 μm mixture, representing the smallest particle diameter ratio of these experiments, was partially resolved at the low eluant conductivity (fractogram B, Figure 11). The peak asymmetry observed here indicates the possibility of particle overloading (the total injection concentration was 2 wt.%) and we suspect that the peaks will become fully resolved using lower sample concentrations. A paper detailing our investigation of sample concentration effects using low conductivity eluants in CHDF is in preparation. When the same mixture was injected into the commercial CHDF, the peak elution time difference doubled; however, the resolution of the two size populations was greatly diminished by the effect of band broadening as shown in fractogram A, Figure 11.

It is important to note that our observations cannot be explained by the electrostatic repulsion between the particles and the tube wall. Rather, the higher resolution obtained using very low conductivity eluants in CHDF has been attributed to the electrokinetic lift on the charged particles as they move along the wall. To emphasize this point in a quantitative manner, we introduce the resolution factor R_s which is defined as:

$$R_s = \frac{\langle t_2 \rangle - \langle t_1 \rangle}{2\left(\sigma_{T,t_1} + \sigma_{T,t_2}\right)} \tag{3}$$

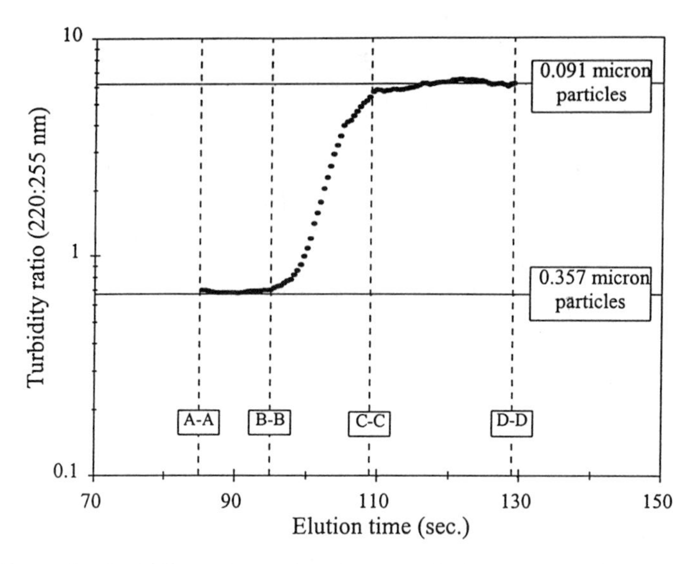

Figure 9: Turbidity ratio-elution time data for fractogram C (Figure 8).

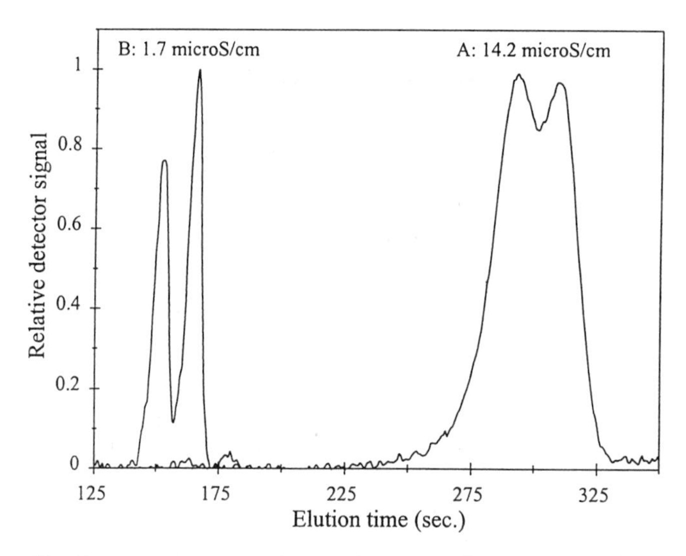

Figure 10: Detector response-elution time data for a 50:50 mixture of PS particles (D_p: 0.176–0.109 μm). Fractogram A: eluant, 0.02% (w/w) BRIJ 35 SP; eluant mean velocity, 2.1 cm/s. Fractogram B: eluant, Matec GR-500 solution.

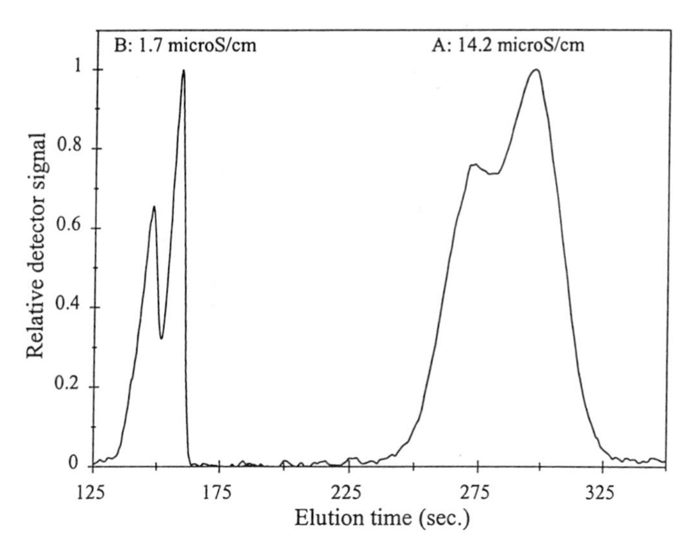

Figure 11: Detector response-elution time data for a 50:50 mixture of PS particles (D_p: 0.234–0.176 μm). Fractogram A: eluant, 0.02% (w/w) BRIJ 35 SP; eluant mean velocity, 2.0 cm/s. Fractogram B: eluant, Matec GR-500 solution.

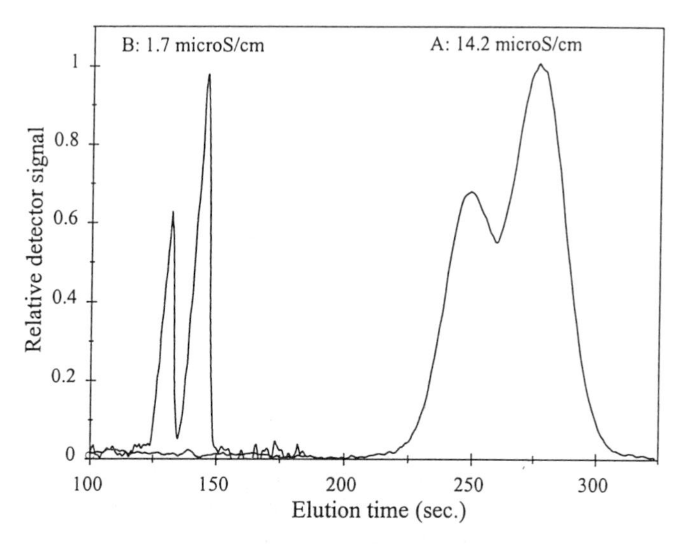

Figure 12: Detector response-elution time data for a 50:50 mixture of PS particles (D_p: 0.357–0.234 μm). Fractogram A: eluant, 0.02% (w/w) BRIJ 35 SP; eluant mean velocity, 2.0 cm/s. Fractogram B: eluant, Matec GR-500 solution.

where $\langle t_i \rangle$ is the average residence time, and $\sigma_{T,ti}$ is the total, time-based standard deviation of the ith size particle population (*23*). In terms of the separation factor and the mean axial dispersivity D^* (*5*), this parameter may be expressed theoretically as:

$$R_s = \frac{\sqrt{v_m L}}{2\sqrt{2}} \frac{\left[R_{f,2}^{-1} - R_{f,1}^{-1}\right]}{\left[R_{f,1}^{-3/2}\sqrt{D_1^*} + R_{f,2}^{-3/2}\sqrt{D_2^*}\right]} \tag{4}$$

where L is the length of the capillary tube. We have neglected the contribution from the dispersion due to non-idealities since the total variance, $\sigma_{T,t}^2$, greatly exceeded the fractogram variance due to the non-ideal mixing at the injection valve and the detector cell. Thus, the axial dispersion coefficient was related to the total variance by $D^* = \sigma_{T,t}^2 \langle v_{pz} \rangle^2 / 2t$. An R_s value of 1.5 indicates that a pair of Gaussian zones of equal σ_T would be completely separated (baseline resolution).

Figure 13 presents the influence of eluant conductivity on R_s for a binary mixture of submicron size particles using the 25.6 μm internal diameter tube. The solid lines represent the theoretical results obtained using equations 1, 2, and 4, together with the appropriate expression for D^* given by Silebi and DosRamos (*5*); the symbols are based on experimental measurements of $\langle t_i \rangle$ and $\sigma_{T,ti}$ obtained using a surfactant-free eluant. In the absence of electrokinetic lift, the predicted R_s values increased when the ionic strength of the eluant decreased because the reduced counterion concentration enhances the electrostatic repulsion between the particles and the wall. Fluid inertia caused the reversal of this trend when v_m increased to 7.2 cm/s; at higher flow rates, the inertial force will dominate the colloidal force of interaction causing R_s to increase gradually for both ionic strength conditions. At small eluant average velocities, R_s decreased because the degree of axial dispersion increases sharply as v_m increases until the hydrodynamic force exerted on the particles by the fluid becomes significant. When the electrokinetic lift force (equations I2 and I3a,b) was included in the simulation, a minimum in the R_s-v_m profile was observed at 0.9 cm/s. Above this eluant average velocity, the predicted resolution increased monotonically with increasing flow rate. Qualitative agreement with the data was observed in this case.

The particle transport behavior observed here bears some similarity to that which has been proposed for the hyperlayer field-flow fractionation technique (*24*). In that elution separation method, an *applied* field (*e.g.*, sedimentation) is oriented in a direction normal to the flow axis forcing particles against one wall of a narrow channel, while a background density gradient keeps the particle layer away from the wall. The steady-state layer which is formed centers about a coordinate position at which the particle density is matched by the density of the background fluid. Because the zones corresponding to different size particles displace differentially along the flow axis, hyperlayer FFF should, in principle, improve the degree of resolution realized by conventional FFF methods. In the low conductivity CHDF process, we suspect that an induced streaming potential in the particle-wall gap produces a significant energy gradient which focuses particles towards the center of the tube.

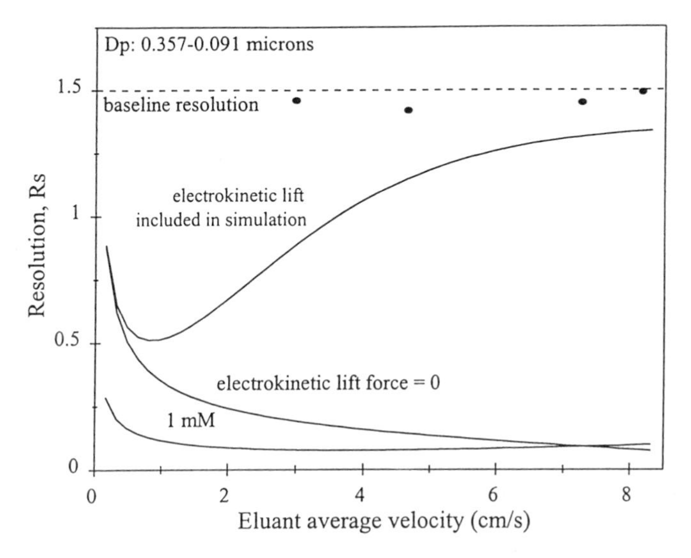

Figure 13: Comparison between theoretical and experimental results of the resolution as a function of eluant average velocity. Capillary length, 1220 cm; ionic strength, 3 x 10^{-6} M; $K = 0.6$ μS/cm; $R_o = 12.8$ μm.

Since the gradient profile depends upon particle size (*e.g.*, $F_{ek}^{t} \propto R_p^{2}$), so too does the boundary defining the corresponding core region in which the particles tend to reside. The particle exclusion zone (in which $E(\beta,\lambda)$ exceeds some multiple of kT energy) would extend further towards the tube axis for larger λ, resulting in the relative displacement of different size populations. Thus the use of very low conductivity eluants in CHDF enhances resolution well above the values possible at higher electrolyte concentrations. The combined effect of *two* lift forces, *i.e.*, F_{ek} and F_h, may result in the focusing of particles to annular regions near the tube axis. The position of these radial "layers" would be centered about those values of β corresponding to the minimum in the respective particle-wall interaction energy profiles.

Conclusions

At very low eluant conductivity ($K < 10$ μS/cm), the average residence time and axial dispersion of colloidal particles suspended in laminar flow through a microcapillary tube appear to be strongly affected by a transverse force of electrokinetic origin. The use of an expression representing the *total* electrokinetic lift force in a mathematical model developed from first principles improved the prediction of separation factors in CHDF, as compared with the case where only the translational variant of F_{ek} was considered. At a conductivity of 3 μS/cm, for example, quantitative agreement with the experimental data was observed; below this value, the model tended to underestimate the data. A more general theory of electrokinetic lift which is valid for thick double layers at arbitrary particle-wall separation and for arbitrary Pe_{ion} is needed to improve the prediction of separation factors in CHDF. Particle separation experiments indicated that the use of very low conductivity eluants in CHDF will provide better resolution of colloidal particles than that presently obtained using higher electrolyte concentrations. Sample concentration should be reduced to avoid particle overloading effects associated with the very low conductivity eluants.

Appendix I

The electrokinetic lift acting on a charged sphere, freely rotating and translating in a slow linear shear flow along a plane wall was determined by Bike and Prieve (*14*). This theoretical result applies to arbitrary particle-wall separation distances provided that $\kappa\delta \gg 1$ and $\kappa R_p \gg 1$. Here, κ^{-1} and δ are the Debye length and gap thickness, respectively. The total lift force is the linear superposition of six component forces including three cross terms (arising from the nonlinearity of the electrical stress):

$$F_z = F_z^{\mathrm{t}} + F_z^{\mathrm{r}} + F_z^{\mathrm{S}} + 2\left(F_z^{\mathrm{tr}} + F_z^{\mathrm{tS}} + F_z^{\mathrm{rS}}\right) \tag{I1}$$

with the superscripts *t*, *r*, and *S* denoting translation, rotation, and linear shear flow. We have chosen the notation F_{ek} to represent the total electrokinetic lift force in our analysis. In cylindrical coordinates, the force is given by:

$$F_{ek}(\beta,\lambda) = \pi\left(\varepsilon/4\pi\right)^3\left(\frac{G(\beta)\zeta}{K}\right)^2 \overline{F}_{ek}(\beta,\lambda) \tag{I2}$$

Here, $G(\beta,\lambda) = 4v_m\beta/R_o$ is the local fluid shear rate for tube flow, K is the fluid conductivity, ζ is the zeta potential ($\zeta_{particle} = \zeta_{wall}$), ε is the eluant dielectric constant, and v_m is the eluant mean velocity. Because the streaming potential profile around a sphere near a wall was expressed as an infinite series of Legendre polynomials, the complete analysis requires the numerical evaluation of each of the terms appearing on the right-hand side of equation I1. To simplify our calculations, the total (nondimensional) electrokinetic lift force which appears in Bike and Prieve's Figure 5 (*14*) was replotted as $\ln \overline{F}_{ek}$ versus $\ln(\delta/R_p)$ and fitted with the following 4th-order polynomial:

$$\overline{F}_{ek}(\beta,\lambda) = \exp\left[C_1 + C_2\, y(\beta,\lambda) + C_3\, y(\beta,\lambda)^2 + C_4\, y(\beta,\lambda)^3 + C_5\, y(\beta,\lambda)^4\right] \tag{I3a}$$

$$y(\beta,\lambda) = \ln\left[\tfrac{1}{\lambda}\left(1-\beta\right)-1\right] \tag{I3b}$$

where the coefficients C_i (i = 1-5) are 1.303, –0.9059, –0.0568, –0.0739, and 0.0109, respectively. Here, the dimensionless particle-wall separation distance, δ/R_p, was expressed in cylindrical coordinates as $\lambda^{-1}(1 - \beta) - 1$.

We have effectively equated $G(\beta)$ with S, the shear rate for *linear* shear flow, and assumed that λ is small enough that G is locally constant at $\beta \pm \lambda$. Note that the first, fourth and fifth terms appearing on the right-hand side of equation I1 were normalized with the reduced translational speed calculated by Goldman *et al.* (*25*) which is appropriate for an electrically neutral sphere and a flat plate. Our calculated particle average velocities are thus understated by a small factor since the appropriate reduced particle velocity for tube flow (and thin double layers) is $v_{pz}(\beta,\lambda)/GR_p$. Also note that the Bike-Prieve analysis requires that the double layer thickness be sufficiently small so that the space charge density will not be perturbed significantly by the flow. In terms of dimensionless numbers, this constraint may be expressed as:

$$Pe_{ion} << 1; \quad Pe_{ion} \equiv \frac{G}{\kappa^2 \omega\, kT} \tag{I4}$$

where $\omega = (6\pi\mu a_i)^{-1}$ is the ionic mobility based on a_i, the mean ion radius, and μ is the fluid viscosity. The characteristic Peclet number represents the strength of the convective forces on the ions, leading to the distortion of the electrical double layer, compared with Brownian forces which tend to restore its equilibrium shape. Here, the characteristic length scale is κ^{-1}, and $\kappa^{-1}G$ is a typical velocity.

Appendix II

Consider a finite, non-dispersing slug of length δl moving in plug flow through a capillary tube of length L^T which is filled with a fluid of dissimilar conductivity. All fluid travels through the tube at a mean eluant velocity v_m. The rate of transport of the particles dispersed in the slug relative to the fluid surrounding them is expressed as the separation factor R_f^H. We have assumed that the particle transport behavior is the same as that which would occur in the three-dimensional Poiseuille flow. Provided that the capillary is sufficiently long and $R_f^H > 1$, the particles will eventually pass from the high conductivity region into the lower conductivity eluant. In this part of the tube, the particles travel at a larger relative average velocity, R_f^L, owing to, among other factors, an electrokinetic lift force. Let L^L and L^H denote the average lengths of the capillary across which the particles travel in high and low conductivity fluid, respectively. Here, the particle axial displacements are additive so that $L^L + L^H = L^T$. At time $t = 0$, the slug containing the particles is centered at the entrance of the capillary, $z = 0$. In the absence of axial dispersion and for $t > 0$, L^H is given by:

$$L^H = \frac{\delta l}{2}\left[1-\left(R_f{}^H\right)^{-1}\right]^{-1} \tag{II1}$$

Using the standard definition of R_f, the overall separation factor becomes the weighted harmonic mean of R_f^L and R_f^H:

$$R_f = \left[\frac{\theta}{R_f{}^L} + \frac{1-\theta}{R_f{}^H}\right]^{-1} \tag{II2}$$

Here, $\theta \equiv L^L/L^T$ is the fractional tube length over which particles travel in the low conductivity environment. Knowing the sample injector flow rate, the sample loop volume, and v_m, one can calculate the slug length at the entrance to the tube. Theoretical values of R_f corresponding to a high and low fluid conductivity, together with equations II1 and II2, can then be used to estimate the effect of sample conductivity on the overall separation factor. Of course, the mean axial dispersivity (band broadening phenomenon) will also be affected by a conductivity difference or gradient; however, no attempt was made to estimate this phenomenological coefficient.

Literature Cited

1. Silebi, C. A.; DosRamos, J. G. *J. Colloid Interface Sci.* **1989**, *130*, 14.
2. DosRamos, J. G.; Silebi, C. A. *J. Colloid Interface Sci.* **1990**, *135*, 165.
3. DosRamos, J. G.; Silebi, C. A. In *Particle Size Distribution II: Assessment and Characterization*; Provder, T., Ed.; ACS Symp. Series No. 472; American Chemical Society: Washington, DC, 1991; pp 292-307.

4. DosRamos, J. G.; Silebi, C. A. *J. Colloid Interface Sci.* **1989**, *133*, 302.
5. Silebi, C. A.; DosRamos, J. G. *AIChE J.* **1989**, *35*, 1351.
6. Venkatesan, J. *Particle Size Sensor Design and Application* Ph.D. Dissertation, Lehigh University, 1992.
7. Hollingsworth, A. D.; Silebi, C. A. *Langmuir* **1996**, *12*, 613.
8. Segré, G.; Silberberg, A. *Nature (London)* **1961**, *189*, 209.
9. Venkatesan, J. In *Particle Size Distribution II: Assessment and Characterization;* Provder, T., Ed.; ACS Symp. Series No. 472; American Chemical Society: Washington, DC, 1991; pp 279-291.
10. Prieve, D. C.; Bike, S. G. *Chem. Eng. Comm.* **1987**, *55*, 149.
11. Bike, S. G.; Prieve, D. C. *J. Colloid Interface Sci.* **1990**, *136*, 95.
12. Bike, S. G.; Prieve, D. C. *J. Colloid Interface Sci.* **1992**, *154*, 87.
13. van de Ven, T. G. M.; Warszynski, P.; Dukhin, S. S. *J. Colloid Interface Sci.* **1993**, *157*, 328.
14. Bike, S. G.; Prieve, D. C. *J. Colloid Interface Sci.* **1995**, *175*, 422.
15. Alexander, B. M.; Prieve, D. C. *Langmuir* **1987**, *3*, 788.
16. Bike, S. G.; Lazarro, L.; Prieve, D. C. *J. Colloid Interface Sci.* **1995**, *175*, 411.
17. Wu, X.; Warszynski, P.; van de Ven, T. G. M. *J. Colloid Interface Sci.* **1996**, *180*, 61.
18. Warszynski, P.; van de Ven, T. G. M., McGill University, unpublished manuscript.
19. Cox, R. G. *J. Fluid Mech.*, in press.
20. Hlatshwayo, A. B.; *Analytical Separation of Colloidal Particles Using Capillary Electrophoresis* Ph.D. Dissertation, Lehigh University, 1994.
21. Ottewill, R. H.; Shaw, J. N. *J. Electroanal. Chem.* **1972**, 37, 133.
22. Wiersema, P. H.; Loeb, A. L.; Overbeek, J. Th. G. *J. Colloid Interface Sci.* **1966**, *22*, 78.
23. Giddings, J. C. *Unified Separation Science*; J. Wiley & Sons, Inc.: New York, NY, 1991; pp 101-105.
24. Giddings, J. C. *Sep. Sci. Technol.* **1983**, *18*, 765.
25. Goldman, A. J.; Cox, R. G.; Brenner, H. *Chem. Eng. Sci.* **1967**, *22*, 653.

Chapter 18

Particle Size Characterization During Emulsion Polymerization

J. Venkatesan and Cesar A. Silebi

Department of Chemical Engineering and Emulsion Polymers Institute, Lehigh University, Bethlehem, PA 18015–4735

Particle size distribution (PSD) information in an emulsion polymerization reactor is critical to understanding the complex kinetics therein and to develop control strategies to produce latexes with pre-specified PSDs. These important benefits of implementing particle size as a quality control parameter in the manufacture of latexes has resulted in extensive research in recent years. The ability to precisely monitor the PSD is critically dependent on the sensor technology employed. Accordingly, the present study demonstrates the use of Capillary Hydrodynamic Fractionation (CHDF) to monitor an emulsion polymerization process using styrene as the model system. In the first part of the paper attention is focussed on implementing a *hybrid* sensor by coupling the high resolution fractionation capability of CHDF with turbidimetric methods. This is achieved by passing the fractionated species through a photo-diode array detector which measures the turbidity at several wavelengths, instantaneously, thereby enabling the use of turbidimetric methods for the final particle size analysis. The distinct advantages offered by this absolute procedure to measure PSDs is found to be particularly advantageous to the monitoring scheme. The sensor is then used to monitor an emulsion polymerization process and specifically focus on two experiments. The first experiment shows the validity of the sensor as an accurate, sensitive and robust tool to track uni-modal growth patterns. The presence of monomer swollen particles does not hinder the particle size measurements. The second experiment demonstrates the sensitivity of the sensor to changes in the monomer feed rate. The results of these experiments provide a sound foundation to monitor a more complex series of polymerizations in the future.

Characterizing, monitoring and controlling particle growth in an emulsion polymerization process enable the user to gain a handle on an important parameter that defines the molecular architecture of the end product; namely, the Particle Size Distribution (PSD). Important application properties of a latex such as opacity, colorability, viscosity, film forming ability etc. are intimately tied to the particle growth process and the final size distribution achieved. However, despite these far reaching benefits offered by potential solutions to this problem of particle size control, the literature is surprisingly thin on research conducted in this area. A careful examination of the literature suggests that a crucial factor that has prevented the implementation of effective control of PSD in a reactor is the lack of a sensor technology that is accurate and reproducible to make PSD information available fast enough to enable appropriate control action.

Dynamic Light Scattering. Recently Kourti et al. [1,2] have demonstrated the utility of using dynamic light scattering (DLS) to measure on-line particle growth during the production of poly(vinyl acetate) latex in a continuous stirred tand reactor. The data showed good reproducibility in the measured average diameters (intensity average and volume average) derived from turbidimetric measurements. This technique is found to work very well for monitoring reactions where growth of one size population is of interest. However, size distribution is assessed using the difference between the weight and intensity average diameters [3]. If these vary significantly, the distribution can be described as relatively broad; but the exact nature of the distribution viz. partial or complete bimodal, broadly distributed etc. cannot be quantitatively described.

Fiber-Optic Dynamic Light Scattering (FODLS). Dynamic light scattering (DLS) measurements on concentrated samples, called Fiber Optic Dynamic Light Scattering (FODLS) [4,5] has an edge over regular dynamic light scattering techniques because it eliminates the need for dilution thereby making this technique more robust for industrial reactors. Additionally, FODLS systems are more compact and have cheaper optical systems. They operate on a similar principle to the DLS with the exception that the DLS measurements are made on the backscattered light. This technique is susceptible to erroneous results from particle-particle interactions in a concentrated dispersion and the fiber optic probe fouling due to an impermeable coating of impurities. In general, the interpretation of the results from FODLS is not straightforward and since the technique is essentially based on DLS it is susceptible to the same problems experienced by DLS.

Hydrodynamic Chromatography (HDC). Hydrodynamic chromatography (HDC) and liquid exclusion chromatography (LEC) have also been reported as viable techniques to track in-situ particle growth. Studies conducted using HDC were successful in monitoring the uniform growth of particles and tracking phenomena like electrolyte induced agglomeration and secondary nucleation during a seeded polymerization of Styrene-Butadiene systems [6]. Sampling and analysis times, however, allowed the measurement of three or four samples during the course of the reaction. This often becomes essential when a certain control action is desired over relatively short interval of time. Issues such as reproducibility and sensitivity to manipulated variables remain unaddressed.

Application of Capillary Hydrodynamic Fractionation (CHDF) for Particle Size Distribution Analysis.

In CHDF, fractionation according to size is achieved by pumping the sample to be analyzed under laminar flow conditions through micro-bore capillary tubes resulting in the larger particles eluting the column before the smaller ones. Separations of various mixtures have been demonstrated using this principle and have been extensively discussed by Silebi *et al* [7,8]. Subsequent to these studies, optimization of the fractionation process resulted in higher resolution separations [9,10] which in turn facilitated shorter residence times and hence shorter overall particle size analysis times.

Absolute Particle Sizing technique. Despite these improvements in the separation efficiency, one shortcoming of the technique, however, still involved the use of a calibration curve for the final calculation of the distribution. The calibration curve in general is influenced by three major parameters: (i) eluant velocity, (ii) eluant composition and (iii) sample concentration. Practically speaking, during a period of prolonged operation, changes in these parameters alters the calibration curve. Failure to account for concomitant changes in the calibration curve correctly will result in erroneous particle size assessments. The process of continuously accounting for these changes will render the procedure tedious and time consuming - two factors that cannot be compromised upon during monitoring of particle growth during polymerization. This was the motivation to attempt a particle sizing methodology that does not rely on a previously calibrated system, namely, an absolute particle sizing procedure. The evolutionary development of the CHDF therefore continued with the incorporation of the multi-wavelength detector as the light sensing device to detect the particulate species after fractionation. This facilitated turbidity measurement of the fractionated species at several wavelengths instantaneously, thereby enabling the use of turbidimetric methods for the final particle size analysis. This methodology therefore couples the advantages offered by each of the individual techniques (fractionation by CHDF and turbidimetric measurements by light scattering) while simultaneously eliminating their inherent deficiencies. For instance, while turbidimetric methods by themselves, perform well for narrow PSDs; broad distributions have been difficult to analyze. The ability of CHDF to fractionate broad samples into narrower ones before resorting to turbidimetric analysis, helps overcome this deficiency. Moreover, it does not rely on a calibrated system thereby making the information absolute in nature.

These new developments were tested to examine if the sensor is independent of variations in the three manipulated parameters (alluded to in the previous paragraph) that have been shown to have a tremendous influence on the calibration curve. The results of these tests are summarized below, details on which can be found elsewhere [8,9]. The particle size calculation procedure was found to be independent of changes in the eluant velocity which becomes an important factor when considering polymerization process monitoring. Frequently, during the course of a reaction, the dynamics of evolution of particles may require a faster monitoring policy. This demand can be met by operating the CHDF at the higher velocity, thereby allowing a quicker response by the sensor. Moreover, the absolute calculation procedure eliminates the need to re-calibrate the system at this higher velocity. Similarly, the procedure was found to be independent of eluant composition variations. Again in a process monitoring situation, components such as initiators, monomers, chain transfer agents and in particular buffer electrolytes can influence the ionic composition of the eluant. The absolute procedure eliminates the tedious process of having to accurately monitor the eluant ionic strength at all times during the course of the polymerization process. Finally, the independence of the absolute technique to the detrimental effects of sample concentration eliminates the need to establish an accurate and cumbersome dilution procedure to ensure that the sample concentration being injected in the column is constant at all times.

Experimental

Recipe Considerations and Materials. Styrene was chosen as the model system for study. Polymerizations were performed with a seeded system where the reactor is furnished with a previously prepared latex referred to as the *seed latex*. The average number of particles introduced through the seed latex is determined from its average particle size. Monomer and initiator are added to this seed latex allowing the particles to grow to a larger size. Details on the exact proportions of the various components used in the recipe are included in the sections describing the individual experiments. The seeded polymerizations were conducted in semi-batch mode where the monomer was fed continuously throughout the course of the reaction. The monomer feed rates were designed to ensure *monomer starved* conditions throughout the course of the reaction which was accomplished by maintaining the monomer feed rate lower than the maximum rate of polymerization. Unimodal particle growth is ensured by limiting the free emulsifier concentration to a value below the critical micelle concentration while simultaneously maintaining a sufficient concentration of emulsifier to prevent coagulation. The batch times for the growth stage of the experiments are estimated from the monomer feed policies employed. The reaction is stopped at the end of this batch time and the system is shut down. Styrene was distilled twice under reduced nitrogen pressure. The distilled monomer was stored at - 2 °C and was used no more than three weeks after distillation and storage. Sodium lauryl sulfate used as the surfactant (SLS - Aldrich Chemicals) and sodium bicarbonate used as the buffer (ACS certified grade) were used as received. Potassium persulfate was recrystallized from distilled deionized water and then dried at room temperature. The water used was distilled and deionized (DDI).

Reactor Configuration. Polymerizations were carried out in a 1000 ml glass flask. The schematic of this laboratory scale reactor and associated components are presented in Figure 1. The reactor is equipped with baffles and an agitator to ensure proper homogeneous mixture of the reactants in the vessel. The glass reactor is jacketed for heating which is accomplished by recirculating hot water through the jacket using an immersion circulator. An internal cooling coil is provided to maintain the temperature constant. A gear pump is used to circulate cold water through the cooling loop. The system is equipped with a nitrogen line for blanketing and oxygen removal for start-up. The same line is used as the inlet port for initiator injection. Two stainless-steel lines are used for monomer addition. The monomer feed rates can

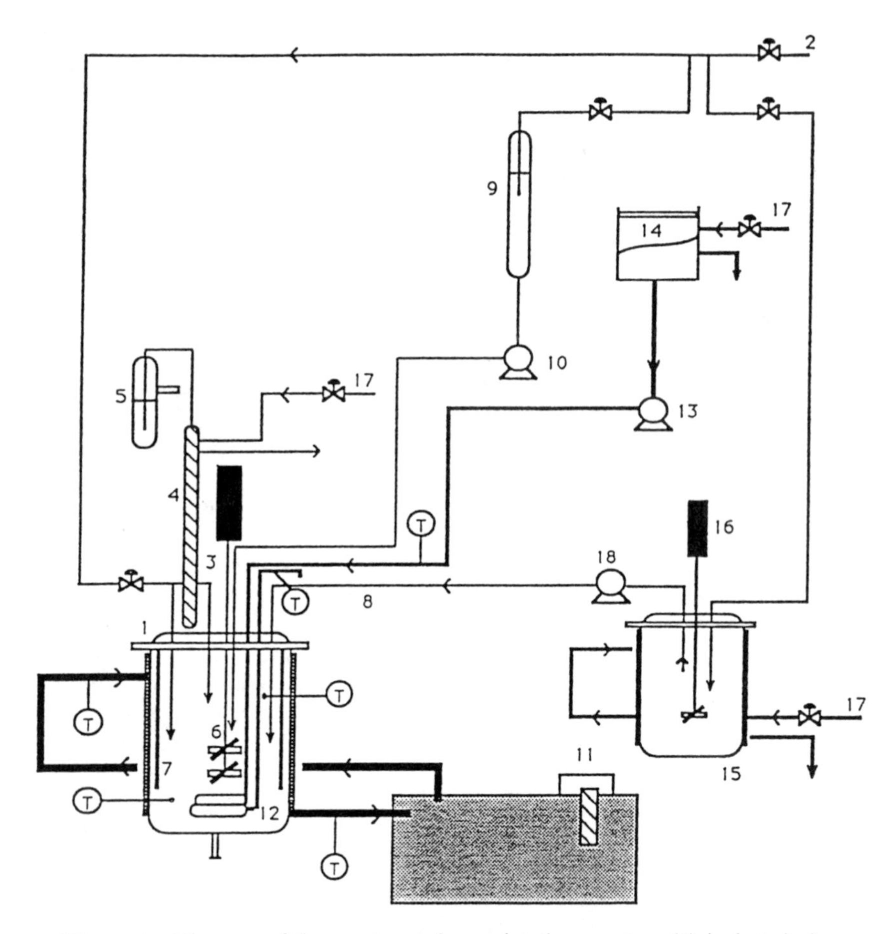

Figure 1: Diagram of the reactor and associated apparatus: (1) jacketed glass reactor; (2) N_2 supply; (3) injection port; (4) reflux condenser; (5) cold trap; (6) double impeller agitator; (7) baffles; (8) emulsifying charge inlet; (9) catalyst addition funnel; (10) catalyst addition pump; (11) jacket water heater and circulator; (12) internal cooling coil; (13) cooling water pump; (14) water buffer and overflow; (15) jacketed glass emulsifying tank; (16) impeller agitator; (17) cold water supply; (18) emulsifier addition pump.

be manipulated very accurately using FMI fluid metering pumps that are designed to deliver very precise amounts of liquid at low flow rates. Thermocouples are introduced at three different spatial locations in the reactor; two to measure the liquid phase temperature and one for the gas phase temperature. Data acquisition and signal conditioning tasks are performed through a control interface using the μMAC communications system.

Monitoring Scheme. The monitoring scheme involves sampling at regular intervals during the course of the reaction. The amount of sample withdrawn is a negligible fraction of the reactor volume and hence does not interfere with the progress of the polymerization. The sample withdrawn is split into two parts; one part is used for the gravimetric analysis and the other is used for the particle size analysis.

Gravimetric Analysis. The samples for the gravimetric analysis are quenched using a 1% solution of hydroquinone as the short-stop. The samples are then dried in an oven at 60 C following which the final weight of the samples is recorded. The overall conversion is calculated using the following formula:

$$Conversion = \frac{W_{polymer}}{W_{seed} + W_{monomer}} \quad (1)$$

The instantaneous conversion at any point in time is calculated by replacing the total monomer added $W_{monomer}$ in equation (1) by the amount of monomer fed up to that particular instant in time. A coded algorithm developed to calculate the overall and instantaneous conversions at each sampling point is compiled to accurately account for the fractional changes in the concentration of the components in the reactor due to intermittent sampling.

Particle Size Analysis. Particle size analyses are performed using both the absolute particle sizing methodology discussed earlier and the conventional calibration based method. The fraction of the total sample reserved for the particle size analysis is first diluted to approximately 2% (weight) using DDI water. This dilution process in itself acts as a quenching step and freezes the reaction thereby eliminating the need for the addition of a short-stop. Addition of hydroquinone as a short-stop would interfere with the light scattering results because it is a strong UV absorber. The CHDF is operated at 4000 psi with an average eluant velocity of 2.0 cm/sec and a residence time of 185 sec. in the capillary. The ability to achieve the fractionation and obtain the particle size analysis results within 7 - 10 min. allows enough time for analyses to be performed twice in between any two successive samplings from the reactor which are typically 20 min. apart. In this fashion the monitoring scheme essentially mimics the ideal case of implementing an on-line monitoring and analysis system.

The sections below discuss two case studies identified to demonstrate the ability of the above-mentioned sensor to monitor the growth of particles in a seeded emulsion polymerization process. The first case study demonstrates the validity of the sensor to monitor changes in the distributions and the average diameter during unimodal particle growth. The second case study evaluates the response of the sensor to a change in a manipulated input variable; namely, monomer feed rate.

Monitoring Unimodal Particle Growth.

The following paragraphs report experimental approaches to monitor the growth of a previously prepared polystyrene seed latex with a measured diameter of 92 nm (details of the analysis presented in the next section). This seed latex, denoted Dow LS-1030-B, was cleaned using the standard serum replacement cell which ensures the removal of surface active agents adsorbed onto the surface of the particles. This seeded polymerization was performed in the semi-batch mode at 50°C. The exact proportions of the various components used in the recipe are shown in the Table I.

Table I. Recipe used for the semi-batch polymerization run SB01 at 50°C.

Ingredients	SB01
Dow latex Seed: LS-1039-B (92 nm)	20 gm.
Styrene (feed)	24.01 gm.
Emulsifier: (Sodium Lauryl Sulfate)	1.00 gm.
Initiator: $K_2S_2O_8$	1.0061 gm.
Buffer: $NaHCO_3$	1.0005 gm.
DDI water	480.00 gm.

Particle Size Analysis of Seed. Figure 2 shows a comprehensive view of the basis for particle size analysis using the turbidity ratio approach for the seed latex used in this experiment. The column of graphs on the left hand side show the fractograms measured at two wavelengths (indicated on top of each graph, in nm) and the colum of graphs on the right hand side represent the intensity ratios taken with respect to the corresponding wavelengths. The solid lines in the graphs on the right hand side are the experimental values for the turbidity ratio and the dashed lines are the average of the turbidity ratio; the average taken across 80% of the fractogram. These average values for the turbidity ratio are now compared with the theoretically calculated value in order to ascertain the average particle size. The average particle size of this seed latex is calculated to be 92 nm. The figure shows four independent turbidity ratios primarily computed in the absorption range. Among the four ratios it is evident that τ_r computed at 220:254 yields the most stable value and is consequently chosen as the preferred τ_r to perform the particle size analysis during the monitoring scheme.

Profile of Reaction. Figure 3 shows the profiles of some of the key variables monitored during the semi-batch polymerization process. The plots are experimental values obtained using the gravimetric analysis discussed earlier. Figure 3a shows three variables: (i) dashed line showing the cumulative increase of the monomer (fed at a constant rate of 0.16 g/min for a total feed time of 145 min.); (ii) dotted line showing the instantaneous amount of polymer formed; and (iii) the solid line showing

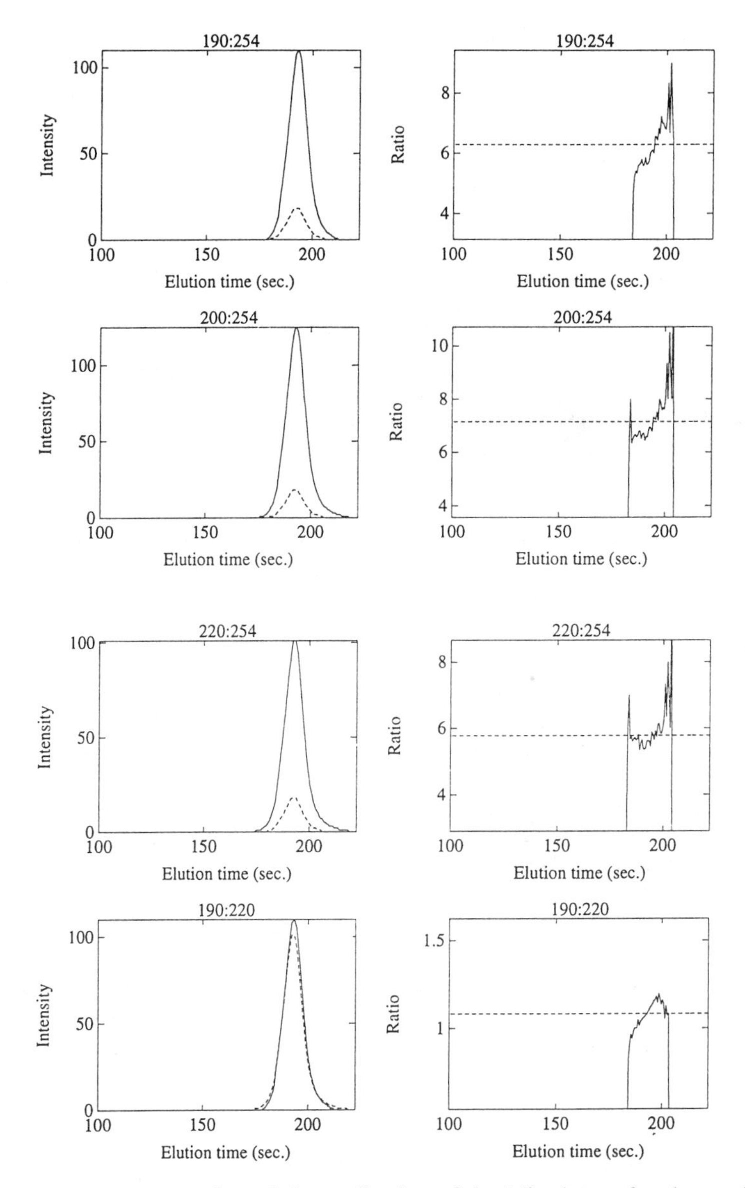

Figure 2: Demonstration of the application of the Mie theory for the particle size analysis of a monodisperse polystyrene standard (average particle size = 92 nm) at wavelengths of 190:254, 200:254, 220:254, and 190:220 nm. Left-hand column: optical density (intensity) versus elution time; right-hand column: turbidity ratio versus elution time.

the difference between (i) and (ii) which translates to the amount of residual monomer at any instant of time. Figure (3b) plots the amount of polymer produced during the reaction. Three distinct regimes are evident from this %solids profile: (a) regime 1 with a minimal slope extending up to 45 min. of the reaction time; (b) regime 2 corresponding to the maximum rate of polymer formation, ranging from 45 min. to about 145 min.; and (c) regime 3 which takes effect when the monomer feed is exhausted (145 min.), corresponds to the final stages of the reaction. Figure 4 shows the measured overall conversion curve, using Equation 1, for reaction SB01 (asterisks). As expected, the overall conversion curve also reflects the description of the three regimes identified above. The instantaneous conversion curve shown by the triangles in Figure 4 decreases initially and then attains a plateau value around 75%. This indicates that some fraction of the monomer exists as a finite amount of residual monomer within the growing polymer particles and another monomer fraction is lost as fugitive emissions. In other words, the monomer fed to the reactor is not completely and instantaneously. Hence although the theoretical calculation for the monomer feed rate was originally designed to ensure completely starved conditions in the reactor; in reality there always exists some finite residual monomer in the system.

Particle Size Distribution Development (PSD). The evolution of the size distribution during the course of the polymerization is discussed in detail in the following paragraphs. Figures 5-8 plot the particle size distributions (by weight) during the course of the reaction described above at various sampling periods. Details on the calculation of the distributions can be found elsewhere [10]. The analysis of the evolution of PSD begins with the first stage of the reaction comprising the first 50 minutes (described earlier as regime 1) which corresponds to the lowest rate of polymerization. Figure 5 shows the weight distributions corresponding to the seed (solid line) and those for samples taken at 25 minutes (dashed) and 50 minutes (dotted); all of which are plotted against a diameter axis. As dictated by the progress of the polymerization, the distributions show a gradual increase towards larger diameters. An interesting point is evident upon a closer look at Figure 5. The shift in the distribution from the seed stage (solid curve) to the 25 min. sample (dashed curve) appears to be more drastic than the corresponding growth from the 25 min. sample (dashed curve) to the 50 min. sample (dotted curve). This suggests that the rate at which particle diameter grows or the rate of polymerization is faster in the first 25 min. of the reaction compared to the next 25 minutes. However, the overall conversion curve, Figure 4, reflects a situation quite to the contrary; i.e. the conversion in the first 25 min. is seen to be negligible and significant conversion of monomer into polymer is observed only following the first 25 min. of reaction. This apparent paradox can be explained by the fact that during the first 25 minutes of the reaction the monomer that is fed into the reactor primarily gets distributed throughout the reaction volume and is absorbed by the seed particles resulting in monomer swollen polymer particles. In the case of styrene, the density of the monomer (0.906 gm/cc) is significantly lower than the density of its polymer (1.11 gm/cc). Therefore by the conservation of mass, the volume of a given mass of a monomer droplet is larger than its equivalent polymer particle. Along the same lines, the volume of a polymer particle swollen by a certain amount of styrene monomer is larger than its equivalent fully converted polymer particle. Since, physically the first

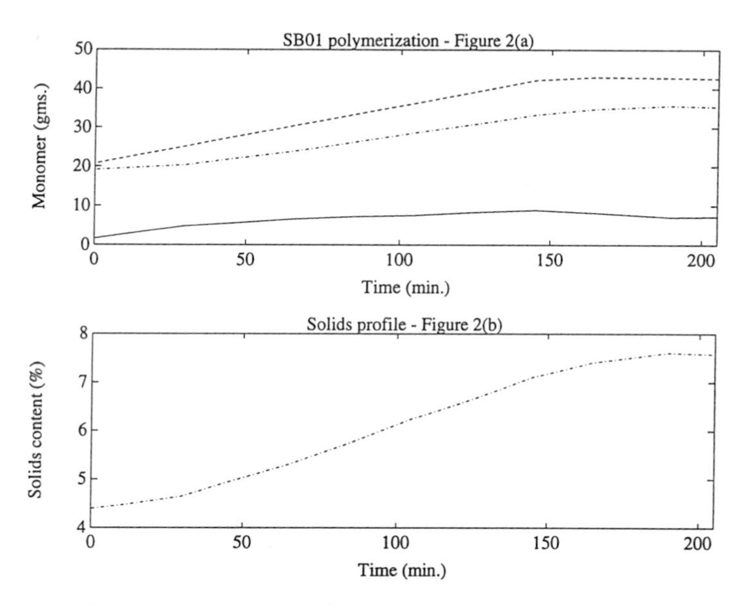

Figure 3: Semi-batch polymerization data for the reaction denoted SB01: (a) monomer feed rate (dashed line), polymer formation rate (dotted line), and residual monomer amount (solid line); (b) solids content profile during the reaction.

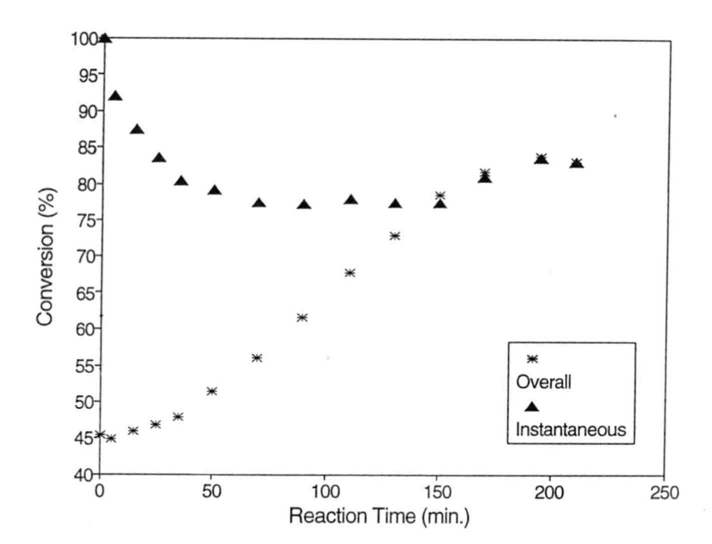

Figure 4: Overall and instantaneous conversion profiles for the reaction denoted SB01.

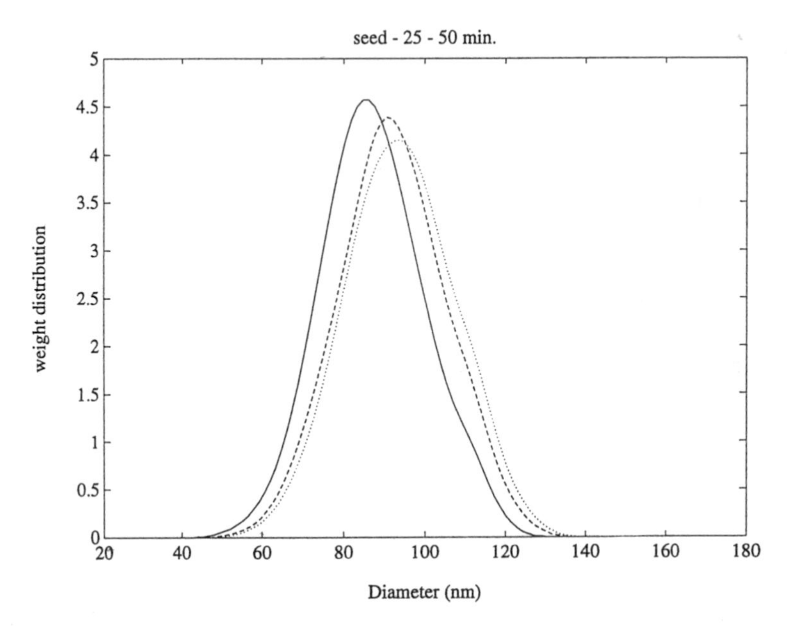

Figure 5: Particle size weight distributions corresponding to the first regime: seed latex particles (solid line), 25 minutes reaction time (dashed line), and 50 minutes reaction time (dotted line).

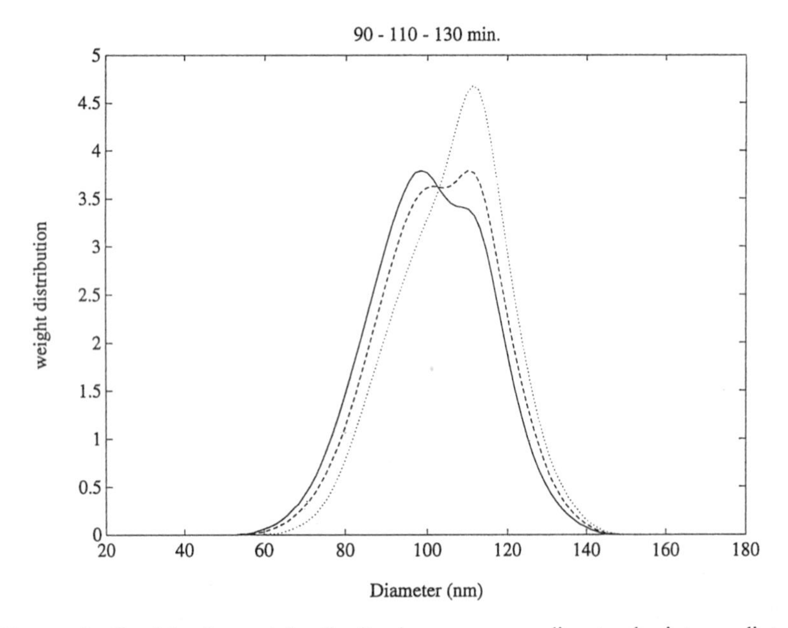

Figure 6: Particle size weight distributions corresponding to the intermediate regime: 90 minutes (solid line), 110 minutes (dashed line), and 130 minutes reaction time (dotted line).

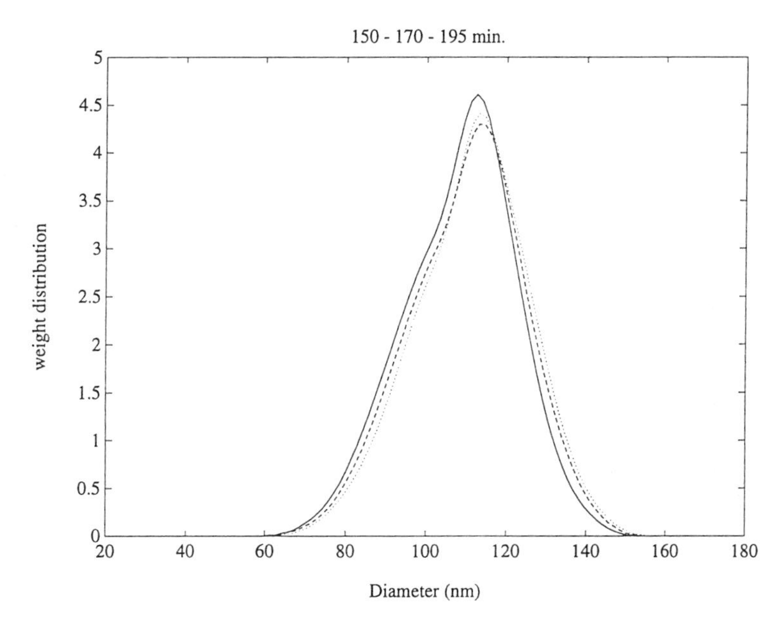

Figure 7: Particle size weight distributions corresponding to the intermediate regime: 150 minutes (solid line), 170 minutes (dashed line), and 195 minutes reaction time (dotted line).

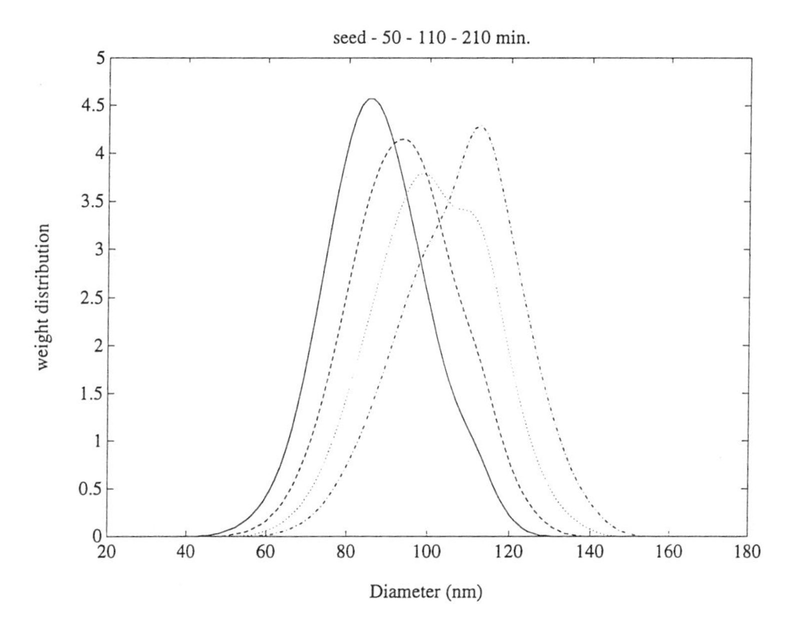

Figure 8: Particle size weight distributions during the reaction denoted SB01: seed latex particles (solid line); 50 minutes (dashed line), 110 minutes (dotted line), and 210 minutes reaction time (dot-dashed line).

25 min. essentially represent the swelling stage followed by the actual conversion of monomer, the corresponding size distribution during the first 25 min. appears to have an exaggerated diameter followed by a more steady growth pattern in the following 25 min. of the reaction. This effect is particularly prominent in monomer-polymer systems where the density difference between a monomer and its corresponding polymer is significant, which is indeed the case for styrene. It should be noted here that the onset of significant conversion of the monomer does not necessarily imply complete disappearance of monomer swollen polymer particles. During seeded emulsion polymerization, in-situ growth of particles occurs continuously as long as the monomer feed is sustained and is the only mode of propagation of the polymerization reaction. Hence unless the monomer that is being fed is instantaneously reacted, the seed particles can always be expected to exhibit a certain degree of swelling.

Figure 6 shows the development of the size distributions corresponding to regime 2 which represents the highest rate of polymerization. Sampling times of 90 min. (solid), 110 min. (dashed) and 130 min. (dotted) are chosen to represent this interval of interest. During this period the monomer swollen particles are polymerized resulting in the overall growth of the polymer particles to newer sizes. The evolution of the larger sized particles is evident from the distribution curves shown in this figure. The final stage of the polymerization is characterized in Figure 7, which plots the distributions corresponding to 150 min. (solid), 170 min. (dashed) and 195 min. (dotted). The monomer being fed to sustain particle growth is exhausted at 145 min. and therefore in this final stage of the reaction the last residual monomer in the system gets converted into polymer. The shape of the distributions in this regime, therefore, do not exhibit any dramatic changes with the exception of a gradual refinement of the distribution to its final form at 195 min. The summary of the characterization of the reaction with respect to the particle size distribution development is illustrated in the distributions in Figure 8. The four distributions plotted are representative of the various stages monitored during particle growth.

Data validation using measured average diameter. The description of the development of particle growth can be further analyzed by plotting the average diameters as a function of reaction time as shown in Figure 9. The points plotted on the graph are experimentally measured values; the weight average diameter (asterisks) are expectedly larger than the number average diameters (triangles). The solid curves (A and B) are the calculated values of the average diameter; computed using stoichiometric considerations. The experimentally measured points show a monotonic increase in particle diameter over the course of the reaction. The top solid curve (A) is the stoichiometrically calculated diameter assuming that all the monomer fed is instantaneously reacted and since this is known not to be the case, the calculated diameter overpredicts the measured diameter. The solid curve (B), on the other hand, is calculated using the instantaneous conversion data and appropriately accounts for the fraction of monomer that is not instantaneously converted. This curve as expected shows better agreement with the experimental data. The calculated diameter is the volume average diameter and therefore lies between the measured number and weight average diameters. Additionally, the diameter calculated is the *dry* particle diameter and does not take into effect the swollen state of the particles.

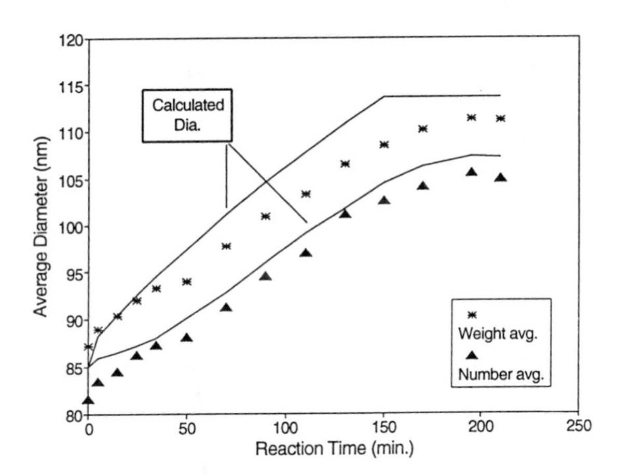

Figure 9: Particle growth profiles based on the number and weight average sizes: experimental data (symbols); calculated diameters (solid lines).

This explains the fact that the calculated diameter (B) slightly underpredicts the weight average diameter.

In order to ensure that the particle growth monitored is indeed unimodal and is devoid of newly generated particle sizes, the Polydispersity Index (PDI) is calculated from the average diameter data. The PDI is the ratio of the weight to number average particle size and is indicative of the extent of polydispersity of the distributions. Values closer to 1 imply more monodisperse distributions. We found that throughout the reaction, the PDI was esentially constant with a value of 1.05. This demonstrates that the growth patterns monitored were indeed unimodal without generation of new particles and/or multimodal species. The diameters were also measured using an alternate sensor; i.e. NICOMP (a dynamic light scattering instrument). Table II shows the measured and calculated values for the average

Table II. Measured and theoretical final weight average diameters.

Method	Initial Diameter (nm)	Final Diameter (nm)	Theoretical Dia. (nm)	% error
CHDF	87.2	109.3	109.8	-0.46
NICOMP	90.3	112.0	113.7	-1.50

diameter. The first two columns compare the measured initial and final diameters using the CHDF and the NICOMP submicron analyzer. The third column shows stoichiometric calculations for the final theoretical diameter assuming the starting diameter respectively measured by each technique. The error between the theoretical and calculated final diameters is found to be minimal. It should be noted here that the particle size analysis using the NICOMP took 30 min. (approx.) before a stable value was obtained and hence could not be used as a valid tool to monitor the growth of the seed particles during the course of the polymerization.

Effect of Monomer Feed Rate. Having established the feasibility of the technique to monitor unimodal particle growth, the next course of action is to test the sensitivity of the measurements to variations in manipulated input variables. This paragraph focuses on studying and demonstrating the ability to monitor the effect of a change in the monomer feed rate. Accordingly, the seed (SD01) is first synthesized in batch
mode. The recipe followed for this stage of the reaction is shown in Table III. The final diameter of the seed latex was 76 nm. The seed latex SD01 was then apportioned equally and used in two separate semi-batch polymerizations, operated at two different monomer feed rates. The recipe used for these two polymerizations, termed SB03 and SB04, are shown in Table IV. It is evident from the table that the two recipes have nearly identical proportions of the various components. This is done with the purpose of ensuring that the only difference in the two recipes is the monomer feed rates. Table IV also shows the differences in the operating conditions between SB03 and SB04. Figure 10 shows the cumulative profile of the amount of monomer fed to the reactor at any instant of time. The feed rate policy for SB04 is

Table III. Recipe used for the semi-batch polymerization run SD01 at 50°C.

Ingredients	SB01
Styrene (feed)	20.14 gm.
Emulsifier: (Sodium Lauryl Sulfate)	2.004 gm.
Initiator: $K_2S_2O_8$	1.003 gm.
Buffer: $NaHCO_3$	1.003 gm.
DDI water	500.0 gm.

Table IV. Recipe used for the semi-batch polymerization run SB03 and SB04 at 50°C.

Ingredients and Operating Conditions	SB03	SB04
Seed latex:SD01 (76 nm)	200.00 gm.	200.00 gm.
Styrene (feed)	21.13 gm.	21.14 gm.
Emulsifier:C_{12} H_{25} SO_4 Na	0.5002 gm.	0.5042 gm.
Initiator: $K_2S_2O_8$	1.013 gm.	1.007 gm.
Buffer: $NaHCO_3$	1.0095 gm.	1.0098 gm.
DDI water	308.03 gm.	308.00 gm.
Feed rate	0.20 (gm/min)	0.10 (gm/min)
Feed time (min)	110 min.	195 min.
Reaction time (min)	210 min.	250 min.

half that of SB03. Consequently the instantaneous monomer concentration at any time in the reactor is higher for SB03 and hence the monomer concentration per polymer particle is also higher. In emulsion polymerization the rate of polymerization is given by:

$$R_{p_{\max}} = \frac{k_p N_s \bar{n} [M]_p}{N_A} \tag{2}$$

where k_p is the propagation rate constant, N_s is the number of particles, n is the average number of radicals per particle, $[M]_p$ is the monomer concentration in the polymer particle and N_A is the Avogadro number. It is clear from this equation that the rate of polymerization is directly proportional to the monomer concentration in the polymer particles. Therefore, the rate of reaction for SB03 should be higher than SB04. Figure 11 plots the overall conversion history from the gravimetric analysis for the two cases. The slope of the overall conversion curve yields the rate of polymerization. Since the slope of the curve for SB03 (asterisks) is higher than SB04 (triangles), it can be concluded that the experiment is consistent with the arguments stated above. Figure 12 shows the development of the average diameter as a function of reaction time. The faster monomer feed rate is represented by asterisks and the

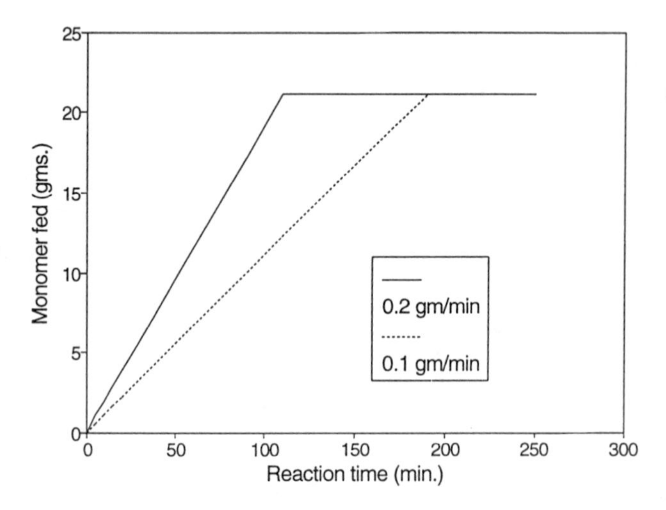

Figure 11: Overall conversion history for the reactions denoted SB03 and SB04.

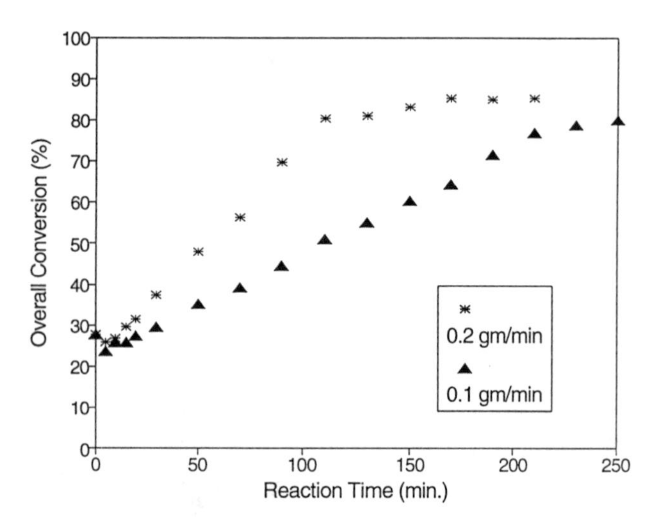

Figure 10: Monomer feed profiles (cumulative) for the reactions denoted SB03 (monomer feed rate = 0.20 g/min) and SB04 (monomer feed rate = 0.10 g/min).

slower feed rate by triangles. As mentioned in the previous paragraph, when the monomer feed rate is increased, at any given instant of time the concentration of the monomer in the reactor is higher than the slower monomer feed rate. Therefore the rate of polymerization will be higher and hence the seed particles will grow more rapidly at faster monomer feed rates. The growth of the polymer particles is also directly proportional to the monomer concentration in the polymer particles. When a slower monomer feed rate is employed (SB04), the average diameter monitored yields a profile (triangles in Figure 12) that is lower than that obtained for SB03. The sensor therefore accurately monitors the effect on the particle growth profile due to the change in the monomer feed rate.

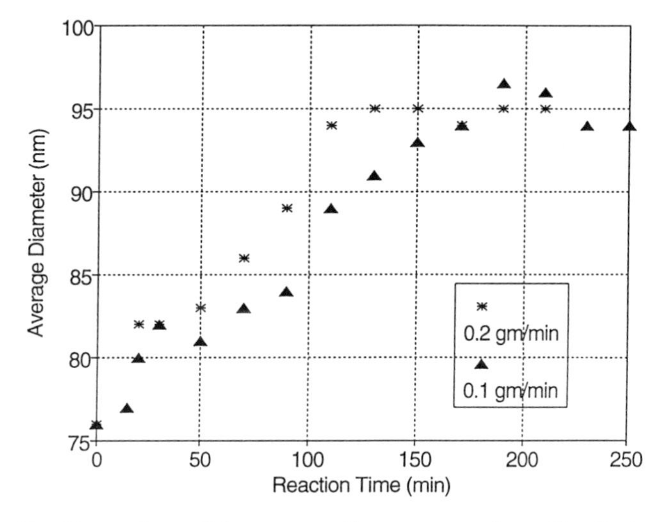

Figure 12: Comparison of the average diameter growth profiles for two semi-batch emulsion polymerizations (reactions denoted SB03 and SB04).

Literature Cited

1. Kourti, T. and Macgregor J. F. In *Particle Size Distributions II: Assessment and Characterization*; Provder, T. Ed.; ACS Symposium Series No. 472; American Chemical Society: Washington, DC, 1991; pp 34-63.
2. Nicoli, D. F., Kourti, T., Wu, J. S., Chang, Y. J., and MacGregor, J. F. In *Particle Size Distributions II: Assessment and Characterization*; Provder, T. Ed.; ACS Symposium Series No. 472; American Chemical Society: Washington, DC, 1991; pp 86-97.
3. Husain, A., J. Vlachopolous and A. E. Hamielec, *J. of Liquid Chromatography,* ***1979****, 2, 517.*
4. Thomas, J. C. In *Particle Size Distributions II: Assessment and Characterization*; Provder, T. Ed.; ACS Symposium Series No. 472; American Chemical Society: Washington, DC, 1991; pp 98-105.

5. Thomas, J. C. and Dimonie, V. *Applied Optics*, **1990**, 36, 5332.
6. Van Gilder, R. L. and Langhorst, M. A. In *Particle Size Distributions: Assessment and Characterization*; Provder, T. Ed.; ACS Symposium Series No. 464; American Chemical Society: Washington, DC, 1987; pp 272-292.
7. Silebi, C. A.; DosRamos, J. G. *J. Colloid and Interface Sci.*, **1989**,130, 14.
8. DosRamos, J. G.; Silebi, C. A. *J. Colloid and Interface Sci.*, **1990**, 135,165.
9. Venkatesan, J.; DosRamos, J. G.; Silebi, C. A. In *Particle Size Distributions II: Assessment and Characterization*; Provder, T. Ed.; ACS Symposium Series No. 472; American Chemical Society: Washington, DC, 1991; pp279-291.
10. Venkatesan, J., *Particle Size Sensor Design and Application*, Ph.D. Dissertation, Lehigh University, 1992.

Chapter 19

A New Method for Size Analysis of Low-Density Particles Using Differential Centrifugal Sedimentation

S. T. Fitzpatrick

Chemical Process Specialists, 7349 S.E. Seagate Lane, Stuart, FL 34997

A new method has been developed to extend particle size analysis using differential centrifugal sedimentation to particles that are lower in density than the fluid in which they are suspended. Low density particles are deposited at the bottom of a centrifuge chamber at the start of the analysis, rather than at the top. In aqueous systems, substituting deuterium oxide for water within the centrifuge allows measurement of particles with densities close, or equal, to the density of water. Virtually any dispersion of particles in an aqueous or non-aqueous fluid can now be accurately characterized for particle size distribution using differential centrifugal sedimentation.

Centrifugal sedimentation of particles suspended in a fluid, according to Stokes' Law (*1*), is a well known technique (*2, 3*) to measure the size distribution of particles in the range of about 0.01 micron to about 30 microns. There are two conventional methods for analysis: **integral** (sometimes called homogeneous) sedimentation and **differential** (sometimes called two-layer) sedimentation. These two methods have inherent advantages and disadvantages.

Integral Sedimentation. In this less commonly used method, each analysis starts with a homogeneous suspension of particles within a centrifuge. The concentration of particles remaining in the suspension is measured during the analysis, usually with a light beam or x-ray beam that passes through the centrifuge. The result of the integral analysis method is a cumulative representation of the particle size distribution. The integral method has several operational disadvantages, the most important of which are inaccuracy due to rapidly changing (and difficult to characterize) conditions at the start of the analysis, possible inaccuracy due to thermal convection within the sample during the analysis, and the need to stop, empty, and clean the centrifuge after each sample. The single important advantage of the integral method is the ability to measure particles that are either higher or lower in density than the fluid in which they are suspended.

Differential Sedimentation. In this more commonly used method, a small sample of dispersed particles is placed on top of clear fluid and subjected to centrifugal acceleration. (Figure 1). All particles start sedimentation at the same distance from the detector beam, and at the same time. Particle size is calculated from arrival time at the detector. The result of the analysis is a differential particle size distribution. An integral distribution may be generated by integrating the differential distribution with respect to particle size.

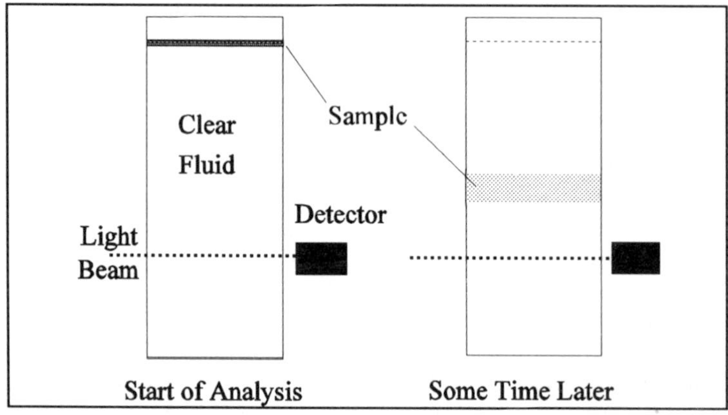

Figure 1. Conventional Differential Sedimentation Method

In actual practice, differential sedimentation requires a very slight density gradient within the fluid inside the centrifuge to insure that no instability develops during sedimentation. This instability is sometimes called "streaming". If the net density of the fluid that contains the sample (particles plus fluid) is greater than the density of the fluid immediately below, then the sedimentation process may become unstable. Stability is assured if the following condition is satisfied:

$$\frac{\delta\rho}{\delta R} \geq 0$$

Where ρ is net fluid density, including particles
R is the distance from the center of rotation

The density gradient in the centrifuge can be quite small; an increase of 0.01 g/cc or less per centimeter of fluid height usually sufficient to insure stable sedimentation, so long as the concentration of particles in the sample is low. The presence of a density gradient tends to eliminate thermal convection currents in the fluid, so the results are not disrupted by modest temperature changes during the analysis. For aqueous fluids, sucrose is usually a good choice to form a gradient, but a wide variety of water soluble compounds may be used. A suitable gradient can be produced manually, by sequential addition of fluids to the centrifuge in order of decreasing density, by an automatic gradient producing machine, or by other methods (*4,5*). The design of the centrifuge varies depending on manufacturer, but the most common design is based upon an optically clear rotating hollow disc (Figure 2).

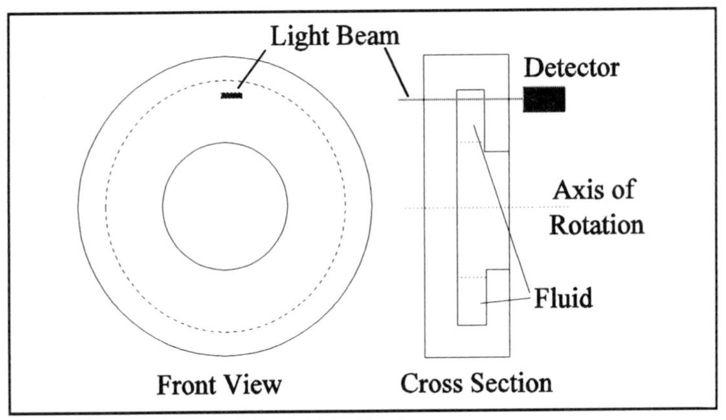

Figure 2. Hollow Disc Centrifuge Design

Differential centrifugal analysis normally yields high resolution, accurate, and reproducible particle size distributions. The principle advantage of the differential method over the integral method is that many samples may be run in series without having to stop and empty the centrifuge. Continuous operation reduces the overall time needed to run an analysis, reduces operator labor, and makes automation of sample injection and data collection straightforward. The most important historical disadvantage of the differential method is the requirement that the particles be more dense than the fluid in which they are suspended. This requirement has made analysis of many low density particles difficult or impossible. Some examples of difficult samples are: butadiene styrene copolymer latexes (densities of 0.92 to 1.03, depending on monomer ratio), polybutadiene latexes (*6*) (density of 0.89g/cc), ground polyethylene resins (0.91 to 0.96 g/cc), nitrile rubber latexes, acrylic adhesive latexes, and many others.

A New Differential Method

A new method (*7*), hereafter called the "low density method", has been developed for differential sedimentation of low density materials. The low density method uses a centrifuge design that deposits a sample of low density particles at the bottom of a spinning centrifuge chamber, rather than at the surface of the fluid in the chamber. The low density method requires that the particles be lower in density than the fluid in which they are suspended; the particles move from the bottom of the chamber toward the top during the analysis. The implementation of the low density method in a centrifuge of the hollow disc design is shown in Figure 3. The cross-section of the centrifuge disc shows how samples are transported to the bottom of the hollow disc centrifuge chamber. A "V" shaped groove is machined into the front face of the hollow disc, and four or more small capillary channels go radially from the base of the "V" groove to

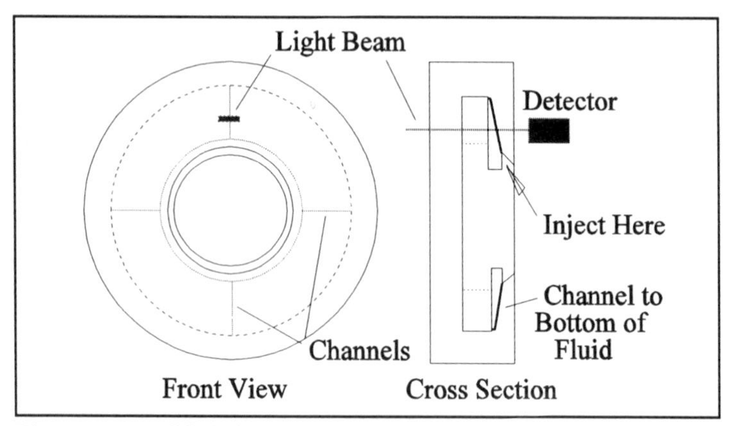

Figure 3. Modified disc design for the low density method

connect with the bottom of the centrifuge chamber. The level of the base of the "V" must be at least slightly above the level of fluid in the centrifuge (a lesser distance from the center of rotation) to keep the groove free of liquid. A sample is injected into the groove at the start of an analysis. Typical injection volume is in the range of 20 to 50 μ liters.

When a sample is injected into the "V" shaped groove, it is quickly (<0.1 second) carried by centrifugal force to the bottom of the centrifuge chamber via the small radial channels. The combined volume of the channels can be less than 10 μ liters, so even a small sample volume is sufficient to displace the liquid in the channels. Any sample that remains in the channels may be flushed to the bottom of the centrifuge by immediately following the sample with a small volume (10 to 20 μ liters) of the same fluid that was used to prepare the sample for injection. The easiest way to do this is to load a single syringe with both rinse fluid and sample. The sample and rinse fluid can be separated within the syringe by a small air bubble to keep them from mixing. This technique is shown in Figure 4.

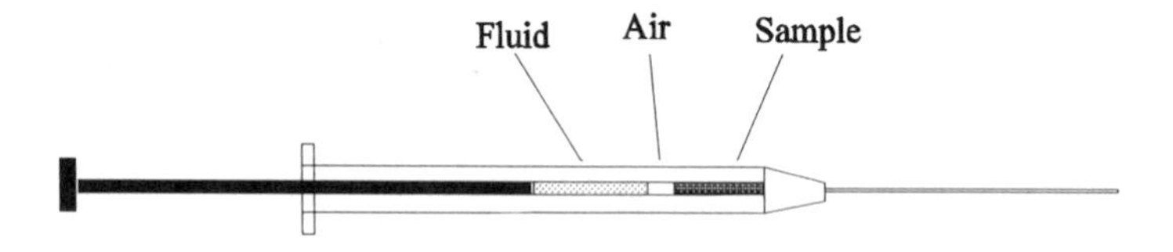

Figure 4. Injection setup with sample and rinse fluid

Samples are prepared for analysis by dilution in a fluid which is more dense than the fluid at the bottom of the centrifuge chamber. The net density of the sample dispersion (average of particles and fluid) must be higher than the density of the fluid at the bottom of the centrifuge chamber, so that the dispersion of particles quickly spreads to form a thin layer at the bottom of the chamber. Sedimentation of the particles proceeds in the normal fashion, except that the particles move toward the surface of the fluid rather than toward the bottom of the centrifuge chamber. Multiple analyses can be run without stopping the centrifuge, and it is even possible to alternate analyses between samples that are higher in density than the fluid, which are injected onto the surface, and samples that are lower in density than the fluid, which are injected into the "V" shaped groove. The low density method can be extended to centrifuges of nearly any design, and to many non-aqueous solvent systems as well, so long as the fluid within the centrifuge has a density gradient and so long as the samples are prepared in a fluid that is both higher in density than the fluid at the bottom of the chamber and miscible with the fluid at the bottom of the centrifuge chamber.

Example Analyses. Figure 5 shows duplicate analyses of a polybutadiene latex (polymer density of 0.89 g/cc) that were run using the low density method. A density gradient was produced by filling the centrifuge with a series of sucrose in water solutions (4% to 0%, 1.0139 g/cc to 0.9989 g/cc). The total fluid height in the centrifuge was about 1 cm, and a detector beam (450 nanometer light) passed through the disc at ~6 millimeters from the outside of the centrifuge chamber. The sample was prepared for injection by dilution to 0.2% active in an aqueous solution of 6% sucrose and 0.05% sodium lauryl sulfate emulsifier. The net density of the prepared sample was 1.0216 g/cc, and 50 μ liters were injected for each analysis. The disc speed was 8,600 RPM and the analysis time (to 0.15 micron) was approximately 17 minutes.

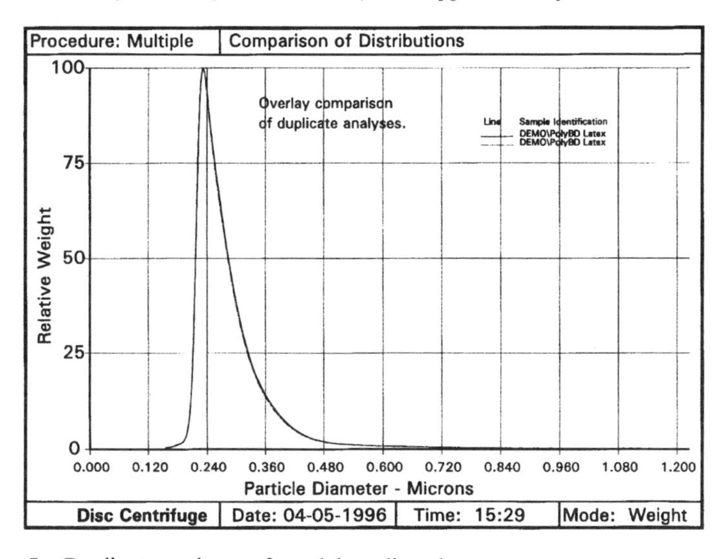

Figure 5. Replicate analyses of a polybutadiene latex

The low density method can be easily extended to measure size distributions for particles in aqueous suspension with densities ranging from slightly below the density of water to slightly above the density of water. Figure 6 shows an overlay comparison of three replicate analyses of a narrow, 0.40 micron polystyrene latex. The density of

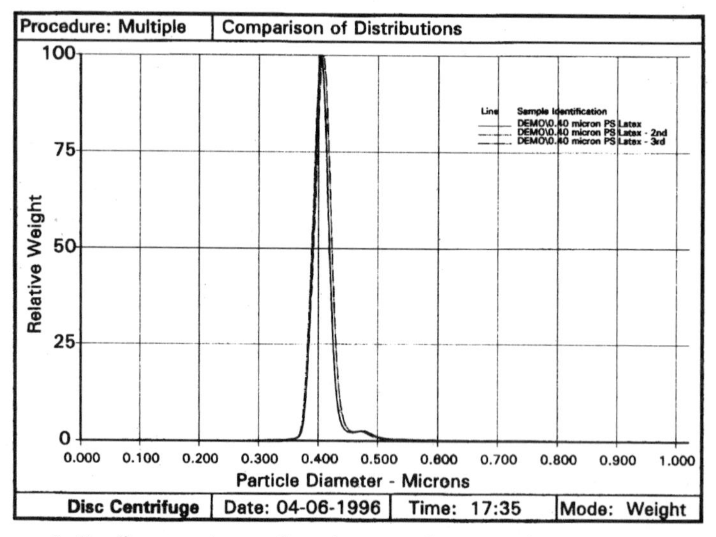

Figure 6. Replicate analyses of a polystyrene latex; run in D_2O

polystyrene (1.050 g/cc) precludes analysis by the low density method in an aqueous system, because the particles are more dense than water. However, these analyses were run using deuterium oxide (>99% D_2O, 1.107 g/cc) in place of water, to provide the required buoyancy for the analysis. A density gradient was produced by filling the centrifuge with a series of sucrose in deuterium oxide solutions that ranged from 4% to 0% sucrose. The sample dilution contained 6% sucrose (in deuterium oxide), and 0.03% polystyrene, to give a prepared sample dispersion with an overall density of about 1.137 g/cc. A diluted sample of 30 μ liters was injected in each analysis. The centrifuge speed was 8,600 RPM, and the analysis time (to 0.25μ) was 10 minutes. Analysis of particles with a density of 1.00 g/cc in deuterium oxide is significantly faster than polystyrene because the net buoyancy of 1.00 g/cc particles in deuterium oxide is higher than polystyrene. The ability to measure polystyrene particles using the low density method allows the many widely available polystyrene latex calibration standards to be used to verify the accuracy of analyses. The near perfect replication of the two analyses in Figure 5 and the three analyses in Figure 6 demonstrate the excellent repeatability of the low density method. The narrow peak widths in Figure 6 show that the resolving power of the method is quite good.

By using either the conventional differential centrifugal method, or the low density method reported here, virtually any sample which is an aqueous dispersion can be measured by differential sedimentation. If the particles are significantly higher in density than water, they can be analyzed using the conventional differential method.

If the particles are significantly lower in density than water, they can be analyzed using the low density method with water in the centrifuge. If the particles are near the density of water, then the low density method can be used with deuterium oxide partially or totally substituted for water in the centrifuge.

Injection Artifacts

The low density method yields distributions that may include small injection artifacts. These artifacts have no connection the actual particle size distribution; they are seen even when a blank (particle free) sample is analyzed. For samples of relatively high particle concentration, where the change in optical density during the analysis is high, the injection artifacts can be insignificant. For samples of relatively low concentration, where the change in optical density during the analysis is small, the injection artifacts are more important, and need to be either minimized or accounted for.

The injection artifacts are of two types: 1) relatively large diameter air bubbles that are entrained when a sample is injected, and 2) a relatively broad baseline deflection that comes from a slight change in optical density of the fluid in the centrifuge when a sample is injected. The injection artifacts can be minimized or eliminated using one or more of the techniques discussed below.

Entrained Air Artifact. Entrained air bubbles appear as large particles because they rise rapidly through the fluid; they are both relatively large in size and much lower in density than the fluid within the centrifuge. Entrained air is mainly the air located in the capillary channels between the base of the V shaped groove and the liquid level within the capillary. When a sample is injected, this air is trapped and transported to the bottom of the chamber by the flow of the sample fluid down the capillary. The volume of entrained air can be minimized by having the level of the fluid within the centrifuge close to the top of the capillary channels (please refer to Figure 3), and by using capillary channels that are of the smallest practical diameter. If the distance between the top of the fluid and the top of the capillary channel is small and the diameter of the capillary channel is also small, then the artifact from entrained air bubbles is minimized.

Optical Density Artifact. The mechanism for production of the second type of injection artifact is not obvious. Before a sample is injected, the fluid in the centrifuge chamber is moving at the same rotational speed as the centrifuge. When a small sample is injected at the bottom of the centrifuge chamber, the total volume of the fluid in the chamber increases very slightly. All of the fluid in the chamber is **raised** slightly when a sample is injected, because the sample is higher in density than the fluid in the chamber and enters at the bottom of the chamber. When the fluid is raised, it rotates at a very slightly smaller radius than it rotated before the injection. The absolute linear velocity of all of the fluid in the chamber **is not** immediately changed when a sample is injected, but the radius of rotation for the fluid in the chamber **is** suddenly (very slightly) reduced when a sample is injected. This means that the rotational speed of all of the fluid in the chamber increases slightly relative to the rotational speed of the centrifuge at the moment a sample is injected. The physical effect of the injection is similar to a small, instantaneous, reduction in centrifuge speed. Inside a hollow disc type centrifuge, the fluid cannot suddenly change in speed, it must gradually return to the rotational speed of the centrifuge disc.

This difference in speed between the centrifuge and fluid causes slight mixing to take place within the fluid as its rotational velocity recovers to match the rotational velocity of the centrifuge. The fluid in the chamber is not uniform in composition; its composition changes due to the presence of the density gradient. If the refractive index of the fluid also changes as the composition changes, then mixing caused by injection of a sample will cause some of the detector light beam to be scattered: the optical transmission of the fluid in the chamber is slightly reduced due to optical inhomogeneity during mixing. As the homogeneity of the fluid gradually recovers over 1 to 2 minutes (due to diffusion of the components that make up the gradient), the optical transmission returns to the original level. The decay of the injection artifact is similar to that shown in Figure 7.

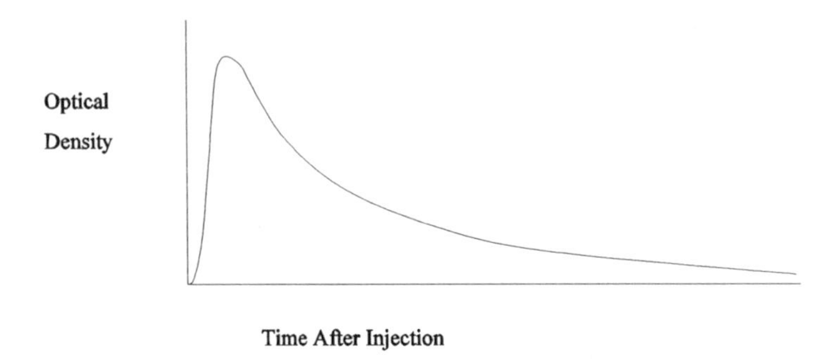

Figure 7. Decay of injection artifact. Return to the original optical density requires ~ 60 - 120 seconds.

The artifact due to changing rotational speed can be minimized in two ways. First, the smallest practical sample volume can be used. The smaller the sample volume (relative to the volume of the centrifuge chamber) the smaller the effect of the injection. Second, a density gradient can be prepared that is constant in refractive index. If all of the fluid in the chamber has the same refractive index (even though the composition and density do change), then mixing does not cause the optical transmission of the fluid to change. Density gradients with virtually constant refractive index can be formed using mixtures of three components. For example, an aqueous density gradient that goes from 2% to 0% (by weight) sucrose while at the same time from 0% to 5% ethanol has nearly constant refractive index over the entire composition range.

Both types of injection artifact can be mathematically subtracted from a particle size distribution. A blank sample (free of particles) can be run to record only the injection artifacts, and then subtracted from the distribution of an unknown sample. The distribution that remains after the subtraction is the distribution for the unknown, free of injection artifacts. Mathematical removal of artifacts is shown in Figures 8 through 10. Figure 8 is the distribution for a sample blank, showing only the artifact.

Figure 9 is the original distribution (including artifact) for a mixture of three narrow polystyrene calibration standards.

Figure 10 is the distribution for the mixture of calibration standards after the injection artifact was subtracted (Figure 9 minus Figure 8).

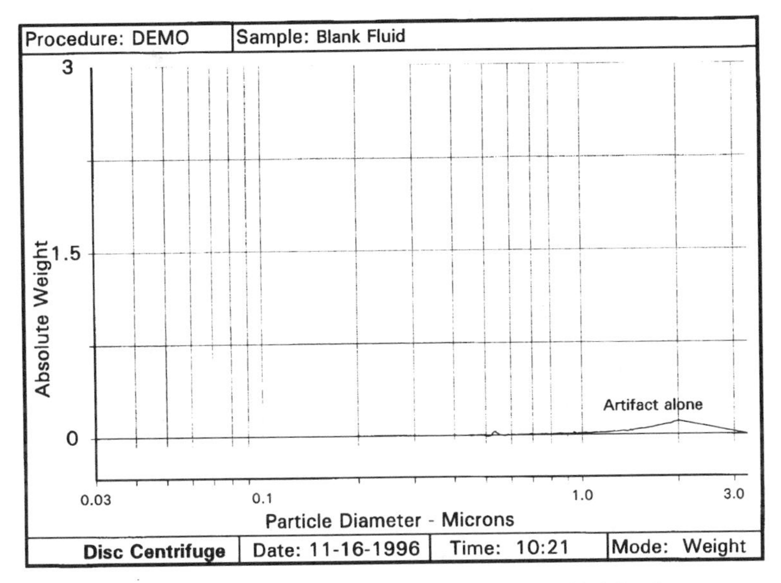

Figure 8. Injection artifact from change in optical density

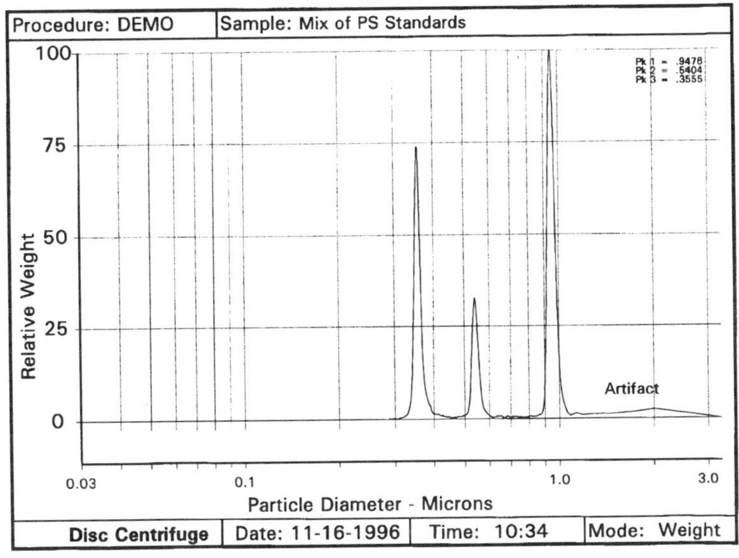

Figure 9. Distribution for mixed polystyrene latexes; artifact included.

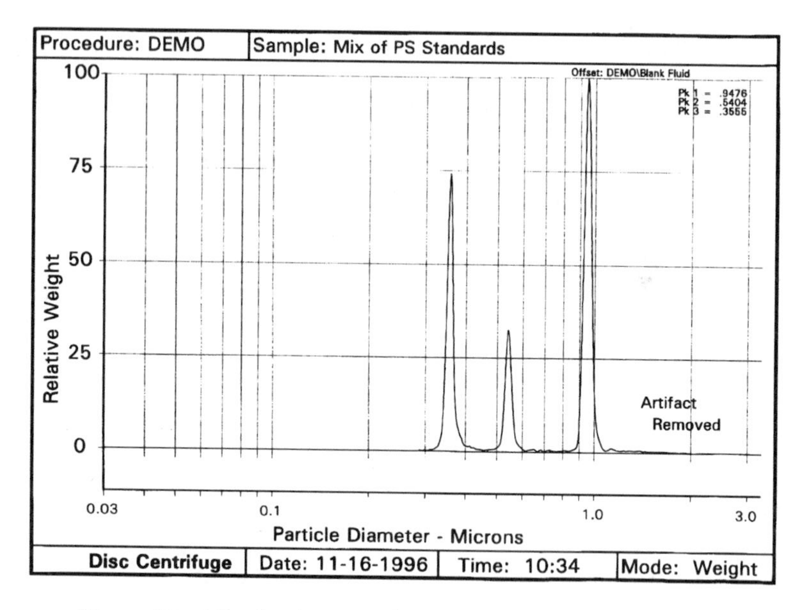

Figure 10. Mixed polystyrene latexes; injection artifact removed.

Conclusion

By using the low density method reported here, differential centrifugal sedimentation can be applied to measure the size distributions of materials that are lower in density than the fluid in which they are suspended. This eliminates the single most important limitation of the differential method, while maintaining the high resolution, accuracy, and operation advantages of the differential method.

Acknowledgment

The author wishes to thank Ronald M. Ellis for construction of the prototype centrifuge disc used for the analysis examples in this chapter.

Literature Cited

1. G.G. Stokes, *Mathematical and Physical Papers*, 11
2. Terence Allen, *Particle Size Measurement* (Chapman and Hall, London, 1968)
3. R.R. Irani, and C.E. Callis, *Particle Size Measurement* (Wiley, New York, 1963)
4. H. Puhk, US Patent 4,699,015, October 13, 1987
5. M.H. Jones, US Patent 3,475,968, November 4, 1969
6. E.M. Verdermen, J.G. Alhas, and A.L. Germanis, *Colloid and Polymer Science,* **1994,** Vol. 272, pp 57-63
7. S.T. Fitzpatrick, US Patent Pending

Electrophoretic and Electroacoustic Separation and Analysis of Particles in Dilute and Concentrated Regimes

Chapter 20

Analytical Separation of Colloidal Particles by Capillary Electrophoresis

H. B. Hlatshwayo and Cesar A. Silebi

Department of Chemical Engineering and Emulsion Polymers Institute, Lehigh University, Bethlehem, PA 18015–4732

Analytical separations of mixtures of seven negatively charged polystyrene latexes ranging in diameter from 90 to 1100 nm have been successfully achieved by capillary electrophoresis. Under the experimental conditions used this work, an electroosmotic flow (in the opposite direction of the electrophoretic migration of the particles) was always present. The electroosmotic mobility was found to decrease as the concentration of sodium lauryl sulfate increased. The direction of migration of the particles was determined by the relative magnitude of the electrophoretic and electroosmotic mobilities. At sodium lauryl sulfate concentrations of 10 mM or more, above the critical micellar concentration (cmc) of SLS, the particles migrated toward the anode, with the bigger particles passing through the detector ahead of the smaller particles. On the other hand, at SLS concentrations of 5 mM or less, below the cmc of SLS, the particles migrated toward the cathode with the smaller particles passing through the detector before the bigger particles. At an SLS concentration of 7.5 mM, a concentration near the cmc of SLS, the particles with diameters ranging from 91-176 nm migrated toward the cathode while the bigger particles migrated toward the anode. The effects of surfactant concentration and the intensity of the applied electric field on the resolution of the separation will be discussed in this chapter.

Introduction

An electric potential applied to a dispersion of charged particles in an electrolyte solution filling a conduit induces two different types of electrokinetic phenomena: electrophoresis and electroosmosis. Electrophoresis is the movement of the dispersed particles toward the electrode of opposite charge, and is the result of the electrical double-layer at the particle/liquid interface. Electroosmosis is the backward flow of the liquid medium along the walls of the electrophoresis chamber. The driving force for electroosmosis is the same as for electrophoresis, except that, in the case of electroosmosis, the solid is fixed and the liquid moves with respect to the solid. Since the discovery of electrophoresis and electroosmosis by Reuss (1) in 1808, these phenomena have been used for the characterization of solid/liquid interfaces and the separation of charged species such as colloidal particles, biological cells, proteins, or ions in solution. Analytical particle electrophoresis has been variously described as microcataphoresis (2), microelectrophoresis (3), cell electrophoresis (4), cellular electrophoresis (5), microscopic electrophoresis (6), capillary electrophoresis or often simply electrophoresis. The technique may be used purely for analytical purposes, for example to elucidate the nature, number and distribution of charge groups in the peripheral zone of biological cells, or alternatively to survey the distribution of electrophoretic mobilities in a mixture of cells or particles with the intent of devising appropriate means to separate subpopulations. Although an analytical separation or a preparative separation of particles or cells by definition involves the electrophoretic movement of particles in an electric field, electroosmosis is always involved to a greater or lesser degree because the experimental conditions require the presence of a static wall in contact with the liquid phase, which invariably has an electrical double layer associated with it. The actual movement of particles under these experimental conditions, therefore, is the summation of the electrophoretic velocity of the particles and the electroosmotic velocity of the eluant.

Electrophoretic separations primarily depend on the differences in the electrophoretic mobilities of the particles. According to the Overbeek-Booth theory, particles in a polydisperse sample having the same zeta potential can be separated according to size by choosing the eluant ionic strength such that the ratio of particle radius to the Debye length has a magnitude between 0.1 and 100. This separation takes advantage of the electrophoretic retardation and relaxation effect, providing the diffuse layer is weakly polarized and the surface conductance is negligible. Based on this theoretical prediction, McCann et al.(7) demonstrated that negatively charged polystyrene (PS) latex particles could be separated according to size using a continuous particle electrophoresis unit. Unfortunately, the complexity of the detection method and the difficulty in relating the migration times to the electrophoretic mobilites made this method difficult to use and implement as an analytical tool. Recently, VanOrman and McIntire (8) demonstrated the use of capillary electrophoresis (CE) to separate negatively charged PS latex particles according to size using a 50 μm ID capillary and 1 mM (millimolar) N-[2-acetamido]-2 aminoethane sulfonic acid as an eluant. These investigators observed that the particles migrated toward the cathode with smaller particles moving faster than larger particles, and postulated that the separation process was controlled by the particle-capillary wall interactions rather than electrophoretic mechanisms. Jones and Ballou (9) also investigated the separation of PS latex particles using a 75 μm capillary and a phosphate buffer as the eluant. The order of elution and the direction of migration of the particles were similar to those observed by VanOrman and McIntire; that is, the particles eluted in order of increasing particle diameter. An attempt was also made by these investigators to use smaller capillary diameters but they were unable to detect the particles. The ionic strength of the eluant used in these CE experiments was relatively low ranging from 1 to 5 mM.

The work described below focuses on the separations of latex particles of different sizes by capillary electrophoresis using 25 μm capillaries and sodium laury sulfate surfactant solutions in water with ionic strengths ranging from 10^{-4} to 10^{-1} M.

Experimental

Equipment. A schematic diagram of the CE equipment is shown in Figure 1. The components include a capillary, detector, high voltage power supply, and a microcomputer for data acquisition. The fused silica capillaries (Polymicro Technologies Inc., Phoenix, AZ) used in this work had an exterior polyimide coating. The tubes had an inner diameter of 25 μm and a length of 60 cm. A 5 mm length of polyimide coating was burnt off at a point halfway between the two ends of the capillary to provide a window for detection. The electric field along the capillary is applied using a Bertran Series 230 direct current voltage power supply (Bertran Associates Inc., Hicksville, NY) with reverse polarity, which supplies constant voltages up to 30 kV. The particles were detected using an on-column Spectra 200 UV-vis absorbance photodiode array detector (Spectra Physics, San Jose, CA) with forward optical scanning which has a wavelength detection range between 190 and 365 nm in the UV range and 366-800 nm in the VIS range. The ability to monitor the fractograms at several wavelengths simultaneously is useful for particle size analysis by turbidimetry. The detector was interfaced with a microcomputer for data acquisition and analysis. The electrophoretic mobility of the latex particles was also measured with a commercially available Pen Kem System 3000 Automated Electrokinetics Analyzer which has been described in some detail elsewhere (10,11).

Materials. The latex particle standards used in this study were a set of polystyrene standards with sulfate functional groups manufactured by the Dow Chemical Company. The average particle diameter and standard deviation as determined by electron microscopy for these standards are shown in Table I. These latices were diluted before injection to approximately a 0.3-0.6% weight fraction with a solution having the same composition as the eluant used for the electrophoretic separation. Sodium lauryl sulfate (SLS), 98 % pure (Stephan Chemical Company), was purified by recrystallization from boiling ethanol, followed by extraction with anhydrous ethyl ether (Fisher, certified grade) and then dried in a vacuum oven. Acetone (High Purity Chemical, Portland, OR) was used as the neutral marker used for measuring the electroosmotic migration times.

Procedure. Prior to use, the capillaries were conditioned by flushing them with 20 mM hydrochloric acid, followed by 20 mM sodium hydroxide and finally distilled deionized water. The flushing was done using a hand syringe equipped with a connector to hold the capillary. A conditioning period of 6 hours is allowed between each treatment with the different solutions. After the capillary has been conditioned, it is filled with the eluant. The capillary ends are then immersed into two 1.5 ml. test tubes filled with the eluant. The eluant in the test tubes was maintained at the same liquid levels to avoid gravitational flow through the capillary.

Electroosmotic backflow. The value of the electroosmotic backflow caused by the negatively charged fused silica surface of the microcapillary was determined by electrokinetically injecting acetone in the capillary, in the same eluant used for the separation, and recording the migration time of the acetone to the detector (located 33cm from the anodic end) under the applied electric field. The electrokinetic injection is made by dipping the anodic end of the capillary into a test tube containing a dilute

TABLE I
Polystyrene Standards

Latex	Dow Batch designation	Diameter (nm)	Standard deviation (nm)
A	LS-1040-A	91	8.0
B	LS-1044-E	109	2.7
C	LS-1045-E	176	2.3
D	LS-1047-E	234	2.6
E	LS-1010-E	357	5.6
F	LS-1115-B	610	4.8
G	LS-1166-B	1101	5.5

solution of acetone in the eluant and applying an electric field for 5 seconds, after which the anodic end of the capillary is returned to the test tube containing the eluant and the electric field is again applied. The electroosmotic mobility is calculated from:

$$\mu_{eo} = \frac{L\lambda}{V t_n}$$

where μ_{eo} is the electroosmotic mobility, L is the total length of capillary over which the potential V is applied, λ is the distance between point of injection and the on-column detector, t_n is the migration time for the neutral marker acetone which migrates at the electroosmotic flow rate.

Particle Electrophoresis. The latex particles were also electrokinetically injected. Since the electrokinetic injection involves the application of an electric field which will induce an electroosmotic backflow in addition to the electrophoretic motion of the particles, the electrokinetic injection of the particles is made by dipping either the anodic or the cathodic end of the capillary into a test tube containing a dilute dispersion. The appropriate end of injection is determined by the magnitude of the electrophoretic velocity relative to the electroosmotic velocities. For samples having an electrophoretic velocity smaller than the electroosmotic velocity, injection is made by dipping the anodic end of the capillary into the test tube containing the latex particles and applying the electric field for 5 seconds, after which the anodic end is returned to the test tube containing the eluant and the electric field is again applied. If the electroosmotic velocity is smaller than the electrophoretic velocity, the injection is then made by dipping the cathodic end into the test tube containing the dispersion. Thus, the electrophoretic mobility of the particles is given by the difference between the apparent electrophoretic mobility and the electroosmotic mobility:

$$\mu_{ep} = \mu_{app} - \mu_{eo}$$

where μ_{ep} is the particle electrophoretic mobility and μ_{app} is the apparent electrophoretic mobility determined experimentally from the following expression:

$$\mu_{app} = \pm \frac{L\lambda}{Vt_p}$$

where t_p is the migration time for a given latex particle and the sign depends on the direction of migration, (+) if the migration is toward the cathode and (-) if the migration is toward the anode.

Results and Discussion

Figure 2 shows the effect of SLS concentration on the electroosmotic mobility. As can be seen, as the SLS concentration increases the electroosmotic flow in the direction of the cathode decreases. Experiments in which sodium chloride was used as the eluant show similar results for the electroosmotic mobility. These results can be attributed to surface charge neutralization of the wall of the capillary by the increased concentration of the sodium counterions, which will decrease the zeta potential of the capillary wall.

Figure 3 shows the electrophoretic mobility of the polystyrene standards used in this study at a sodium lauryl sulfate concentration of 100 mM. The electrophoretic mobilities determined by CE are sohon with those determined using a PenKem 3000 at the same SLS concentraiton. The difference between the two techniques can be attributed to the difference in the intensity of the electric fields used. In the PenKem the electric field is significantly lower than the 254 vots/cm use in the CE instrument. This effect of the intensity of the electric field on the electrophoretic mobility can be attributed to a significant deformation and expansion of the cloud of counterions that surround the charged particle, resulting in a reduce drag of counterions which move in the opposite direction. It can also be seen in Figure 3 that the absolute value of the electrophoretic mobilities for these standards increases with particle diameter. This suggests that at this concentration of SLS these particles can be separated by size with the larger diameter particles moving faster toward the cathode if their electrophoretic mobilities are greater than the electroosmotic mobility.

Figure 4 shows the fractogram obtained for a mixture of the negatively charged PS standards electrokinetically injected in the cathode into a 100 mM SLS eluant. All the particles have migrated to the detection zone within 14 minutes and have been analytically separated. The order of elution was such that the larger particles migrated faster than smaller particles. A comparison between the electrophoretic mobilities shown in Figure 3 and the electroosmotic mobility at the same SLS concentration from Figure 2 shows that all the particles except the two smaller standards had an electrophoretic mobility greater than the electroosmotic flow . This suggests that the two smaller particle sizes should have been dragged to the cathode by the electroosmotic flow; however this was not observed and all the seven polystyrene standards migrated in the direction of the anode. This indicates that under our CE experimental conditions all of these particles have an electrophoretic mobility greater than the electroosmotic mobility.

Migration towards the anode occurs if the electrophoretic mobility of the particles is

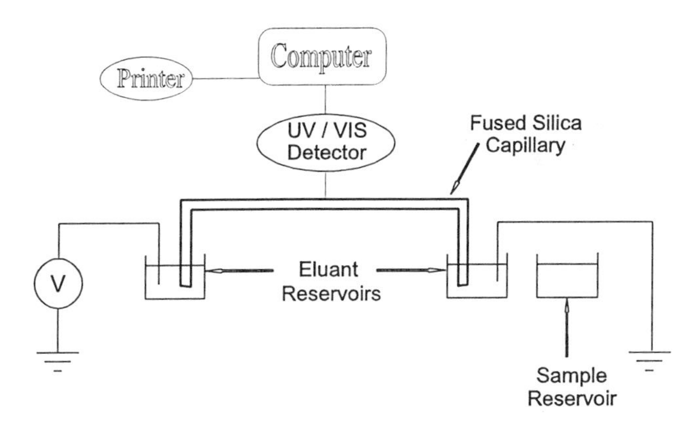

Figure 1. Schematic diagram of a capillary electrophoresis system.

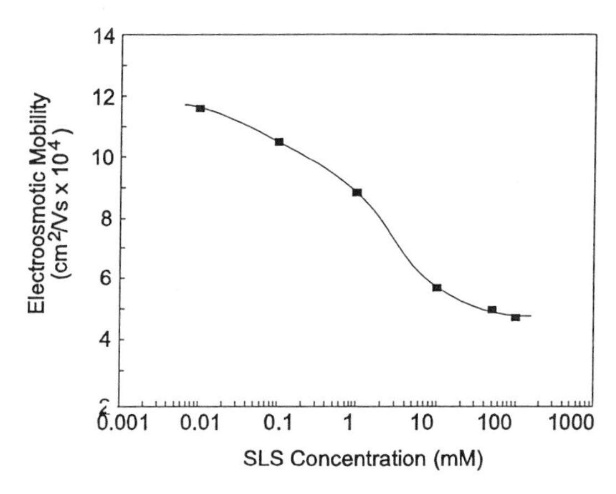

Figure 2. Electrophoretic mobilities of PS latex particles in a 100 mM SLS eluent.

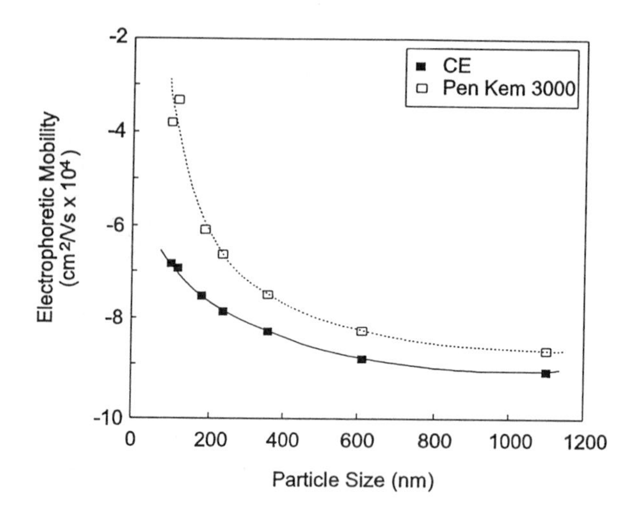

Figure 3. Comparison of the electrophoretic mobilities of PS latex particles obtained from CE and the Pen Kem 3000 in a 100 mM SLS solution as eluant.

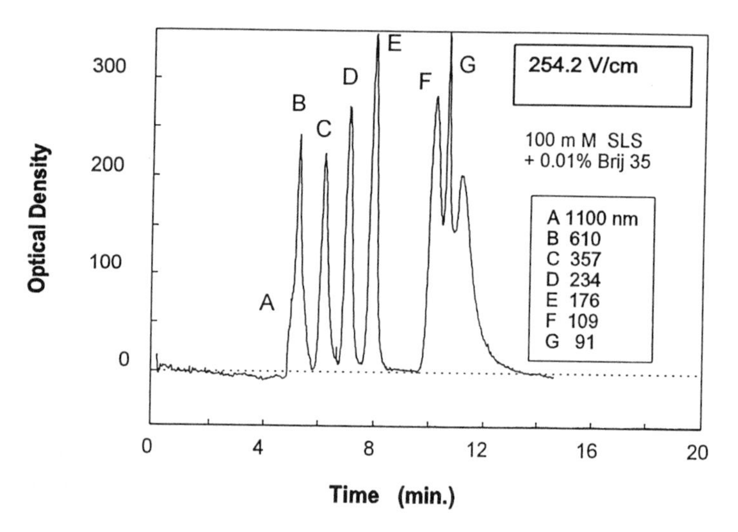

Figure 4. Electropherogram of a mixture of seven PS latex particles after migrating along 29 cm of a 25 μm ID capillary towards the anode in a 100 mM SLS eluent at 254.2 V/cm.

greater than the electroosmotic velocity of the eluant. Thus, the observed order of migration is due to the greater absolute value of the electrophoretic mobility of the larger particles relative to the smaller particles. The apparent discrepancy with the two smaller particles can be attributed to the effect of the intensity of the electric field on the electrophoretic mobility of the particles. Like Jones and Ballou (9), we found that at very high voltage gradients the electrophoretic mobility increased as the electric field increased. This effect on the electrophoretic mobility can be attributed to a significant deformation and expansion of the cloud of counterions surrounding the particles, resulting in greater net charge of the particles and as a consequence a greater electrophoretic mobility. Also illustrated in Figure 4 is the effect of the electric field. At 254.2 V/cm the 1100 nm particles are not well resolved from the 610 nm particles, however at 423.7 V/cm a distinct peak corresponding to the 1100 nm particles is clearly seen. The improved resolution can be attributed to a decrease in peak spreading due to shorter residence times at the greater electric field as well as an increase in electrophoretic mobilities at higher voltage gradients. At the higher voltage gradient, all the particle populations have migrated to the detector in less than 8 minutes. This order of migration, where large particles are followed by small particles, was also observed with concentrations of SLS greater than 10 mM as illustrated in Figure 5 where the apparent electrophoretic mobilities are shown for SLS concentrations ranging from 0.003 to 100 mM. Because of the increase in the electroosmotic mobility the particles are slowed down in their migration toward the anode and consequently for a given electric field, the migration times increases as the SLS concentration decreases to 10 mM. For SLS concentrations of 5 mM or less, the migration times decrease as the concentration of SLS is decreased.

As shown in Figure 2, at SLS concentrations of 5mM or less, the electroosmotic velocity of the eluant dramatically increases, becoming greater than the electrophoretic mobilities of the particles, and consequently the particles need to be injected in the cathode. Thus, in this range of concentration of SLS, which is below the cmc of SLS (8.1 mM), the apparent electrophoretic mobility is positive, as shown in Figure 5, and consequently the particles need to be injected in the cathode. Under these conditions, the negatively charged particles are dragged by the eluant toward the anode and the order of migration towards the detection point is now reversed with smaller particles moving faster than the larger ones. Figure 6 shows the electrophoretic separation in a 1 mM SLS eluant. At this concentration of SLS, the electroosmotic velocity of the fluid in the capillary is greater than the electrophoretic mobility of the particles. Thus, under this condition, the particles were injected at the anode and migrated toward the cathode.

At an SLS concentration of 7.5 mM, which is close to the cmc of SLS, when the particles were injected at the cathode, only the particle populations with sizes between 234 nm and 1100 nm were detected as shown in Figure 7. The order of migration to the detector was the same as the one observed at the higher SLS concentrations, where the bigger particles reached the detector first followed by the smaller particles. In order to detect the other three particle populations present in the mixture, the sample had to be injected at the anode. In this case the three peaks corresponding to the particles with sizes ranging from 91 to 176 nm with the smaller particles passing through the detector first followed by the bigger particles as illustrated in Figure 8. These result clearly indicates that at the cmc of SLS both the electrophoretic and electroosmotic mobilities vary in a way that the direction of particle migration is reversed. The slope of the

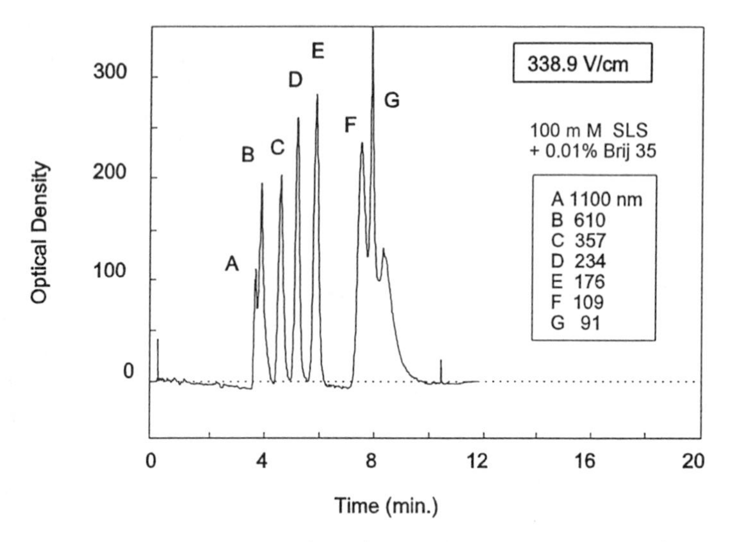

Figure 5. Electropherogram of a mixture of seven PS latex particles after migrating along 29 cm of a 25 μm ID capillary towards the anode in a 100 mM SLS eluent at 338.9 V/cm.

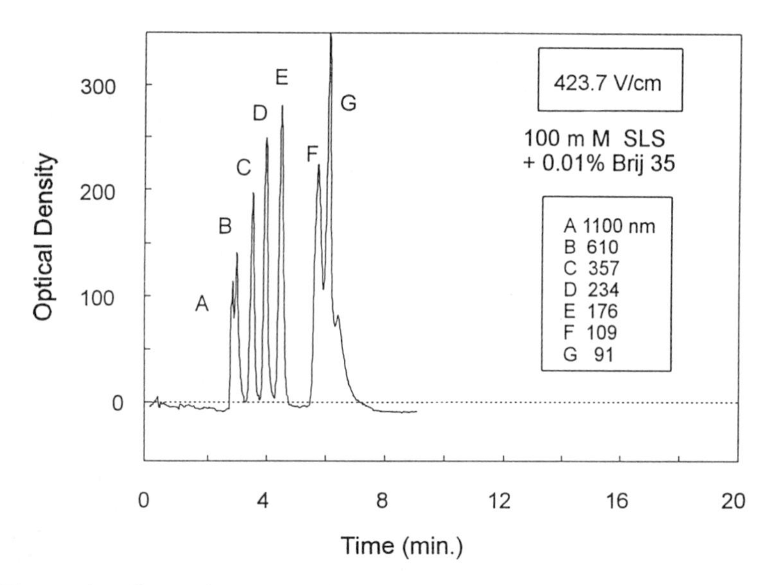

Figure 6. Electropherogram of a mixture of seven PS latex particles after migrating along 29 cm of a 25 μm ID capillary towards the anode in a 100 mM SLS eluent at 423.7V/cm.

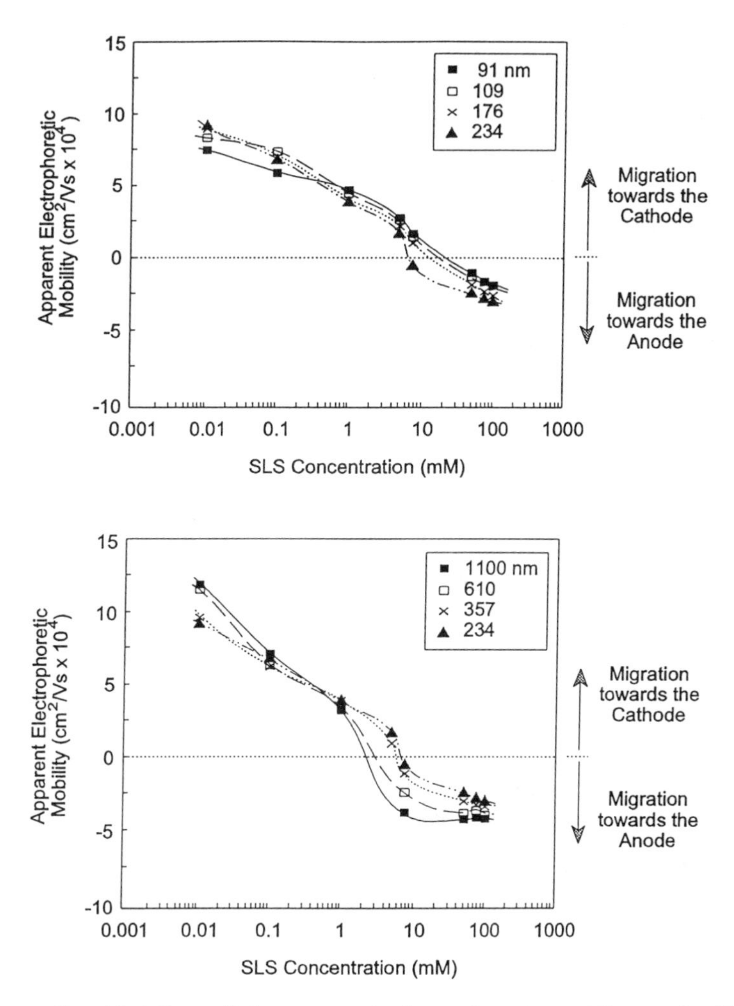

Figure 7. Variation of the apparent electrophoretic mobility with SLS concentration for the seven polystyrene standards in a 25 μm ID capillary.

change in the electroosmotic mobility is more pronounced than that of the electrophoretic mobility of the latex particles, this leads to a crossover of the absolute values of these electrokinetic velocities. However, we should point out that this crossover of the absolute values of electroomotic and electrophoretic mobilities was not observed with polymethyl-methacrylate latexes.

In all of these electrophoretic separations, the peaks were identified either from the peak elution times obtained when the latexes were injected individually or from the turbidity (optical density) ratios at two wavelengths which are characteristic of the particle size. The turbidity ratios at wavelengths of 210 and 254 nm for the different particle sizes are shown in Figure 9. The curve in Figure 9 was obtained from application of the Mie theory of light scattering using the complex refractive index values obtained by Venkatesan (12) for polystyrene particles. This figure also shows that particles with diameters greater than 400 nm cannot be identified uniquely by the turbidity ratio because of the multiplicity of this ratio for particle diameters above this treshold. Thus, the particle size of the fractionated species can be determined using the ratio of the turbidity at two wavelengths. Figures 10 and 11 show the electropherogram of the 91 and 176 nm particles at two wavelengths, 210 and 254 nm. Also included in these figures are the turbidity ratios at these wavelengths. These ratios suggest that the 176 nm particles have a narrower size distribution than the 91 nm particle standards in agreement with the standard deviations of the particle sizes measured by electron microscopy reported in Table I.

Conclusions

The applicability of capillary electrophoresis to the separation and electrokinetic characterization of particles with diameters ranging from 90 to 1100 nm has been demonstrated. Fast and effective separations of polystyrene particles by capillary electrophoresis has been successfully carried out.

As the concentration of SLS decreased, the electroosmotic mobility of the eluant increased. At SLS concentrations greater than 10 mM, the electroosmotic mobility is smaller than the absolute value of the electrophoretic mobility of the particles, while at concentrations of SLS smaller than 6mM the opposite is true. At SLS concentrations of 10 mM or more, the negatively charged latex particles migrated toward the anode with the larger particles eluting ahead of the smaller particles. We have also shown that at SLS concentrations of 5 mM or less, the faster electrosmotic flow dragged the negatively charged polystyrene particles toward the cathode with the smaller particles eluting ahead of the larger particles. Finally we also demonstrated that at an SLS concentration of 7.5 mM the particles smaller than 200 nm migrated toward the cathode while the particles bigger than 200 nm migrated toward the anode. In addition to the electrophoretic mobility of the particles, their size can also be obtained by measuring their turbidity ratios as the fractionated particles pass through the photodiode array detector.

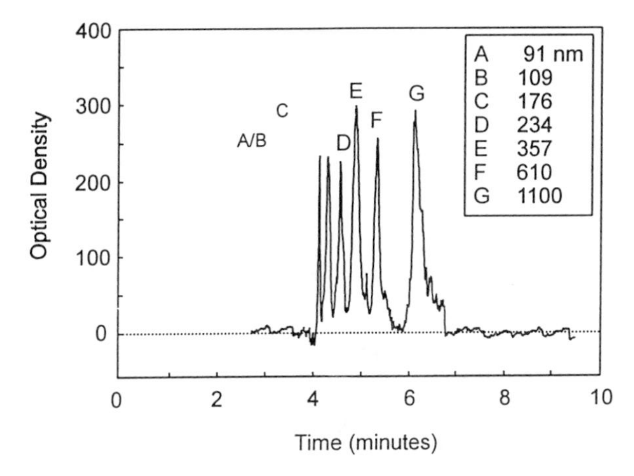

Figure 8. Electropherogram of a mixture of seven PS latex particles after migrating along 30 cm of a 25 μm ID capillary towards the cathode in a 5 mM SLS.

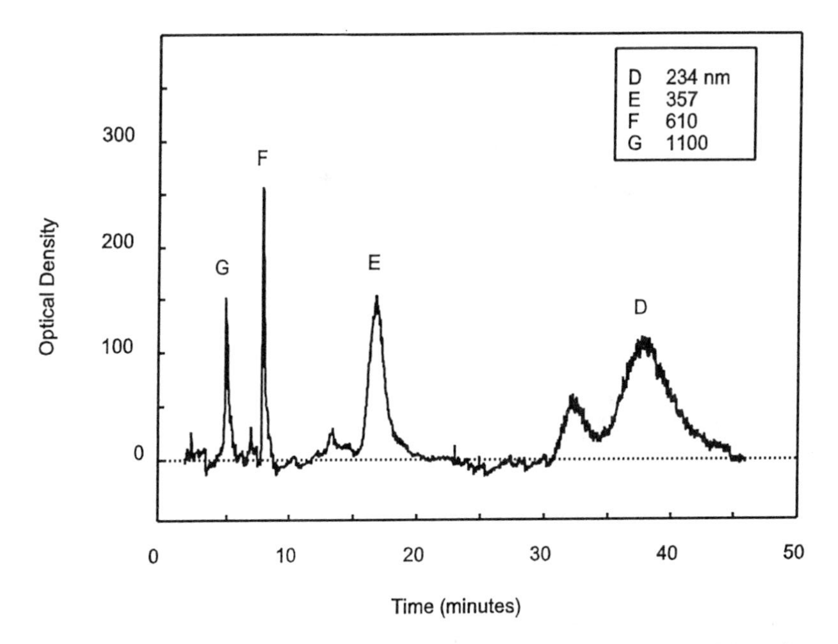

Figure 9. Electropherogram of a mixture of seven PS latex particles after migrating along 29 cm of a 25 μm ID capillary towards the cathode in a 7.5 mM SLS eluent.

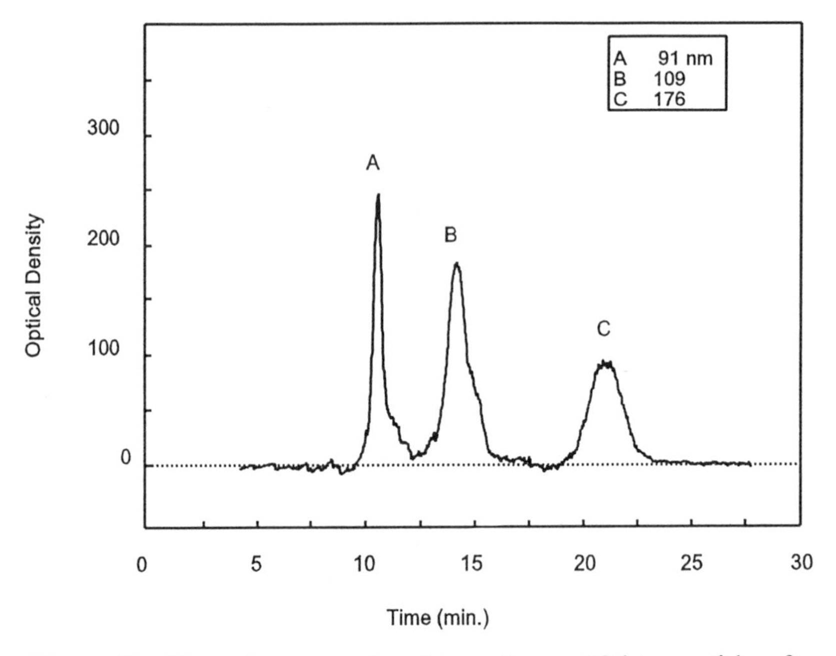

Figure 10. Electropherogram of a mixture of seven PS latex particles after migrating along 29 cm of a 25 μm ID capillary towards the anode in a 7.5 mM SLS eluent at 423.7V/cm.

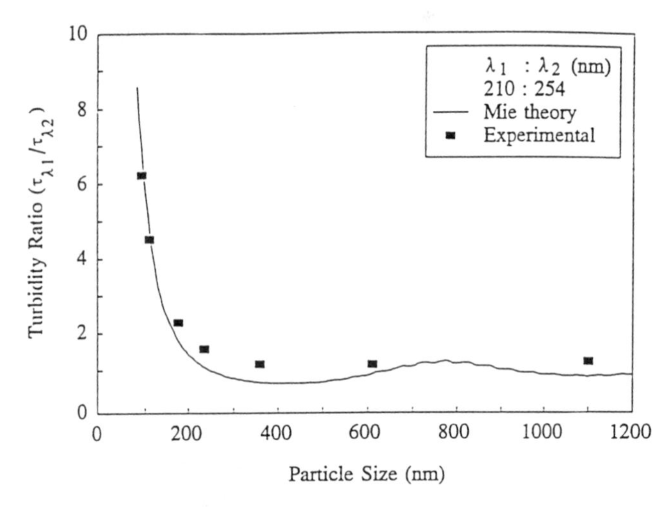

Figure 11. Comparison between the theoretical and experimental turbidity ratios for PS latex particles at wavelengths of 210 and 254 nm.

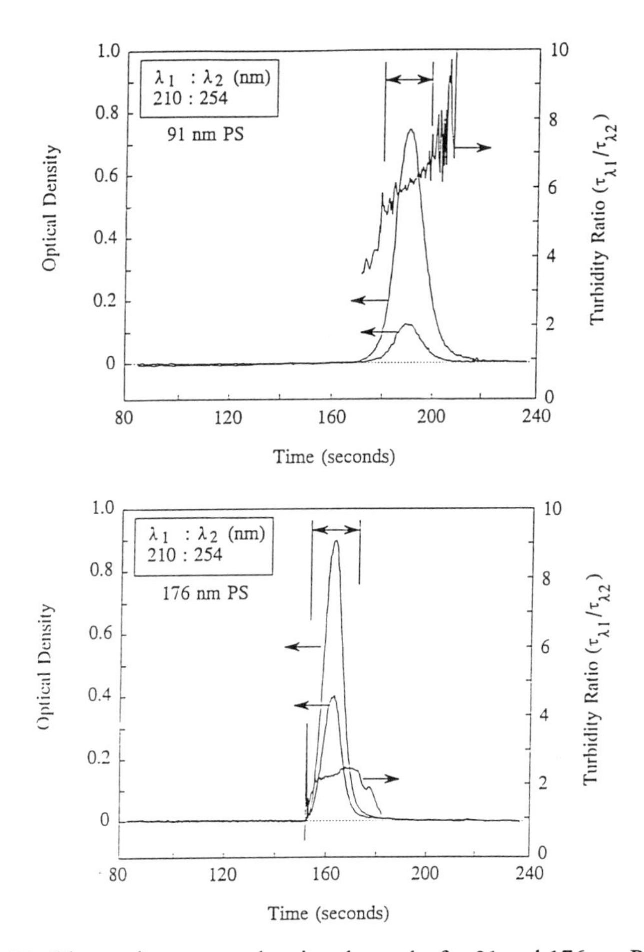

Figure 12. Electropherograms showing the peaks for 91 and 176 nm PS latex particles monitored at wavelengths of 210 and 254 nm and the corresponding turbidity ratios.

Literature Cited

1. Reuss, F. F., Memoirs de la Societe Imperiale des Naturalistes de Mskou. **1809**, 2, 324.
2. Brown, H.C. and Broom, J.C. , Proc. Roy. Soc. B. **1936**, 119, 231.
3. Seaman, G.V.F., Microelectrophoresis of Red Blood Cells, Ph.D. Dissertation, Univ. of Cambridge, **1958**.
4. Ambrose, E.J., and A. Churchill, Cell Electrophoresis, London, **1965**.
5. Weiss, L. J. Natl. Cancer Inst. **1966**, 36, 837.
6. Abramson, H. A. Electrokinetic Phenomena and Their Application to Biology of Medicine, The Chemical Catalog Cl. Inc., New York **1934**.
7. McCann, G.D., Vanderhoff, J.W., Strickler, A., and Sacks, T.I., Sep. Purif. Methods, **1979**, 2, 153.
8. VanOrman, B.B., McIntire, G.L., J. Microcol. Sep. **1990**, 1, 289.
9. Jones, H.K., and Ballou, N.E., Anal. Chem. **1990**, 62, 2482.
10. Goetz, P. J. U.S. Patent 4154669.
11. Morfesis, A. A.; Rowell, R. L. Langmuir **1990**, 6, 1088.
12. Venkatesan, J. Ph. D. Thesis, Lehigh University, **1993**.

Chapter 21

Electroacoustic Determination of Droplet Size and Zeta Potential in Concentrated Emulsions

R. W. O'Brien[1], T. A. Wade[2], M. L. Carasso[2,3], R. J. Hunter[2], W. N. Rowlands[1], and J. K. Beattie[2]

[1]Colloidal Dynamics Pty. Ltd., Unit 125 National Innovation Centre, Australian Technology Park, Eveleigh, New South Wales 1430, Australia
[2]School of Chemistry, University of Sydney, New South Wales 2006, Australia

Recent developments in electroacoustic instrumentation have allowed the dynamic (or high frequency) mobilities of colloidal particles in concentrated suspensions to be measured for the first time. From the frequency dependence of the dynamic mobility it is possible to determine the zeta potential and the size distribution of the suspended particles, provided that a suitable electroacoustic theory is applied. An analytical equation has been derived relating the dynamic mobility to the particle properties for near neutrally buoyant particles in concentrated suspensions.

O'Brien's thin double-layer theory may be used to determine zeta and size from the dynamic mobility spectrum for dilute suspensions of solid particles with thin electrical double layers (*1*). Here, the term 'dilute' means suspensions having a particle volume fraction less than 0.02, or, typically 4-10% by weight. This is still a highly concentrated suspension in terms of optical techniques, but in electroacoustic terms it marks the limit above which particle-particle interactions significantly affect the dynamic mobility at Megahertz frequencies. O'Brien has recently developed an approximate theory which allows the measured dynamic mobility to be interpreted in terms of zeta and size for suspensions of near-neutrally buoyant particles with volume fractions up to around 0.6.

Some of the most commonly encountered examples of such suspensions are oil-in-water-emulsions. In such systems the density difference between the oil droplets and the suspension medium is typically around 10%. In this paper, we present the results of electroacoustic measurements of zeta potentials and droplet size distributions on concentrated industrial emulsions, obtained using O'Brien's concentrated theory. Although this theory assumes solid particles, experimental measurements in this laboratory on dilute oil-in-water emulsions have shown that, electroacoustically, emulsion droplets behave as solid particles under most conditions (*2*). The industrial 'emulsion' is, in any case, a bitumen oil of extremely high viscosity.

[3]Current address: Bell Laboratories Lucent Technologies, 600 Mountain Avenue, Murray Hill, NJ 07974–0636.

Background Theory

When an isolated colloidal particle is placed in a spatially uniform ambient electric field, the field lines become distorted because they have to bend around the particle. Thus the particle creates a disturbance in the field, and that disturbance extends out several particle radii. If the particle is charged it will be set in motion by the field, and that will set up a motion in the surrounding liquid that will also extend several radii from the particle. Thus if two particles are placed within a few radii of each other, they will interact and their dynamic mobility will be different from that of an isolated sphere. To develop a theory for the dynamic mobility in a concentrated suspension it is necessary to take into account these interactions between neighbouring spheres.

Unfortunately the exact solution of the multiple particle problem is not feasible at present, so we must fall back on some simplifying assumptions. In this paper we will present an approximate formula for the dynamic mobility of a suspension of nearly-buoyant particles. The theory arose from the observation that the velocity disturbance for a neutrally buoyant particle is short ranged. [To be more precise, the disturbance is contained within the viscous boundary layer thickness $\sqrt{(\nu/\omega)}$ of the particle surface, where ν is the kinematic velocity of the suspending liquid and ω is the angular frequency of the applied field.] Thus in a suspension of buoyant (or nearly-buoyant) particles, only the near-neighbours interact hydrodynamically. In this paper we make the assumption that the hydrodynamic interactions between the near-neighbours can be calculated using a pair-wise additive approximation. That is, the contribution to the velocity of the test particle from a neighbour is the same as the velocity disturbance that the neighbour would cause if it were alone in an infinite liquid. The disturbance in the electric field due to the other particles is approximated by the usual Clausius-Mossitti approach used in the calculation of dielectric properties in a concentrated material (*3*).

The resulting formulae for the dynamic mobility in a monodisperse concentrated suspension are presented in the appendix to this paper. To apply these results to a polydisperse suspension, we make the additional assumption that each particle moves as if it were alone in a concentrated monodisperse suspension of particles of the same concentration as the true suspension. Thus the measured dynamic mobility, which is an average over the particles, is given by

$$\langle\mu\rangle = \int_0^\infty \mu(a,\phi)\, p(a)\, da \tag{1}$$

where $\mu(a,\phi)$ is the mobility formula given in the appendix for a monodisperse suspension of particles of radius a and volume fraction ϕ. The quantity $\mu(a,\phi)$ also depends on the frequency of the applied field, and on the zeta potential of the particle, but that dependence is not shown explicitly here.

The formula (1) can be used for determining particle zeta potential and size distribution from the measured dynamic mobility spectrum. The procedure involves the determination of the zeta and size distribution that will give the best fit between the measured and theoretical dynamic mobility spectra. The size and zeta results presented here were obtained on the assumption that the particles have a log-normal size distribution.

Experimental

Electroacoustic measurements were made using the AcoustoSizer (Matec Applied Sciences, MA, USA). The instrument was calibrated as recommended by the manufacturers.

Three types of oil-in-water emulsions were measured in this study. Recombined dairy cream was prepared by homogenising anhydrous milk fat (Unilac, Australia) with commercial skim milk as the plasma (or background electrolyte) phase. Dilutions of the recombined cream samples also used commercial skim milk as the plasma phase. Commercial dairy cream was obtained locally. This sample was diluted to the concentration of milk using a skim milk plasma phase prepared by centrifugation of the cream sample at 3330 g for 30 minutes. The viscosities of the plasma phases were measured using an Ostwald viscometer, and their densities were measured using a PAAR DMA-02C densitometer. The density of the milk fat droplets was assumed to be 0.92 g cm^{-3}.

Samples of a parenteral nutrition fat emulsion (Intralipid, Kabi Pharmacia, Sweden) used for intravenous injection were supplied by Baxter Healthcare, NSW, Australia. This product is supplied in two concentrations, at 10 and 20% w/v purified soybean oil. A sample of the 20% w/v emulsion was progressively diluted with a background electrolyte sample obtained by centrifuging a similar sample at 40,000 rpm for one hour. The density of the soybean oil was 0.92 g cm^{-3}.

Bitumen emulsions with volume fractions up to 0.6 were supplied by an Australian bitumen manufacturer. Three batches of one particular cationic formulation (samples A1 to A3) were supplied, together with sedimentation data. A second formulation, from a different plant, was diluted from 60% by volume to 40% with distilled water. The density of the oil phase was taken to be 1.05 g cm^{-3}.

In the case of the dairy emulsions, and also in the case of the more dilute parenteral fat emulsions, the ESA signal from the plasma phase or background electrolyte was significant compared to the signal from the emulsion droplets themselves. These background signals arise because the ions from electrolytes and the casein micelles from skimmed milk generate ESA signals as they move in the applied electric field. In such cases, the background signal should be subtracted vectorially from the measured signal to obtain the true ESA signal from the emulsion droplets. This was achieved by measuring the appropriate plasma phase or background electrolyte separately and subtracting the signals using the instrument software.

Results and Discussion

Dairy Emulsions. A series of dynamic mobility spectra obtained from the progressive dilution of a 38% by volume recombined cream emulsion with commercial skim milk as the plasma phase is shown in Figure 1. The mobilities are shown as the magnitude and the argument (or phase lag) as a function of frequency. As the emulsions are undergoing only a four-fold dilution with the plasma phase, the particle size distribution

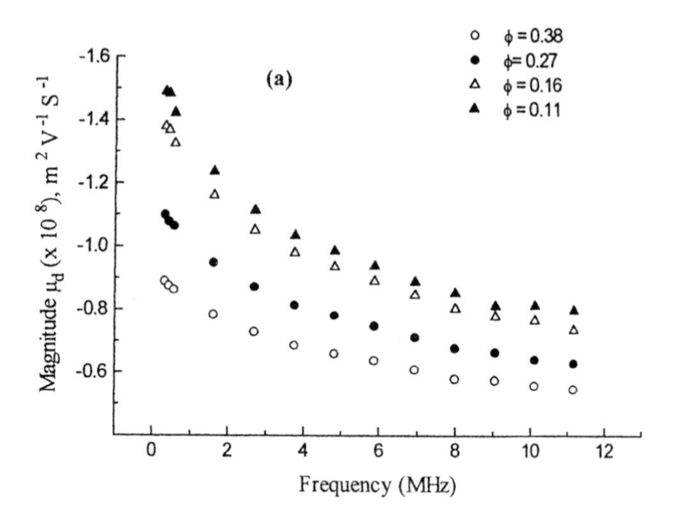

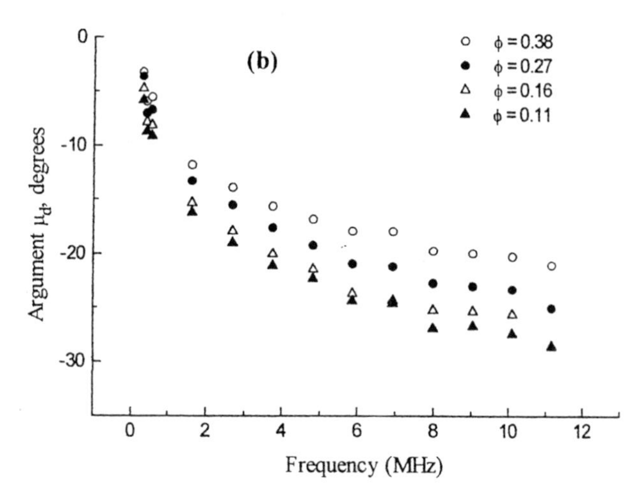

Figure 1. Dynamic mobility of dairy cream emulsion reconstituted with commercial skimmed milk to produce various fat concentrations as indicated. (a) Magnitude in SI units and (b) Argument (or phase angle) in degrees as a function of the frequency of the applied field.

is not expected to change significantly during the dilution process, an expectation confirmed by optical microscopy. Thus, most of the changes in the dynamic mobility spectra in Figure 1 are due to the effects of particle concentration. For the purposes of comparison, we may regard the 10% by volume emulsion as dilute, because the particle interactions are not significantly greater than for a 2% emulsion in this near neutrally buoyant system. Looking first at the mobility arguments (Figure 1b), it can be seen that as the particle concentration increases, the argument of the mobility becomes less negative at a given frequency. Secondly, as the concentration increases, the mobility magnitudes (Figure 1a) drop off less steeply from low frequency to high frequency, i.e., the curves become flatter. The absolute values of the magnitudes at any given frequency also decrease, which is a result predicted by the concentrated theory. However, this last result could also be the result of an increase in the zeta potential of the droplets upon dilution. In a complex natural system such as this it is difficult to be sure that no such change in zeta has occurred, even though the dilution is only four-fold over the experiment and an appropriate diluent is used.

Each of the changes described above is due to the mobility-inhibiting effect of increasing particle concentration. Insofar as determining particle size and zeta potential of the droplets from the dynamic mobility is concerned, the effect of increasing concentration is to make the droplets 'appear' to be smaller, and to have a lower zeta potential, than is truly the case. By taking the particle interactions into account, O'Brien's concentrated theory enables size and zeta to be determined in near-neutrally buoyant suspensions of arbitrary concentration.

The sizing results for the recombined cream emulsions are shown in Table I, along with the results for the other emulsions in this study. For comparison, the mobility spectra have also been sized as though the emulsions were dilute, without compensating for the particle interactions.

Referring to the results obtained using the concentrated theory, it is apparent that the zeta potential of the droplets changes by only about 10% over the concentration range studied and the median diameter of the droplets changes by about the same amount. The small changes in droplet size are consistent with measurements taken from light micrographs. The second set of columns in Table I illustrates the gross underestimates in zeta potential which occur if particle interactions are not taken into account in the analysis of the mobility spectra at high concentrations. The underestimates in the droplet sizes are also quite dramatic, though these do not show quite such a strong concentration-dependence as the zeta potentials.

The commercial dairy cream sample (38% by volume fat) showed a lower zeta potential (-23 mV) and considerably larger droplets (median 3.08 microns) than the recombined cream at the same concentration. This is a consequence of the different membranes covering the droplet surfaces, and also the different preparation techniques. The commercial cream droplets are likely to be covered with a membrane where the charge is imparted mainly by phospholipids. The recombined cream droplets are likely to be covered with a protein membrane from the skimmed milk plasma phase. The recombined sample was homogenised, hence the relatively small droplet sizes and narrow size distributions. The commercial dairy cream was not homogenised.

The commercial dairy cream was also diluted to the fat concentration found in full-fat milk (4% by volume) with a plasma phase obtained by centrifugation. Table I shows that the zeta potentials and droplet sizes again do not change very much on dilution, and that the concentrated and dilute formulae give essentially identical results at this concentration.

Table I

Particle size distributions and zeta potentials calculated from dynamic mobility data according to O'Brien's formulae. Particle sizes are given as diameters in microns, where d_{50}, d_{15}, and d_{85} represent the median, 15th and 85th percentiles of a log-normal distribution. Zeta potentials are in mV.

Emulsion	*Volume Fraction*	*Concentrated Formula* ζ	d_{50}	d_{15}	d_{85}	*Dilute Formula* ζ	d_{50}	d_{15}	d_{85}
1.Reconstit-uted dairy cream[a]	0.38	-36.8	1.49	1.04	2.12	-19.7	0.82	0.74	0.90
	0.27	-37.6	1.38	1.13	1.68	-25.0	0.95	0.86	1.05
	0.16	-40.0	1.33	1.20	1.47	-32.0	1.08	0.98	1.20
	0.11	-40.5	1.32	1.17	1.46	-35.0	1.15	1.05	1.28
2.Dairy cream[b]	0.38	-23.0	3.08	2.79	3.41	-12.9	1.86	1.68	2.06
	0.04	-25.3	3.14	2.84	3.48	-24.2	3.05	2.75	3.37
3. Intravenous fat emulsion (Intralipid 20%w/v)	0.217	-40.8	0.40	0.36	0.44	-29.4	0.28	0.26	0.31
	0.163	-40.3	0.40	0.36	0.44	-31.9	0.31	0.28	0.35
	0.109	-40.2	0.40	0.30	0.53	-34.2	0.35	0.31	0.38
	0.054	-40.6	0.40	0.26	0.60	-37.5	0.39	0.28	0.53
4. Intralipid[c]	0.109	-40.4	0.33	0.23	0.48	-34.4	0.28	0.21	0.38
5. Bitumen A1	0.40	67	5.9	2.6	13	32	2.9	1.5	5.4
A2	0.40	65	5.3	2.5	11	31	2.6	1.5	4.5
A3	0.40	45	6.6	2.0	21	21	3.2	1.2	9.1
6. Bitumen B	0.60	43	3.5	2.6	4.7	12	1.0	0.9	1.1
Diluted B	0.40	55	3.9	2.2	6.7	26	1.9	1.4	2.5

[a] In commercial skimmed milk [b] In a centifugate [c]10% w/v as supplied.

Intravenous Lipid Emulsions. Dilution to the level of 4-5 volume percent, where the emulsions may be regarded as dilute in electroacoustic terms, often introduces significant experimental errors in the complex dairy emulsions (Wade, T.A. and Beattie, J. K. *Colloids and Surfaces Section B* in press). This is because the ESA signals from the emulsion droplets and from the plasma phase approach the same order of magnitude, so that the background correction procedure involves the subtraction of similar (small) numbers. Random noise and systematic errors such as uncertainties in the exact plasma phase composition therefore become important.

For this reason, the intravenous lipid emulsions provide a more suitable examination of the concentrated theory in some respects. The overall signals are stronger here (because the droplets are smaller and have less inertia at high frequency) and the background electrolyte signals are much less significant. The mobility spectra for a series of dilutions of a 20% w/v intravenous emulsion sample with centrifugate are shown in Figure 2. The same general features as described for Figure 1 are seen. In this case it is reasonable to expect the zeta potentials and particle size distributions to remain largely unchanged upon four-fold dilution, and the data given in Table I bear this out. Again, the dilute and concentrated formulae agree at 5% volume fraction. A different batch of Intralipid 10% w/v gave similar results for the zeta potential, with a smaller median droplet size. The zeta potentials and droplet sizes obtained using electroacoustic measurements are in good agreement with literature values established using other techniques (*2*).

The bitumen samples were measured at very high volume fractions. The dramatic differences between the sizes and zeta potentials obtained using the concentrated and dilute algorithms illustrate the importance of accounting for the concentration dependence of the dynamic mobilities. The measured size distributions and zeta potentials for the series of samples A1-A3 were found to correlate well with

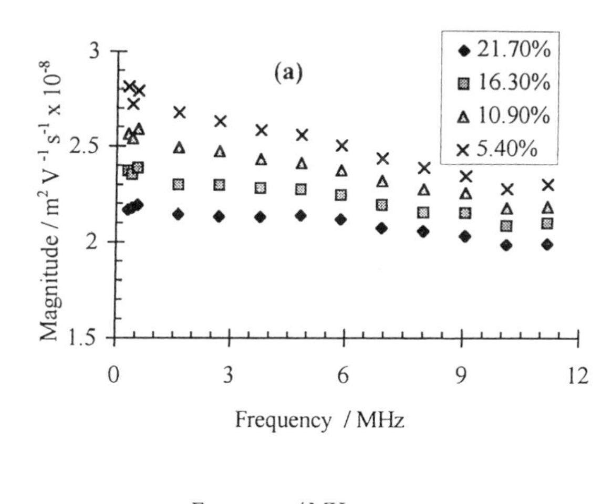

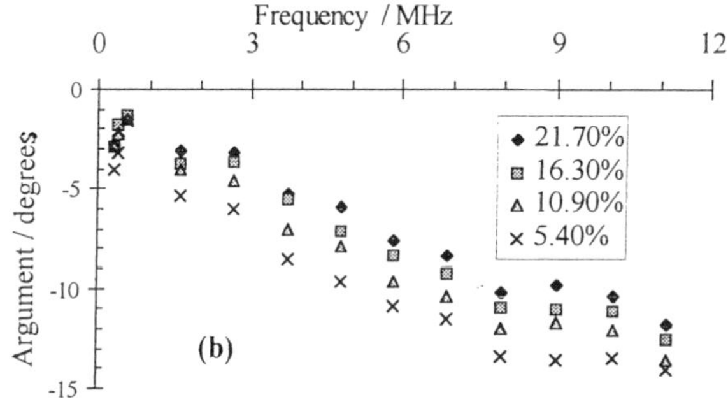

Figure 2. As for Figure 1 but for the intravenous lipid emulsion.

sedimentation results and storage life data supplied by the manufacturer. The dilution of the sample B from 60% by volume to 40% by volume with distilled water presumably causes a substantial change in the background electrolyte concentration and the double-layer thickness. The increase in zeta potential shown in Table I is therefore reasonable.

Conclusions

Electroacoustic measurements of the dynamic mobilities of emulsion systems provide a means of determining particle size and zeta potential without sample dilution. The absence of a need for sample dilution is particularly important for the study of complex natural systems such as dairy foods. It also potentially allows in-line measurements of size and zeta potential in other commercial emulsion systems, for process control and product characterisation. O'Brien's theory for the dynamic mobility of concentrated suspensions of near-neutral buoyancy represents the crucial link between the dynamic mobility and the more familiar quantities of droplet size and zeta potential. As well as covering the majority of oil-in-water emulsion systems, the theory is applicable to concentrated dispersions of other neutrally-buoyant particles such as polymer latices and casein micelles (*4*).

Literature Cited

1. O'Brien, R.W., Cannon, D.W. and Rowlands,W.N., *J. Colloid. Interface Sci.* **1995** *173*, pp.406-418.
2. Carasso, M.L., Rowlands, W.N. and Kennedy,R.A., *J. Colloid Interface Sci.* **1995** *174*, 405-413.
3. Feynman, R.P., Leighton, R.B. and Sands, M. "*The Feynman Lectures on Physics*"; Addison-Wesley; Reading, MA (1972); Vol 2, 11-6 and 32-7.
4. Wade, T.A. Beattie, J. K. Rowlands, W.N. and Augustin,M.A. *J. Dairy Research*, **1996**, *63*, 387-404.

Appendix: the Approximate Formula for the Dynamic Mobility in a Concentrated Suspension

The dynamic mobility in a concentrated suspension of nearly-buoyant particles is given by

$$\mu = \frac{\epsilon\zeta}{\eta} H \left[1 - \frac{2\lambda^2}{(2+\phi)(3+3\lambda+\lambda^2)} - \frac{3\phi}{2+\phi} F \right] \qquad \text{(A1)}$$

where ζ is the zeta potential of the particle, and ϵ and η are the permittivity and viscosity of the solvent. The factors H and F are defined by

$$H = \frac{3+3\lambda+\lambda^2}{3+3\lambda+\frac{1}{3}\lambda^2(1+2\frac{\rho_p}{\rho})} \qquad \text{(A2)}$$

and

$$F = \frac{2}{3}[(4\lambda^2 I + (1+2\lambda)e^{-2\lambda})J + \frac{1}{2}] \tag{A3}$$

Here

$$J = \frac{e^{\lambda}}{1+\lambda+\frac{\lambda^2}{3}} \tag{A4}$$

and

$$I = \int_1^{\infty}(g(r)-1)r\,e^{-2\lambda r}\,dr \tag{A5}$$

where g(r) is the pair distribution. Finally the parameter λ is given by

$$\lambda = (1+i)\sqrt{\frac{\omega a^2}{2\nu}} \tag{A6}$$

where a is the particle radius, ν is the kinematic viscosity of the suspending liquid and ω is the angular frequency of the applied field. The pair distribution function depends on the particle volume fraction ϕ. In our calculations we use the Percus-Yevick formula for g (see R. J. Hunter, in, "*Foundations of Colloid Science*", Volume II, pp. 696-703, Oxford University Press, Oxford, U.K. (1989).

Chapter 22

Acoustic Spectroscopy: Novel Technique for Particle Size Measurement in both High- and Low-Density Contrast Materials over Wide Particle Size and Concentration Ranges

Remi Trottier[1], James Szalanski[1], Charles Dobbs[1], and Felix Alba[2]

[1]Alcoa Technical Center, 100 Technical Drive, Alcoa Center, PA 15069
[2]Felix Alba Consultants, Inc., 5760 S. Ridge Road, Murray, UT 84107

Particle size data are of major importance to virtually all chemical processing industries. To name a few, particle size is directly related to the combustion rate of powdered fuels, the flow properties of granular materials, the hiding power and reflectance of paint pigments, the lubrication of roll milled products, as well as the flame retardant properties of aluminum trihydroxide when used as a filler. The realization of the importance of particle size characterization in such a variety of processes has provided the necessary incentive for the development of several technologies for fast and precise measurements. One of the major weaknesses of most of the conventional technologies, especially for in-process applications is the requirement that highly diluted samples suspended in transparent liquids be used in the measurement. Recent developments in ultrasonic spectroscopy instrumentation are showing great potential for the characterization of opaque emulsions and slurries at process concentrations over particle sizes ranging from 0.01μm to 1000μm.

Measurement

Particle sizing by acoustic spectroscopy is an exercise which can be broken down into three basic steps:

1) Raw Measurement - This step consists of acquiring the attenuation spectrum of ultrasound waves traveling through the suspension over a wide range of frequencies (1 to 150 MHz). The measurement volume is defined by the cylindrical space between the emitting and receiving transducers as illustrated in Figure 1. The attenuation of the signal results from a combination of the various extinction mechanisms: viscous, thermal, scattering, and diffraction.

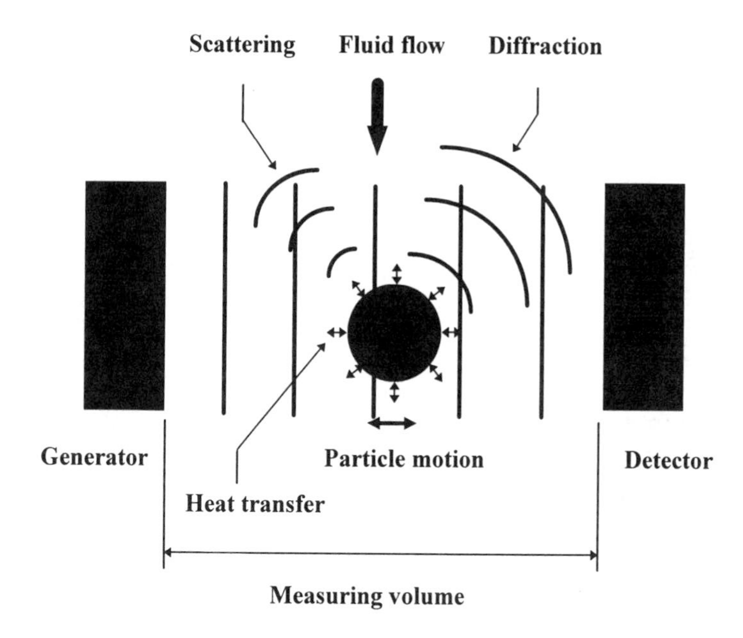

Figure 1 - Basic measurement arrangement and particle-wave interactions

2) Modeling - The modeling part of the experiment consists of a precise mathematical model predicting the attenuation spectrum as a function of the dispersed phase size distribution and concentration. The model used is based on the work of Epstein, Carhart and Allegra who obtained a rigorous solution to the wave equations describing the interactions between acoustic waves and the particles when multiple scattering is negligible (*1, 2*). When multiple scattering is significant the theory becomes much more involved and the instrument requires calibration to preserve the same accuracy (*3*). Multiple scattering phenomena are the most significant when the main attenuation mechanism is viscous-inertial and the particle size is under 10 μm. In the case of emulsions, for which the main attenuation mechanism is thermal diffusion, multiple scattering is negligible regardless of the particle size even at concentrations as high as 70% by volume. The advantage of this fundamental approach (physics first principles) is that in most of the applications, no instrument calibration is necessary and the disadvantage is that the knowledge of several mechanical, thermodynamical and transport properties of both constituent phases is required for modeling low density contrast systems. For the case of high density contrast, however, the required properties reduce to the density of the particles and the viscosity of the suspending medium, though the accuracy of the instrument may deteriorate by some 10%, particularly for coarse particles (>10μm).
3) Data Inversion - The data inversion algorithm step is accomplished using a non-linear optimization algorithm to minimize (in the least squares sense) the difference between the measured spectrum and the predicted one for any conceivable size distribution and concentration. For example, Figure 2 is a

qualitative representation of the variation of sound attenuation as a function of particle size for the case of solid rigid high density contrast particle suspended in water.

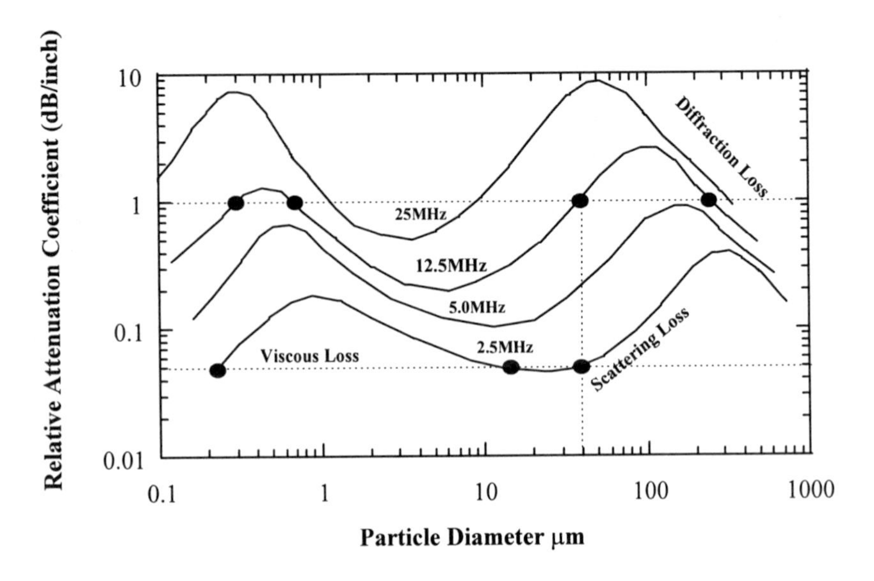

Figure 2 - Variation of attenuation with particle size at different frequencies

Each of these four curves shows the dependence of the attenuation of the sound wave relative to a monosize particle population at a fixed volume concentration. For the simple case of a monodispersed sample, an experimental attenuation of 1.0 dB/inch for a frequency set at 12.5 MHz, as illustrated in figure 2, would narrow down the possible particle size down to 4 possible values. If the attenuation spectrum is then measured at 2.5 MHz, it is possible to quickly converge on the 40 μm particle size, as used in this example. Therefore, in the absence of measurement and modeling errors, probing the sample with only two frequencies would be adequate to zero in on the actual particle size of a monosize population. However, for the simultaneous measurement of the size distribution and volume concentration of a polydispersion, the attenuation spectrum over a large range of discrete frequencies is required. A complete description of the instrument is available from US Patent No. 5,121,629 (*4*).

Size Distribution of Emulsion Systems

The size distribution of the dispersed phase plays an important role in the use, processing and manufacturing of emulsion systems and is of importance to the food, pharmaceutical, cosmetic, petroleum, lubrication, and agricultural industries (*5, 6*). The motivation for the development of ultrasonic technology to characterize emulsions for process applications stems from the lack of reliable conventional techniques to perform the measurement. In the aluminum industry for example, water-based emulsions and dispersions are commonly used in hot and cold rolling mills. These emulsions and dispersions have particle sizes ranging roughly from 0.05 μm to 200 μm, in many cases within a single distribution, and volume concentrations from 1 to 10% oil in water. The sampling and subsequent dilution of these dispersions, which are a requirement of conventional laboratory instrumentation, are problematic since most of these dispersion systems are unstable in the diluted form. Therefore, in order to obtain useful measurements for these types of systems, an instrument capable of on-line, or at-line measurement at process concentration is necessary.

One of the emulsion systems selected to test the instrument (Ultraspec) is referred to as Type "A". The emulsion was prepared by blending the two phases (oil containing the proper surfactants and water) with an industrial blender for typically 2 minutes. Type "A" oil was selected for this test since it is a highly stable emulsion resulting in very little creaming, aggregation and coalescence, even after dilution, and therefore had the added benefit that other well established laboratory techniques could be used to verify the results. Because of the ultrafine narrow distribution nature of the microemulsion prepared, the instruments used had the requirement of being able to measure within a size range from below 0.1μm to 1 μm. Figure 3 shows the experimental, or measured attenuation spectrum as well as

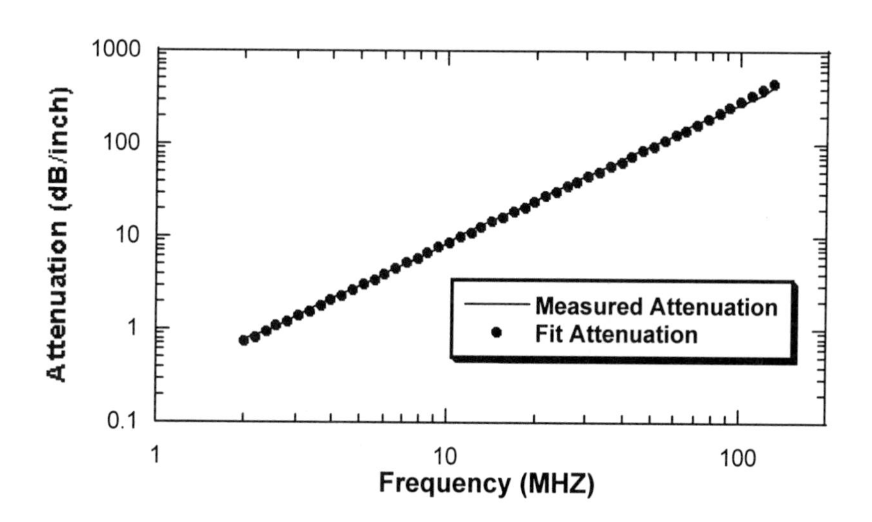

Figure 3 - Measured and fit attenuation spectra for type "A" emulsion.

the fit attenuation which is back calculated from the size distribution and concentration which "best fits" the measured attenuation spectrum
Figure 4 shows the results for the Horiba LA-900 (light scattering), Brookhaven BI-90 (photon correlation spectroscopy), and Ultraspec. The sample analyzed ultrasonically was made to an actual concentration of 5.0% by volume and Ultraspec returned a concentration of 5.25%. The samples analyzed by other techniques were diluted by factors well over 100 in order to obtain the proper levels of obscuration for type "A" oil in water emulsions.

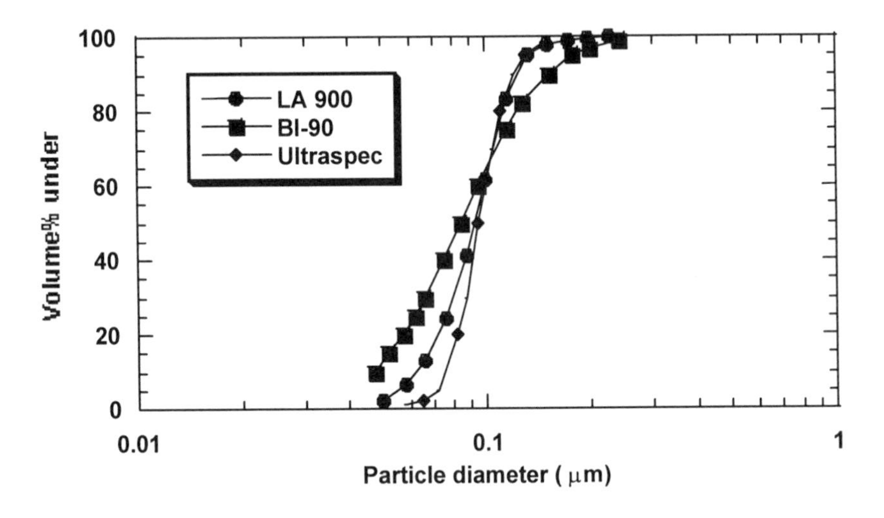

Figure 4 - Particle size distribution as returned by 3 different particle sizing instruments.

Size distribution of High Density Contrast Materials

The first prototype of Ultraspec was designed to measure the size distribution of high concentration TiO_2 slurries, and also was tested for limestone and glass suspensions several years ago (7). Extensive evaluation of the instrument with glass particles and other materials has been conducted by Malvern Instruments which recently acquired the technology. Figure 5 shows the attenuation spectra for the measured and fitted data for 10% volume glass spheres dispersed in water. Figure 6 shows the comparison between the size distribution of the glass sphere dispersion, as measured by Ultraspec and the Malvern Mastersizer X (light scattering). Very good agreement is obtained from the two methods over hundreds of different measurements.

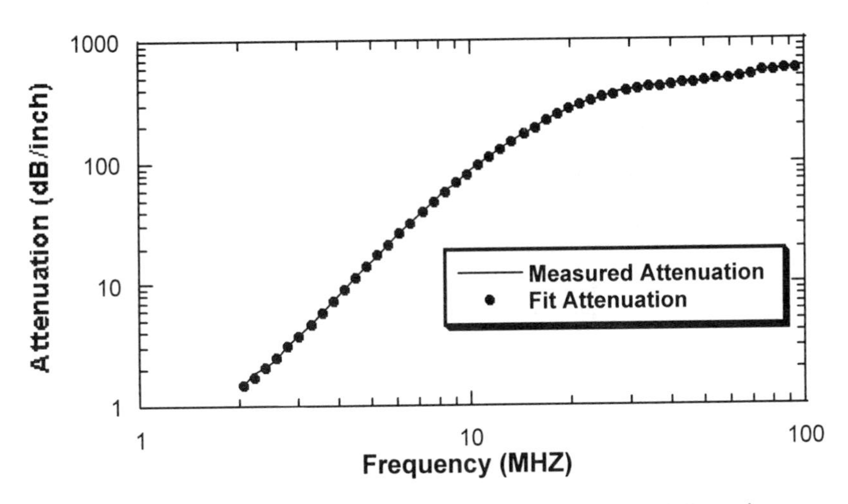

Figure 5 - Measured and fit attenuation spectra for glass bead dispersion.

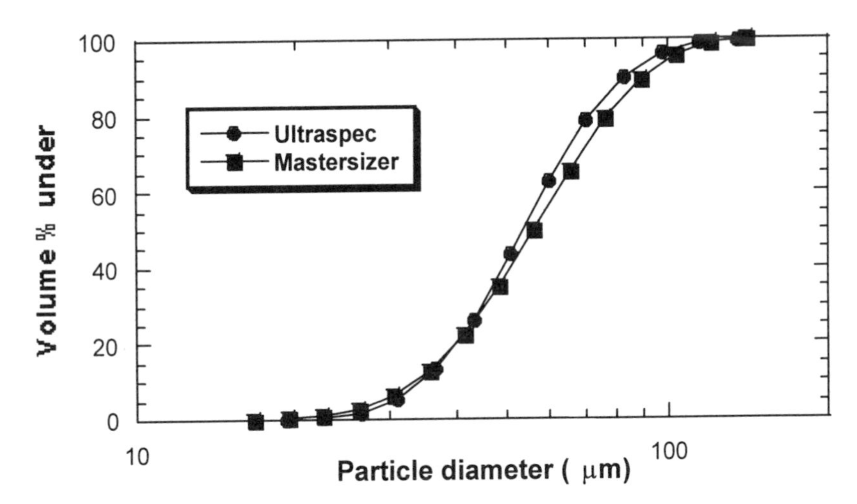

Figure 6 - Particle size distribution of glass spheres as returned by two different particle sizing instruments.

Conclusions

This investigation suggests that the particle size distributions in both high and low density contrast materials can be accurately measured by means of acoustic spectroscopy as performed by Ultraspec. This technique has several distinctions from other size characterization technologies. These are:

1. Since the suspension is probed by ultrasonic energy, as opposed to optical energy, the measurement can be performed through opaque, high concentration suspensions.
2. The technique also exhibits an extremely wide dynamic range (0.01μm to 1000μm) as claimed by the manufacturer (tested here from 0.05μm to 150μm).
3. This technology is also well suited for on-line applications as it is robust enough to withstand fairly harsh industrial environments, can perform measurements up to 70% weight concentration, and therefore eliminates the need for dilution.

Literature Cited

1) Allegra J.R., *Theoretical and Experimental Investigation of the Attenuation of Sound in Suspensions and Emulsions*, Ph.D. Thesis, Harvard University, Cambridge, MA., 1970.
2) Allegra J.R., *J. Acoust. Soc. Am.* **1972**, *51*, pp.1545- 1564.
3) Waterman, P.C., Truell, R., *J. Math. Phys.*, **1961**, 2 (4), , pp 512-537.
4) Alba, F., *Method and Apparatus for Determining Particle Size Distribution and Concentration in Suspension using Ultrasonics*, 1992, U.S. Patent No.5,121,629.
5) Becher, P., Editor, In *Encyclopedia of Emulsion Technology*, Marcel Dekker, inc. New York and Basel, 1995, Vol 1& 2.
6) Schramm,L.L., *Emulsion-Fundamentals and Applications in the Petroleum Industry*, 1992, American Chem. Society.
7) Alba F., Dobbs C.L., and Sparks R.G., *First International Particle Technology Forum*, **1994,** Denver**,** Part 1, pp. 36-46.

Chapter 23

A New Instrument for the Measurement of Very Small Electrophoretic Mobilities Using Phase Analysis Light Scattering

Walther W. Tscharnuter, Fraser McNeil-Watson, and David Fairhurst

Brookhaven Instrument Corporation, Brookhaven Corporate Park, 750 Blue Point Road, Holtsville, NY 11742

A new instrument, the Brookhaven ZetaPALS, based on the principles of phase analysis light scattering (PALS) applied to the measurement of electrophoretic mobilities, has been produced. Such measurements are particularly useful in the study of dispersions in non-polar media, since for a given zeta potential, the mobility is proportional to the dielectric constant. It is also potentially useful where mobilities are low and the use of high electric fields (a traditional remedy) is inappropriate. This is the case when the ionic concentration and hence conductivity of the medium is high. The present PALS configuration has been shown to be able to measure mobilities up to three orders of magnitudes lower than conventional LDE. The device is based on developments reported elsewhere but has a number of new features. In particular all the signal processing is digital and the optical system a reference beam configuration. In this paper we present data showing that on suitable samples both the PALS technique and conventional LDE can be performed on the same instrument, and that the techniques have good agreement.

An understanding of many aqueous dispersions is based on their electrophoretic mobilities, which is of importance in determining, for example, suspension stability, rheological properties and coating behaviour. This arises because the mobility, defined as the velocity the particle attains per unit electric field, can be related to the so-called zeta potential. This potential is defined to be the potential at the surface of shear where the particle with a shell of electrostatically attracted counter-ions moves through the bulk solution. The value of the zeta potential is not the same as the surface potential, due to the presence of the counter ions: however it is the relevant potential for calculating the interaction energy of the dispersion.

The electrophoretic mobility, μ, can be shown to be related to the zeta potential, ζ, by

$$\mu = \frac{\zeta \varepsilon f(\kappa a)}{\eta} \tag{1}$$

where η is the viscosity, ε is the dielectric constant of the dispersing medium, $f(\kappa a)$ is a function of the particle size (radius a) and $1/\kappa$ the thickness of the double layer of counter ions and ions surrounding an individual particle.

In many aqueous based dispersions with a moderate ion concentration and not too low particle size, where κa is 100 or more, $f(\kappa a) = 1.5$, and the equation (1) becomes the so-called Smoluchowski relationship. For particles in non-polar media, where $\kappa a<1$, $f(\kappa a) = 1$ and equation (1) becomes the Hückel relationship. For intermediate values of κa, and for a wide range of zeta potential values, the form of $f(\kappa a)$ can only be evaluated numerically (*1*).

For aqueous systems to which the Smoluchowski relationship applies, at 25° C where the viscosity of water is 0.8905 cP and the dielectric constant is 79, we have a mobility of $1.0\times10^{-8}\ m^2V^{-1}s^{-1}$ corresponding to a zeta potential of 12.6mV. Values of mobility in the range +/- $7\times10^{-8}m^2V^{-1}s^{-1}$ and zeta potentials in the range +/- 90mV are common in aqueous colloids and many biological systems.

Typically, as the surrounding ion concentration increases, the zeta potential and hence the mobility both fall, owing to the shielding effect of the ion atmosphere around the particle. For example, human red blood cells (RBC) in physiological saline conditions (0.145M NaCl, pH 6.8) have a mobility of $-1.07\times10^{-8}\ m^2V^{-1}s^{-1}$ (*2*). In conventional electrophoresis instruments electric fields of the order 1000 Vm^{-1} are used, so that velocities of the order of 10 to 100 microns per second must be measured. This can be done by laser Doppler electrophoresis (LDE). However, for many non-aqueous dispersions (and some other cases) the situation is very different. The change of $f(\kappa a)$ from 1.5 to 1 has a marginal effect: more important are the effects of changes in the viscosity and the dielectric constant. For example, a dispersion in toluene has a viscosity of 0.56 cP, which would tend to raise the mobility for a given zeta potential. Conversely, the dielectric constant for toluene is 2.38, (due to the low polarizability of a non-polar molecule). The resultant effect of both the changes is to reduce the mobility by a factor of about 20. Other examples are given in Table 1.

The stabilization and behaviour of suspensions of solids or liquids in low dielectric media are of considerable interest. Systems of this type are widely used in the paint and coatings industry, in lubrication technology, in pharmaceutical, agrochemical and cosmetic formulations, in reprographic applications and in the development of high performance ceramics and magnetic recording hardware.

Unfortunately, knowledge of the electrophoretic properties of suspensions in non-polar liquids is much less well developed than that in aqueous suspensions. A prime reason is that the measurement of particle mobilities in media of low dielectric constant has previously been experimentally more difficult than the corresponding measurement in aqueous systems (*3*). In addition to lower dielectric constants, the viscosities of many organic liquids are higher, thus reducing electrophoretic mobilities even more. This is clearly illustrated with the examples of some common organic liquids in Table 1 below. A low dielectric constant causes additional problems of cell capacitances and discontinuous electric fields when a driving potential is applied (*4*).

Many applications, particularly in the cosmetic and personal care industries, utilise oils and waxes for the preparation of W/O emulsions and, also, as emollients. While some oils, such as isopropyl myristate, have a moderate value for the dielectric constant and the viscosity, silicone oils are insulators and can have viscosities of many orders of magnitude greater than water making measurement even more difficult, if not impossible, with conventional instrumentation.

Non-aqueous systems are not the only ones which pose a challenge; there are many aqueous examples. We have already mentioned that as the electrolyte concentration of a solution increases, the particle mobility decreases owing to the shielding effect of the ion atmosphere around the particle. Although in the medical and biomedical area organisms such as bacteria and blood cells are suspended in water under physiological conditions, which results in a reduction in mobility of nearly an order of magnitude compared to distilled water (*5*), some body fluids can exceed the electrolyte concentration of normal physiological solutions by an order of magnitude (*6*).

Severe difficulties also arise in obtaining reliable data in environmental applications where the solution conductivity is very large, as in brine and sea water (*7,8*).

Table 1. Mobility ratio for particles with the same zeta potential in various media

Liquid	Viscosity	Dielectric Constant	Mobility Ratio
	η	$\varepsilon/\varepsilon_0$	U_e
Water	00.89	78.0	1.000
Methanol	00.54	33.0	0.700
Toluene	00.56	02.4	0.050
Ethylene Glycol	17.00	40.0	0.030
Glycerol	01.20	43.0	0.400
Oleic Acid	26.00	02.5	0.001
n-Octane	00.54	02.0	0.040
1:4 Dioxane	01.26	02.2	0.020

Finally, particles that are sterically stabilized by adsorbed non-ionic surfactants, macromolecules and synthetic polymers all have electrophoretic mobilities at, or near, to zero. Even electrostatically stabilized suspensions will have little charge and, hence a very low mobilities, if brought close to their iso-electric points (IEP).

Conventional Laser Doppler Electrophoresis (LDE)

Before we describe how the new apparatus addresses the measurement of such dispersions we will briefly review the operation of conventional LDE and its limitations.

Conventional LDE is based on the mixing of scattered light from a sample of a suspension of colloidal particles moving in an electric field, with light directly from the source. The scattered light is frequency shifted by the Doppler effect, and optical mixing of this with the 'unshifted' reference beam light, leads to a beating at a frequency dependent on the speed of the particles. A phase modulation may be applied to the 'reference' beam which generates an additional frequency shift to the spectrum. This allows for either a positive or negative frequency shift due to the movement of the particles to be distinguished (providing this shift is less than the modulation frequency) because the two shifts are additive. Since the direction of the electric field is known, the sign as well as magnitude of the electrophoretic mobility can be determined. The limitation of this method, when a mobility is low, arises from the small displacement of the particles in a given time. If the displacement is less than K^{-1} (K being the scattering vector as defined below) the signal will not produce a complete cycle of 2π and cannot be accurately measured using either spectral analysis or correlation techniques. Note that the addition of the modulator frequency does not materially alter this argument since we are now trying to quantify a shift of less than one cycle added to many. In conventional dynamic light scattering (where we measure the mean square displacement rather than the mean displacement itself) we may still measure small displacements simply by waiting a sufficiently long time for them to become commensurate with K^{-1}. This option is not a good one in an electrophoresis experiment as the continual application of a field in one direction for periods longer than a few seconds can lead to electrode polarization. Similarly increases in the field from the usual 1000 V/m or so lead to excessive Joule heating and other undesirable effects; irregular motions at high fields have been observed in conventional microelectrophoretic instruments (*9*). In addition, a phenomenon known as dielectrophoresis can occur at high fields when there is a large difference in dielectric constant between particle and medium (*10*). With truly non-polar suspensions, where conductive effects should be absent, the existence of small flows due to temperature gradients can mask the very low mobilities ($<10^{-10}$ $m^2V^{-1}s^{-1}$) that are to be measured. Additionally, the presence of Brownian motion added to any electrophoretic motion makes the measurement of a small mobility less precise as this random motion broadens the peak due to any directed motion, and makes its accurate location more difficult.

Phase analysis light scattering (PALS)

The difficulties outlined above may be overcome by recasting the problem somewhat. We apply phase modulation, so that the Doppler frequency of a zero mobility particle, is equal to the modulation frequency ω_0. We may measure the deviation of the actual frequency, present in the scattered light by performing a phase comparison of the detected signal and the imposed modulator frequency. If the mobility is truly zero the relative phase of the two will be constant: if a small mobility is present the relative phase will shift, and small phase shifts can be detected by a phase comparator. The essence of the extra precision is that phase comparison takes place over many cycles of the respective waveforms whereas spectral analysis is sensitive to the period of one cycle. This comparison can be carried out in a variety of ways. In the original PALS development (*11,12*) an analogue lock-in amplifier was used; the instrument described in the present work uses digital signal processing.

The average phase change as a function of time can be shown (*12*) to be

$$\langle Q(t) - Q(0) \rangle = \langle A \rangle K[\langle \mu \rangle E(t) + V_c t] \quad (2)$$

The mobility, μ, that is extracted is the mean mobility for the sample since the average of all phase changes due to the many particles is detected. The scattering vector is defined to be $K = 4\pi n / \lambda_0 \sin(\theta/2)$; λ_0 being the wavelength in vacuum, θ the scattering angle and n the refractive index of the suspending liquid. Q(t) is the amplitude weighted phase at time t, A is the signal amplitude, $\langle\mu\rangle$ is the mean electrophoretic mobility, E(t) is the electric field, and V_c is some collective motion (due to temperature gradients, for example), supposedly constant over the time scale of the field application.

To perform the electrophoresis experiment we apply a rather higher frequency field than the more usual 0.5 to 1.0 Hz used in conventional LDE. The phase change, since the start of the experiment, is then seen to oscillate as shown in the lower curve in Figure 1. If a sinusoidal field is applied a sinusoidal oscillation will be detected. If a square wave is applied the phase changes linearly in each half-cycle of the waveform reversing direction when the field switches. Obviously the total phase excursion depends on the time of application (or inverse frequency) of the field. The signal fluctuates in amplitude due to the relative motion of the particles and fluctuations of concentration. It has been previously shown (*12*) that weighting the phase with the amplitude of the signal was of benefit since the phase cannot be determined accurately when the signal amplitude is small. Since the phase difference function is synchronous with the field it can readily be averaged over a number of separate applications of the field to yield high quality results.

The presence of a collective motion is shown in the upper curve in Figure 1. We see that by suitable data treatment we can separate the collective motion from the electrophoretic motion. Indeed, by adjusting the reference frequency to account for the frequency shift of V_c, this collective term can be virtually eliminated.

Instrumental details

The instrument described in the present work is the first to adapt the PALS methodology to utilize the much more robust and simple optical arrangement of the homodyne method and is based around the configuration of the Brookhaven ZetaPlus in which the method of applying the phase modulation frequency also differed from that used by Miller *et al* (*12*).

Finally, a major objective was to reduce the total size of the original apparatus and design a commercially viable unit. The original PALS system was configured on a large optical bench and used a rack of analogue electronics.

The general arrangement of the Brookhaven Instruments Corp. ZetaPALS is shown in Figure 2. A laser beam is split to produce a scattering and reference beam. The latter can be modulated by a piezo-electric phase modulator at frequencies in the 62.5Hz to 2000Hz range. Scattered light at 15 degrees is combined with this reference beam to produce a homodyne signal, or, with the reference beam interrupted, a self-beat signal for particle size analysis by dynamic light scattering (DLS). The use of the reference beam configuration is also a significant departure from the original PALS device of Miller *et al* and allows considerable ease in alignment and optical stability since flare light from the walls of the cell becomes unimportant as a source of error. This also greatly assists the measurement of weak scatterers such as small particles or dispersions of low refractive index contrast. A fiber optic also allows DLS to be performed at 90 degrees scattering angle. The ratio of scattered to reference signal is optimized by automatic adjustment of a continuous attenuator in the incident beam path. The scattered light is detected by a photomultiplier and passed to both a data buffer, for spectral analysis and an (optional) digital correlator for particle size analysis. The sample is in a 1cm square cuvette thermostatted by a Peltier device. The electrodes plug into the cell and provide a field that can be reversed automatically and is remote from any cell wall. The cell configuration, based on that used in the Brookhaven ZetaPlus, avoids the troubling factor of electro-osmosis and, in addition, no alignment to a stationary level is required; both effects are discussed in detail elsewhere (*13,14*). A further advantage, of the cell configuration, is that there is no necessity to refocus the optics when working with media of differing refractive indices. Electrodes of gold or palladium are available, the latter being preferred when conductive (polar) dispersions are to be measured. In conventional LDE the field current is regulated, and the electric field computed from the current and measured conductance, which is carried out at high frequency.

The whole instrument is controlled by a host PC. For phase analysis we add an additional electronic module containing a slave digital signal processor. This processes the 16 bit photon count signal to produce phase estimates at intervals that are a multiple of the fundamental sampling time, and a small fraction of the electric field period. This measurement can be synchronized with the application of the electric field. The cell can be driven by fields of up to 60 kVm^{-1} at frequencies from DC to >1MHz, with either a square or sine wave format. The frequency is generated by a high precision numerically controlled oscillator and has a wide dynamic range.

Experimental details

A number of samples were studied by both conventional LDE and the new PALS methodology. Since, as yet, no 'standard' types of non-aqueous dispersion are available which have been characterized by LDE, the samples used initially were

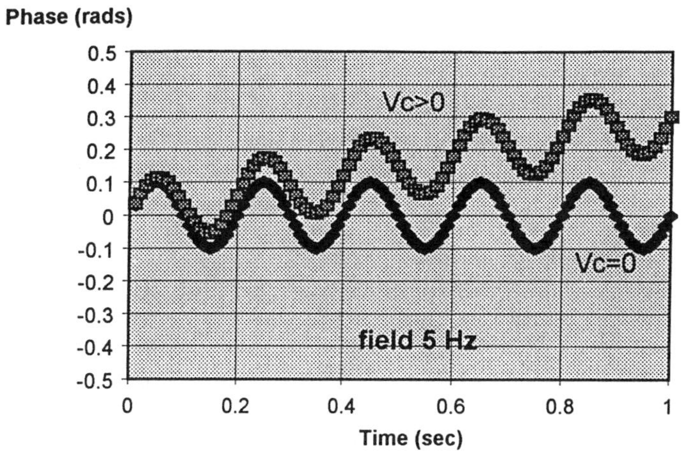

Phase (rads)

0.5 0.4 0.3 0.2 0.1 0 -0.1 -0.2 -0.3 -0.4 -0.5

Vc>0

Vc=0

field 5 Hz

0 0.2 0.4 0.6 0.8 1

Time (sec)

Figure 1
AWPD spectra for Vc=0 and Vc>0

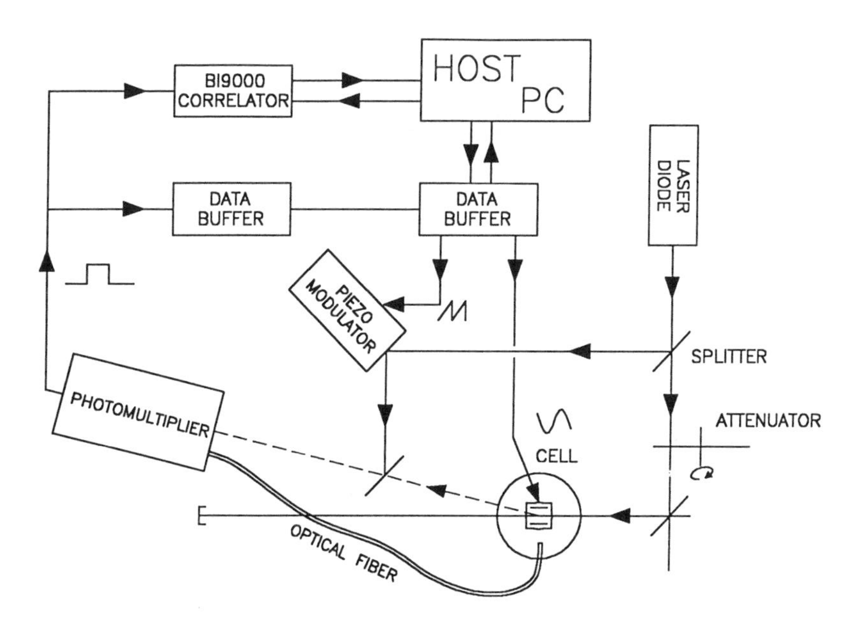

Figure 2
A schematic diagram of the instrument

aqueous based dispersions. Various materials were simply dispersed in a range of different electrolytes between 10^{-4} M and above 1M. The materials and solution conditions were chosen to provide examples of both positively and negatively charged surfaces; a particle size range of 49nm (PS latex) to 7μm (human RBC) was covered by the different samples. Measurements were conducted at a temperature of 25° C using the standard temperature control system fitted to the PALS instrument.
Conventional LDE was performed on a Brookhaven ZetaPlus. On this instrument the electric field is applied for up to about 1 second in each direction and the resultant Doppler shift for the two directions averaged. The instrument checks the apparent mobility with no field applied and waits until the Doppler shift is less than 4 Hz before taking data. This procedure largely eliminates the effects of thermal collective motion of the particles which would otherwise distort the electrophoretic motion. In contrast, the ZetaPALS can directly measure the collective motion which, as we have shown above, can be separated from the electrophoretic motion. The examples given below in Figures 3 and 4 illustrate such an effect. In addition, the ZetaPALS has the ability to adjust the phase demodulation frequency so compensating for any substantial thermal or other drift in the course of the experiment. Measurement times on the ZetaPALS are typically less than 20 seconds or so and in this time some 30 separate phase spectra are collected and averaged.

Results I

The first set of samples, on which we report electrophoretic mobilities, are listed below in Table 2(a) and 2(b). Where an error figure is quoted it is the standard error of the mean for between 5 and 10 repeated measurements.

Sample A is used within Brookhaven Instruments Corporation as a material for initial optical alignment and checking the ZetaPlus; the electrophoretic mobility value of -3.1 $m^2V^{-1}s^{-1}$ has been established by many measurements over more than a year. Agreement between the two techniques is excellent.

Sample B is a uniform polymer microsphere used as a calibrant or control for particle sizing instrumentation (Duke Scientific Corp. Palo Alto, California). Composed of DVB crosslinked PS , it was chosen as an example of a weakly scattering system. In addition to the mean values given in Table 2 above, the experimental data outputs from the two techniques are shown in Figures 4 and 5. The zeta potential value, calculated from the mobility data is approximately -44mV and is not unusual for aqueous polymer lattices electrostatically stabilized by an anionic surfactant.

Sample C is the only 'official' reference material for electrophoretic mobility and is manufactured and distributed by the NIST (Gaithersburg, Maryland). Figure 6 shows the electrophoretic mobility distribution obtained for the NIST Standard Reference Material 1980 using the Brookhaven ZetaPlus. This material is a sample of Goethite (α-FeO(OH)) which, when prepared under standard conditions, has a certified electrophoretic mobility of $+2.53\times10^{-8} \pm 0.12\times10^{-8}$ $m^2V^{-1}s^{-1}$ (*15*). The average mobility found from Figure 6 is $+2.54\times10^{-8}$ $m^2V^{-1}s^{-1}$ and a measurement using the PALS system gave +2.51 $m^2V^{-1}s^{-1}$ — all three values in excellent agreement.

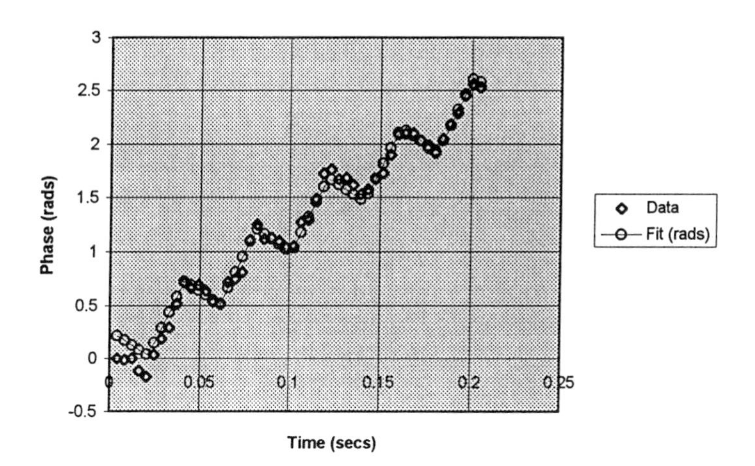

Figure 3
BIC Zeta potential reference sample BI ZR2 at low electric field using PALS (no autotracking applied). The mobility from the fit gave a value -3.07×10^{-8} $m^2/V^{-1}s^{-1}$. The 'accepted' value is -3.1 +/- 0.15×10^{-8}

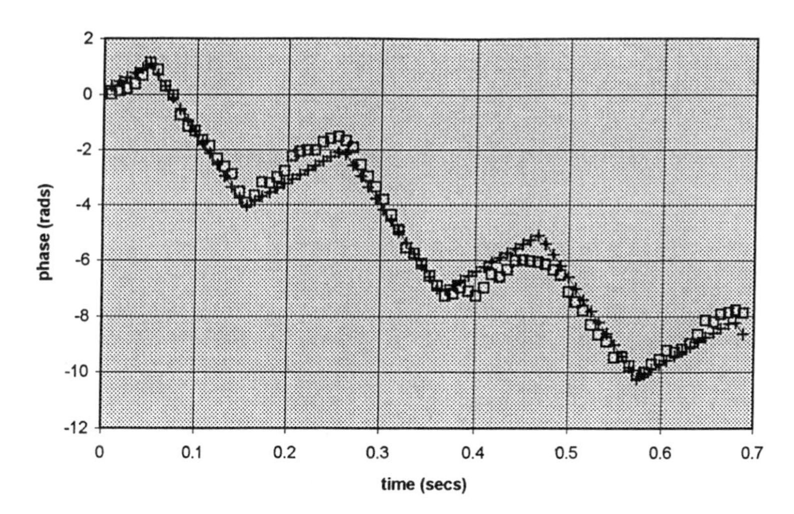

Figure 4
AWPD for 49 nm PS latex in 1mM KCl square field 833 V/m, 5 Hz., No autotracking. Experimental data is shown as open squares, the fitted model as a crossed line. The fit gave a mobility of -2.685×10^{-8} $m^2/V^{-1}s^{-1}$

Table 2. Comparison of Electrophoretic mobilities by conventional and PALS measurements

Sample	Mobility by LDE (or 'accepted' value)	Mobility by PALS $x10^{-8}$ $m^2V^{-1}s^{-1}$
A. Brookhaven Ref. material ZR2 (A blue colored organic dye) dispersed in 1mM KCl.	-3.1+/-0.15	-3.2 +/- 0.1
B. Duke Scientific Corporation 49 nm PS Latex dispersed in 1mM KCl.	-2.65 +/-0.2	-2.68 +/- 0.1
C. NIST Reference material SRM1980 in water adjusted to pH 3.4 using HNO_3.	2.54 +/- 0.10 (NIST)	2.51 +/- 0.11
D. Human RBC in Dulbecco's Phosphate Buffered Saline solution (0.145M NaCl, pH=7.4).	- 1.07 +/- 0.02 (ref 2)	- 1.08 +/- 0.015
E. Ferrite in Dodecane.	?	0.013 +/- 0.0015
F. TiO_2 in Ethanol.	?	-0.503 +/- 0.010
G. The same TiO_2 dispersed in Toluene.	?	0.255 +/- 0.010
H. The same TiO_2 after drying, dispersed in Toluene.	?	0.155 +/- 0.011
I. The same TiO_2 dispersed in Xylene.	?	0.095 +/- .005
J. Casein in Polyethylene Glycol (PEG).	?	-0.025 +/- .002

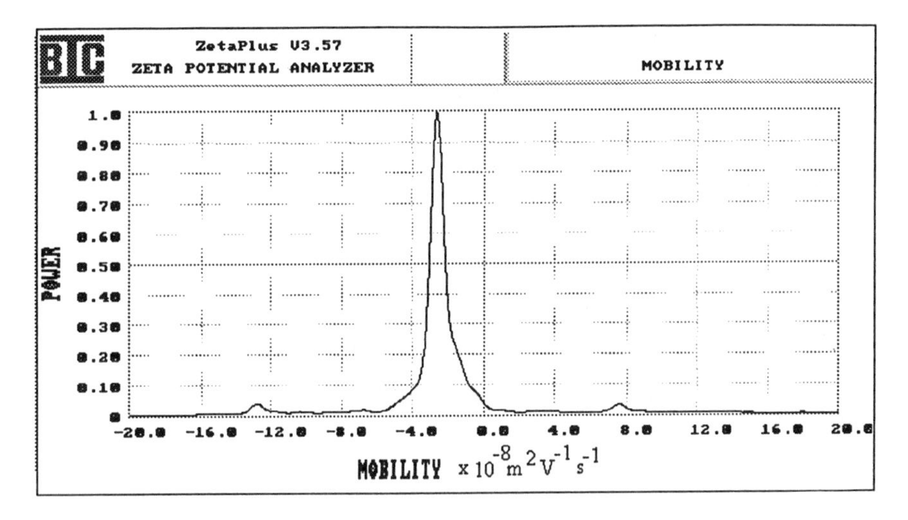

Figure 5
Conventional LDE analysis for the same sample. The field in this case was 1.68 kV/m, the mode of the mobility spectrum is at -2.64×10^{-8} m^2 / $V^{-1}s^{-1}$

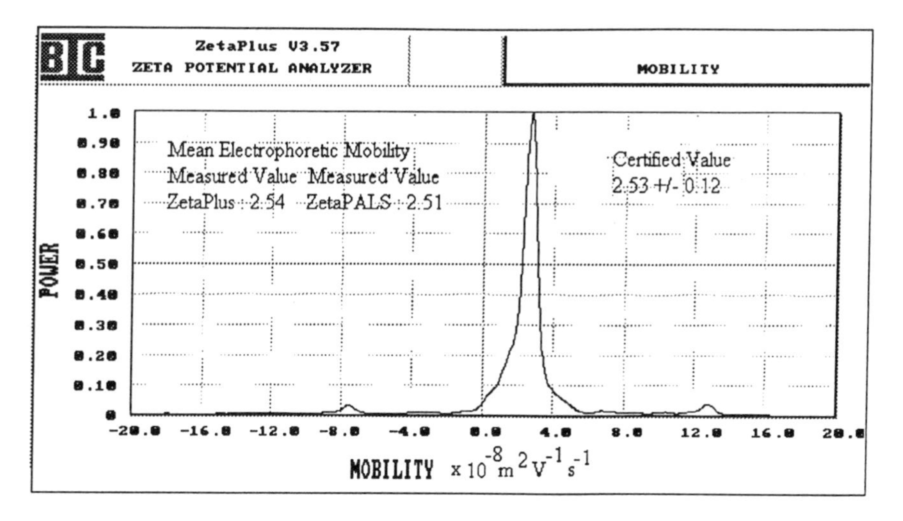

Figure 6
Electrophoretic mobility of NIST Reference Material SRM 1980.

Sample D was prepared using RBC drawn from one of the present authors (WWT). For human RBC, a spread of 2% across the population and depending on the physiological condition of the individual has been routinely observed (*5*).

In summary, the agreement for all of these aqueous based samples between PALS and conventional LDE measurements is very close. It is commonly held (*16*) that, for many reasons, an agreement of 5% between different measurements of the electrophoretic mobility of aqueous suspensions and a reproducibility of perhaps 2% may be expected. The PALS technique easily surpasses this criterion.

The remaining measurements were made on samples that have previously been beyond the range of conventional LDE, at least for routine application.

Sample E was a proprietary commercial suspension of ferrite in silicone oil sent originally to Brookhaven Instruments Corporation for particle size analysis. The stabilizing moiety was believed to be a low molecular weight polymer. The initial material, supplied as a moderately concentrated dispersion, was diluted into dodecane for the PALS measurement. The measured electrophoretic mobility value of $0.013 \times 10^{-8}\ m^2V^{-1}s^{-1}$ was the lowest of all the samples chosen for the present study and was more than two orders of magnitude smaller than any of the aqueous mobility values and is consistent with what might be expected from a "sterically stabilized" system. However, using the Hückel equation, the zeta potential is calculated to be approximately 15mV suggesting an electrostatic contribution to the stability.

Aside from the excellent reproducibility of such small measured values the data clearly highlights the superior sensitivity of the PALS methodology. Indeed, the PALS technique is potentially capable of determining electrophoretic mobilities down to $10^{-12}m^2V^{-1}s^{-1}$ (*17,18*). This would correspond to only one or two charge sites/groups per particle; clearly, at this level, the term "zeta potential" would have no meaning.

Although no 'accepted' value for TiO_2 is known, the reversal of sign between polar ethanol and non-polar toluene or non-polar xylene is expected on electron donor-acceptor arguments. (*19,20*). The reduction in the relatively high value of $0.255x10^{-8}$ $m^2V^{-1}s^{-1}$ to $0.155x10^{-8}$ $m^2V^{-1}s^{-1}$ on drying is also explicable when it is recognized that traces of water will stay on the surface of the particle due to the hydrophobicity of the toluene and largely determine the surface charge. It is well recognized that trace amounts of water can dramatically effect the surface charge of particles in non-aqueous media (*21,22*).

The final result was obtained on another proprietary commercial sample sent for electrophoretic mobility analysis: casein dispersed in a PEG having an average MW of 200. The measurement was made at 37° C. Even at this temperature the viscosity is still 27 cP, so a small value of mobility is to be expected. This measured value of $-0.025x10^{-8}$ $m^2V^{-1}s^{-1}$ is to be contrasted with the typical literature value for casein in water of about $-6.0x10^{-8}$ $m^2V^{-1}s^{-1}$ (*23*).

Results II : Measurements on quartz spheroids over a range of ionic molarities

To investigate the performance of the ZetaPALS over a range of salt conditions, rather wider than so far demonstrated, a set of measurements on the material BCR66 were performed. This material is a certified reference material made of crushed quartz spheroids of size range 0.35 microns to 3.5 microns (*24*). It is used to check the accuracy and precision of particle sizing instrumentation and also as a calibrant. The samples were prepared by simple dispersion of the BCR66 into D.I. water containing different concentrations of KCl using ultrasonics; no additional dispersing agent was employed. The results obtained are tabulated below in Table 3 and plotted in Figure 7.

As expected for quartz in aqueous suspension, all the measured mobility values are negative. In low electrolyte concentration (< 0.01M) the zeta potential is calculated to be about -65mV, typical of literature values for quartz (*16*). The measurements above 0.1M could not be performed by conventional LDE. Data was obtained even in saturated salt (>4M) but the reproducibility was not felt to be acceptable enough, at this time, to justify publication, though the mobility obtained, interestingly, was still negative. The general trend of the effect of increasing electrolyte concentration, shown in Figure 7, including the extremum at 0.001M (probably due to slight specific adsorption of chloride ions) is also in agreement with expectation.

Table 3

BCR66 in KCl Molarity	**Mobility x $10^{-8}m^2V^{-1}s^{-1}$**
0.0001	-4.30 +/- 0.10
0.001	-4.56 +/- 0.05
0.01	-3.50 +/- 0.12
0.1	-2.10 +/ 0.10
0.5	-1.75 +/- 0.20
1.0	-1.33
2.0	-0.73

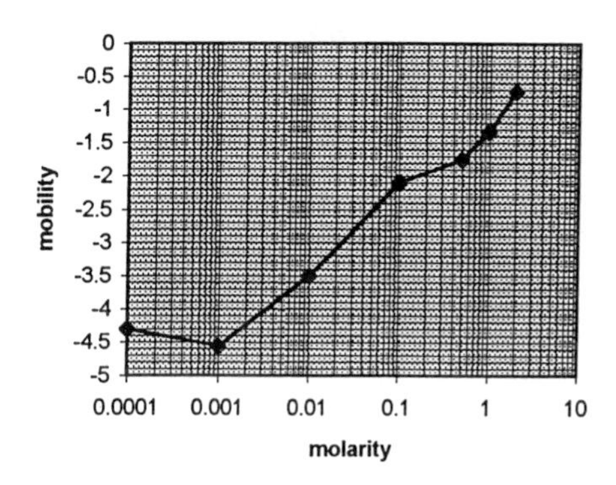

Figure 7
BCR 66 in KCl

Conclusions

The Brookhaven ZetaPALS instrument has been used to measure the electrophoretic mobility of a range of samples some of which are also able to be measured by conventional LDE. The agreement between the two techniques, where appropriate, is shown to be excellent. Examples of measurements on samples beyond the reach of conventional LDE due to low dielectric constants, high viscosities, and high ionic strengths, have also been clearly demonstrated; routine measurements of such systems are now possible. With a current sensitivity of up to three orders of magnitude better than conventional LDE, the PALS methodology represents a quantum leap in the technological development of commercially available instrumentation to measure the electrophoretic mobility of colloidal dispersions.

References

1. O'Brien, R.W.; White, L.R. *J.Chem.Soc.*, Farad.II, **1978**, *74,* 1607
2. Bangham, A. et al. *Nature*, **1958**, *182,* 642
3. Morrison, I.D.; Tarnawskyj, C.J. *Langmuir*, **1991**, *7,* 2358
4. Kornbrekke, R.E. et al. *Langmuir*, **1992**, *8*, 1211
5. Seaman, G.V.F. In *The Red Blood Cell*; Surgenor, D., Ed.; Academic Press: 1975
6. Beretta, D.A.; Pollack S.R. *J.Ortho.Res.*, **1986**, *4,* 337
7. Coffey, M.D.; Lauzon, R.V. *Int.Symp.Oilfield Chem.*, **1975**, SPE *No.5302,* 93
8. Yousef, A.A.; Bibawy, T.I. *Tenside Deterg.*, **1976**, *13,* 316
9. Novotney, V. *J.Appl. Phys.*, **1979**, *50,* 324
10. Pohl, H.A. In *Dielectrophoresis;* Cambridge University Press: 1978
11. Miller, J.F. Ph.*D. Thesis*; University of Bristol 1990
12. Miller, J.F.; Schätzel, K., Vincent, B. *J.Coll.Int.Sci.*, **1991**, *143,* 532
13. Uzgiris, E.E. *Prog.Surf.Sci.*, **1981**, *10* 53
14. McFadyen, P.; Fairhurst, D. *Proc.Brit.Ceram.Soc.*, **1993**, *51*
15. Hackley, V.A. et al. *Coll. and Surf.*, **1995**, *98,* 209
16. Hunter, R.J. In *Zeta Potential in Colloid Science;* Academic Press: 1981
17. Miller, J.F. et al., *Coll. and Surf.*, **1992**, *66,* 197
18. Bradbrook, S. *PhD.Thesis*, University of Bristol 1996
19. Labib, M.E.; Williams, R, *J.Coll. and Int.Sci.*, **1984**, *97,* 356
20. Fowkes, F. et al. In *Colloids and Surfaces in Reprographic Technology;* Hair, M and Croucher, M.D. Ed.; ACS Symposium Series: 1982, 200
21. McGown, D.N. et al. *J.Coll.and Int.Sci.*, **1965**, *20,* 650
22. Kosmulski, M.; Matijevic, M. *Langmuir*, **1991**, *7,* 2066
23. Moore, D.H. In *Electrophoresis: Theory, Methods and Applications*; Bier, M. Ed.; Academic Press: 1959
24. *European Community Bureau of Reference*, 1980, *EUR 6825*

INDEXES

Author Index

Subject Index

C

D

E

M

O

P

W

Z